MW00760358

Tunneling in Complex Systems

PROCEEDINGS FROM THE INSTITUTE FOR NUCLEAR THEORY

Series Editors: Wick C. Haxton *(Univ. of Washington)*
Ernest M. Henley *(Univ. of Washington)*

Published

Vol. 1: Nucleon Resonances and Nucleon Structure
ed. G. A. Miller

Vol. 2: Solar Modeling
eds. A. B. Balantekin and J. N. Bahcall

Vol. 3: Phenomenology and Lattice QCD
eds. G. Kilcup and S. Sharpe

Vol. 4: N* Physics
eds. T.-S. H. Lee and W. Roberts

PREFACE

The six advanced courses contained in this book are representative of the subjects discussed during the spring 1997 INT program, "Tunneling in Complex Systems". The courses do not cover all the topics raised during the program nor does this book make any attempt to cover all the physics that could be represented as tunneling in complex systems. Instead, you will find contained herein two main thrusts. First is a diverse, albeit small in number, sampling of modern physical contexts in which tunneling is playing or may play an important role, and second are the theoretical advances in tunneling driven either by these new physical systems or by the resurgence of interest in the study of dynamical systems more generally.

The book opens with Anthony Leggett's course. Being the first course of the program as well, he begins with very general background material on tunneling problems in a single coordinate coupled to a complex environment. In the second half of his lectures, he turns his attention to tunneling processes involving Bose-condensed atomic alkali gases.

This is followed by Stephen Creagh's course on tunneling in two degrees of freedom. A unique synthesis is presented of the relatively recent realization that all of the dynamical complexity possible with simple, few degree-of-freedom systems may be manifest in tunneling processes in a number of ways. Phenomena such as dynamical tunneling in KAM systems, chaos-assisted tunneling, and tunneling between chaotic states are discussed for the first time in a unified way.

The next two courses are by Philip Stamp and Anupam Garg. They are both motivated by recent experiments in nanomagnetic systems, but approach the subject from differing viewpoints. Dr. Stamp, amongst other things, reviews the effects of various bath environments with an emphasis on the role of a nuclear spin bath as the main progenitor of decoherence. Dr. Garg also discusses the role of nuclear spins and emphasizes its effects on the tunneling spectrum.

Superdeformation in excited, high spin nuclei is the main subject of the next course by Teng Lek Khoo. He summarizes the main physical processes in the experiments and discusses the relevance of tunneling to the sudden decay of nuclei out of the superdeformed bands back into excited states of normal deformation.

The final course in the book is by William Reinhardt. He discusses the appearance and dynamics of solitons within a mean field description of gaseous Bose condensates. He then discusses analogies between the Josephson tunneling effect and the soliton description of zero-energy condensate collisions.

In closing, I thank the authors for their extensive efforts in producing their courses. You will find herein, well organized, illuminating, and pedagogical courses that are accessible to senior graduate students, yet capable of serving experts in the field. Finally, on behalf of all the participants of the program, I gratefully acknowledge the support of the INT and its staff beginning right from the conception of the program back in 1994 all the way through its completion three years later, and the hard work of the other organizers, Oriol Bohigas and Tony Leggett.

December 1997

Steven Tomsovic
Pullman, Washington

CONTENTS

SOME GENERAL ASPECTS OF QUANTUM TUNNELING AND COHERENCE: APPLICATION TO THE BEC ATOMIC GASES

A. J. Leggett
Department of Physics
University of Illinois
1110 West Green Street
Urbana, IL 61801-3080

In these lectures I want to do two separate though related things. The first is to review some general results concerning the phenomena of tunneling and coherence, with particular reference to the case of a macroscopic (or mesoscopic) degree of freedom which is likely to be coupled to a complex environment. The second is to discuss some issues relevant to tunneling in a specific system, namely the recently realized Bose-condensed phases of dilute alkali gases.

The original and perhaps conceptually simplest tunneling problem refers to a "particle" with a single coordinate q, which is originally situated in a metastable well of the corresponding potential V(q) but can penetrate the potential barrier by a quantum tunneling process, or alternatively, at finite temperatures, undergo thermal activation over it: see fig. 1. In this case the obvious question is: what is the rate of escape as a function of temperature T? Examples include tunneling of an α-particle out of a heavy nucleus, many processes in chemical physics, and the escape of a Josephson junction from the metastable zero-voltage state.

A second, related problem is that illustrated in fig. 2; one has a double-well potential, in general asymmetric, such that one cannot treat the quantized levels as a continuum, and the particle is initially located in the upper well. The interesting question here is: how does the transition rate depend on the coincidence or not of the levels in the upper and lower wells (and also on temperature)? Controlled experiments to look at these questions have been done in SQUID rings(1) and in the mesoscopic system Mn^{12}-acetate(2,3).

A third type of situation involves a particle located in a symmetric double-well potential (fig. 3), such that in the absence of any interaction with the environment it would perform oscillations of the type exemplified by the NH_3 inversion resonance. In this case the most interesting questions refer to the coherent oscillation behavior, if any. In particular, if $\sigma_z = +1(-1)$ denotes the groundstate in the left (right) well respectively, we can ask (a) what is $< \sigma_z > (t)$ given that $\sigma_z(0) = +1$? (b) what is the correlation function $< \sigma_z(t)\sigma_z(0) >$? It should be emphasized that these questions, while related, are not identical, and in certain circumstances the answers to them can be

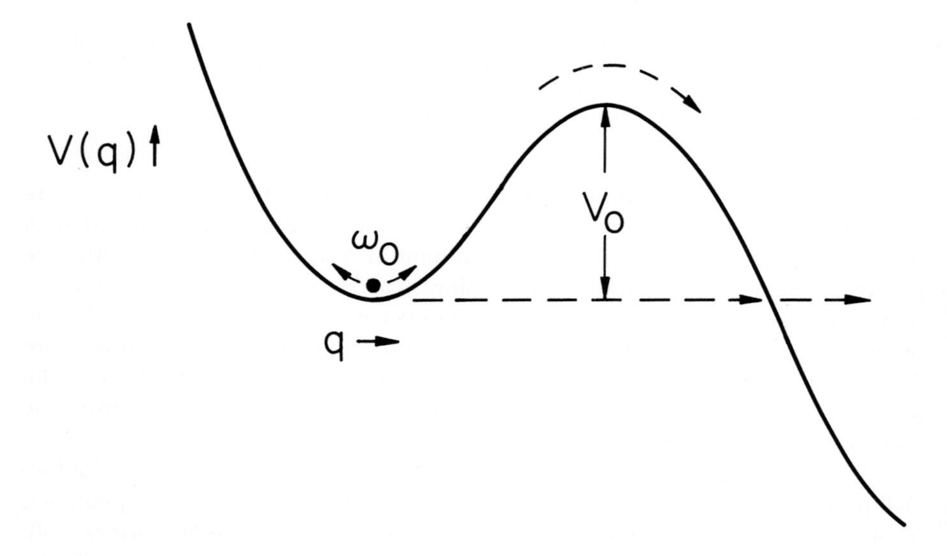

Figure 1: Tunneling out of a metastable well

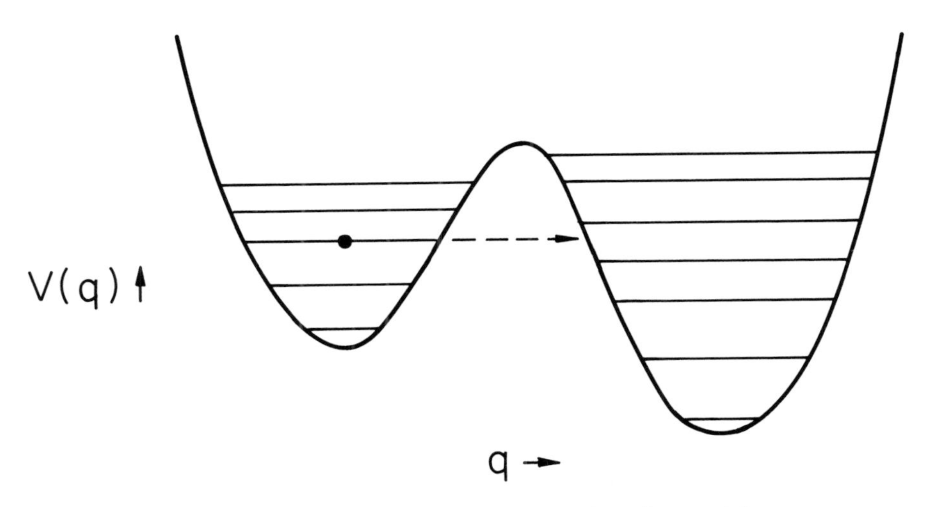

Figure 2: Resonant activation in a double-well potential

qualitatively different.

I'll first briefly discuss these questions for the idealized case of a totally isolated system, then consider how the interaction with the environment may modify the behavior quantitatively or qualitatively in each case.

First, then, consider the process schematically indicated in Fig. 1, namely the decay of a metastable state into the continuum. At very short times, the behavior is liable to depend on the precise preparation of the initial state. At very long times, there is a general theorem(4) that the decay must be a power law in the time. However, over a vast range of intermediate times (and effectively at all times of practical interest for real-life systems) the general belief is that the decay will be exponential in time, i.e. the probability of finding the system still in the well at time t, given that it is known to have been there at time zero, is given by

$$P_{in}(t) = exp - \Gamma t \tag{1}$$

where $\Gamma \equiv \Gamma(T)$ is the decay rate. If ω_o is the frequency of small oscillations around the potential minimum, then at high temperatures ($kT >> \hbar\omega_o$) the decay is dominantly by a classical thermal activation process and the rate $\Gamma(T)$ is given by the Arrhenius-Kramers(5) formula

$$\Gamma_{T\to\infty}(T) = A_{cl}\omega_o exp - V_o/kT \equiv \Gamma_{cl}(T) \tag{2}$$

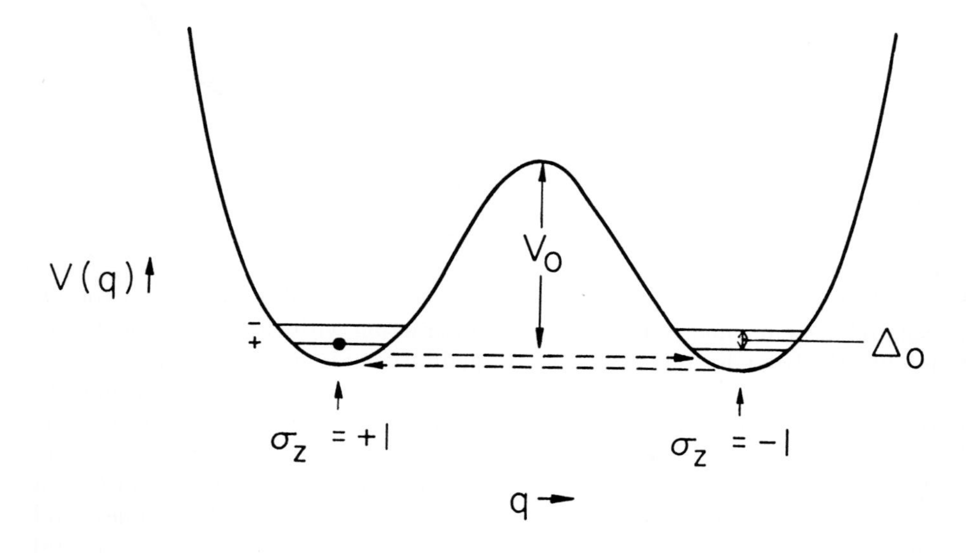

Figure 3: Coherent oscillations in a symmetric double-well potential

where V_o is the height of the barrier and A_{cl} is a constant of order unity which depends on the friction(5). By contrast, at zero temperature the decay is entirely by quantum tunneling through the barrier, and the rate is given by a WKB-type formula

$$\Gamma_{T=0} = A_{QM}\omega_o exp - B_{T=0} \tag{3}$$

where B is the WKB exponent and, again, A_{QM} is a prefactor which is generally of order unity. To obtain the exponent B a number of different methods are available (a particularly good discussion is given by Schmid in ref. (6)). In particular, one can

(a) solve the time-dependent Schrödinger equation explicitly subject to outgoing-wave boundary conditions (this requires us to treat the wave "reflected" at the outer edge of the potential barrier explicitly).

(b) solve the time-independent Schrödinger equation within the well and barrier region and analytically continue the solution beyond the outer edge of the barrier to represent an outgoing wave: or,

(c) use an imaginary-time path integral ("instanton") technique to calculate the imaginary part of the groundstate energy coming formally from a nontrivial trajectory ("bounce") in the inverted potential. Any of these methods, or many others, give the standard textbook result for the value of B at T=0:

$$B_{T=0} = 2\int_0^{TP} (2mV(x)/\hbar^2)^{1/2}dx \tag{4}$$

where "TP" indicates the classical turning point, i.e. the point on the outside of the barrier at which $V(x) = 0$.

An easy mnemonic for the order of magnitude of B is that it can always be written in the form $CV_0/\hbar\omega_0$ where V_0 is the barrier height, and that for all but pathological forms of the potential the constant C is of order 2π (e.g. for the quadratic-plus-cubic form, which is generic for a system near the classical lability limit in the absence of special symmetries, C is 36/5, while for a quadratic-plus-quartic it is 32/3).

The decay by quantum tunneling of a metastable state at finite temperature ($0 < T \lesssim \hbar\omega_0/k_B$) raises some new points. In general, if we consider the "persistence probability" $P_{in}(t)$ for a system which at $t = 0$ is inside the metastable well and, subject to this constraint, described by a thermal equilibrium density matrix, ($\rho_n = Z^{-1}exp - \beta E_n$ where $\beta \equiv 1/kT$ and Z is the partition function), but apart from this is isolated from its environment, then the possibility of quantum tunneling for $t > 0$ leads to the formula

$$P_{in}(t) = Z^{-1} \sum_n exp - \beta E_n exp - \Gamma_n t \tag{5}$$

where Γ_n is the rate of escape from the n-nth excited state. It is clear that in general (5) cannot be written in the simple exponential form (1), so that the rate $\Gamma(T)$ is not even defined. The usual response to this difficulty is to assume that the characteristic time τ for "repopulation" of levels depleted by tunneling, i.e. for re-establishment of thermal equilibrium within the well, is much shorter than the inverse of the dominant $\Gamma_n^{\prime}s$. In that case, we can legitimately define $\Gamma(T)$ by the prescription

$$\Gamma(T) = \frac{d}{dt} lnP(t) \mid_{t=0} \tag{6}$$

since at each instant we as it were start afresh from a thermal distribution. Under these conditions the decay (when averaged as usual over an appropriate ensemble) is once more exponential (eqn. (1)) with a rate given by

$$\Gamma = Z^{-1} \sum_n \Gamma_n exp - \beta E_n \tag{7}$$

If we formally write $E_n \equiv Re\ E_n + i\Gamma_n$ and assume $\Gamma_n << Re\ E_n \approx E_n$, then this leads to the formal relation, often quoted in the literature,

$$\Gamma(T) = -kT(Im(lnZ)) \equiv ImF(T) \tag{8}$$

where F is the free energy of the system in the metastable well.

A general idea of the way in which $\Gamma(T)$ varies with temperature may be obtained from the following argument starting from eqn. (7): Generally speaking, the formula for the decay rate Γ_n is the obvious generalization of (4), namely

$$\Gamma_n \sim \omega_0 exp - \int_A^B 2(2m(V(x) - E_n))^{1/2}/\hbar dx \tag{9}$$

where A and B are the "classical turning points" for a particle with energy E_n (see fig. 4). It is clear that in general Γ_n will increase with E_n. On the other hand, the thermal population ($\sim exp - \beta E_n$) decreases with increasing E_n. Hence, the tunneling will be dominated by energies in the neighborhood of the solution of the equation

$$\beta E + \int_A^B 2(2m(V(x) - E))^{1/2}/\hbar dx = min., \tag{10}$$

$$i.e. \quad \beta \equiv \frac{1}{kT} = -\frac{\partial}{\partial E}\int_A^B 2(2mV(x)-E)^{1/2}/\hbar dx = 2\int_A^B \frac{dx/\hbar}{(2mV(x)-E)^{1/2}} \tag{11}$$

But the last quantity in (11) is nothing but $\hbar^{-1}$ times the period $\tau(E)$ of a complete classical oscillation ("bounce") of a particle of energy -E in the inverted potential $V_{cl}(x) \equiv -V(x)$; thus the criterion (10) reduces to

$$\tau(E) = \hbar/kT \tag{12}$$

and the finite-temperature decay rate due to quantum tunneling can be written in the WKB form, i.e.

$$\Gamma_{QM}(T) = A'_{QM}\omega_0 exp - B(T) \tag{13}$$

provided that the WKB exponent B(T) is given by the formula

$$B(T) = \beta E + B(E) \tag{14}$$

with $B(E)$ the exponent of (9).

Now, the quantity $\tau(E)$ diverges as $|\ln E|$ in the limit $E \to 0$, and it then follows that for $T \to 0$ the quantity B(T) is given by $B_{T=0}$ (eqn. (4)). Moreover for a nonpathological potential $\tau(E)$ cannot be smaller than $2\pi/\omega_{max}$, where ω_{max} is the frequency of "small oscillations" around the maximum of the actual potential (minimum of the inverted potential), i.e. $(-m^{-1}V''(x)\,|_{V_0})^{1/2}$. Thus, for $T > T_o \equiv \hbar\omega_{max}/2\pi k_B$ eqn. (12) has no solution. The physical interpretation is that above this temperature the dominant escape mechanism is by thermal activation to levels which are already above the top of the barrier, and should therefore be described to a first approximation by the Arrhenius-Kramers formula (2). If then we plot B(T) versus 1/T, the plot should be approximately a straight line of slope V_o/k_B for $T > T_o$, crossing over to the constant value $B_{T=0}$ as $T \to 0$. It is important to emphasize that the shape of the crossover may depend quite strongly on the details of the potential $V(x)$. To see this, let's compare the cases of the standard generic quadratic-plus-cubic potential and the quadratic-plus-quartic (relevant to some magnetic tunneling problems). In each case, an important role is played by the ratio of the Arrhenius-Kramers exponent as $T \to T_0$ from above, which according to the above is $2\pi V_0/\hbar\omega_{max}$, with the zerotemperature WKB exponent (4). For the quadratic-plus-cubic case the latter is (36/5) $V_0/\hbar\omega_0$, and moreover for this potential $\omega_{max} = \omega_0$; thus the ratio of the two exponents is a factor $5\pi/18 \approx 0.87$, which is quite close to 1. Hence we expect that in this case the

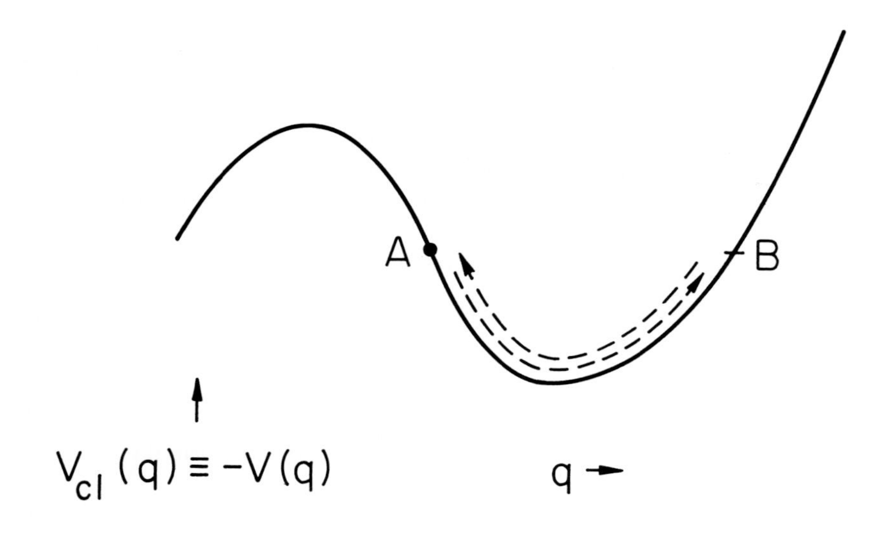

Figure 4: A finite-temperature "bounce"

actual curve of B(T) against 1/T is very sharply kinked at $T \approx T_0$, indicating an almost discontinuous crossover from the AK to the T = 0 WKB regime. On the other hand, for the quadratic-plus-quartic case, the zero-temperature WKB exponent is larger (32/3), and in addition ω_{max} is twice ω_0; thus the ratio is only $3\pi/32 \sim 0.3$, so we expect a rather smooth crossover. The actual details of the crossover region need of course a more quantitative calculation, for which see e.g. ref. (7): in particular, one finds that quantum effects may still be quite appreciable even for $T \sim 2T_o$.

Let's now turn more briefly to the other two problems mentioned above. For the symmetric two-well problem in the absence of coupling to the environment the result is standard: the quantity $< \sigma_z(t) > \equiv P(t)_{\sigma_z(0)=1}$ oscillates at a rate Δ_0, i.e.

$$< \sigma_z(t) > \cong cos\Delta_0 t \tag{15}$$

where the quantity Δ_0 is the splitting between the symmetric and antisymmetric combinations of the groundstates in the two wells. Similarly, the correlation C $(t) \equiv < \sigma_z(t)\sigma_z(0) >$ is given by $cos\Delta_0 t$, so that in this simple (isolated) case the quantities P(t) and C(t) are identical. As to the value of the tunneling splitting Δ_0, it can, again, be calculated by any one of a number of methods; however, the trajectory which occurs in the "instanton" technique is now a genuine instanton, i.e. half of a "bounce", and the formula for the WKB exponent is accordingly halved by comparison with the case of decay from a metastable state. (This is basically because we are now calculating a tunneling amplitude rather than a tunneling probability). In any case, the general expression for the quantity Δ_0 is of the form

$$\Delta_0 = A'\omega_0 exp - \int (2mV(x)/\hbar^2)^{1/2} dx \tag{16}$$

(note no factor of 2 in front of WKB integral!) where the numerical factor A' depends on the details of the potential. For the popular quadratic-plus-quartic form of double-well potential, the WKB exponent may be written in the form $(16/3)(V_0/\hbar\omega_0)$ where V_0 is the height of the barrier.

For an isolated system, finite temperature does not change the zero-temperature formula for P(t) and C(t) provided that $kT << \hbar\omega_0$ (so that none of the excited states within each well are appreciably excited), even though kT may be large compared to the characteristic energy $\hbar\Delta_0$. This is essentially because provided one knows that the system is confined to the 2-dimensional "groundstate manifold spanned by the groundstates in the two wells separately, then the constraint that (e.g.) the system is in the left well at T=0 is sufficient to specify the state completely. (As we shall see below, this feature is qualitatively changed by interaction with an environment). For $kT \geq \hbar\omega_o$ the situation is a good deal more complicated, since one now has to take account of the possibility of tunneling from an excited state in one well to the corresponding state in the other.

Finally, turning to the case of resonant tunneling for an isolated system in an asymmetric double-well potential (fig. 2), it is clear that we should expect the transition rates to show a dramatic peaking whenever a pair of energy levels calculated for the two wells separately "cross", or more precisely when the energy difference between them is comparable to or less than the WKB tunneling amplitude at that energy; then the tunneling process can result in a coherent superposition, just as it does for the symmetric double-well potential. However, it is clear that this problem is somewhat unrealistic since the behavior is likely to be very sensitive to the "width" of one or both levels due to intra-well decay process, and in most real-life cases this is likely to be much larger than the WKB splitting. Thus to make this problem realistic it is essential to take proper account of the coupling to the environment, something which I shall not attempt in these lectures (but cf. the lectures of A. K. Garg and P. C. E. Stamp in this volume).

I now turn, in the context of the metastable-state decay and symmetric two-well problems, to the effects of coupling to a complex environment. Quite generally, we can represent this as a problem of quantum mechanics in a many-dimensional space, with one coordinate (call it q) representing the behavior of the "system" (i.e. the degree of freedom we are primarily interested in) and a collection of other coordinates $\{\xi\}$ describing the behavior of the environment. The Hamiltonian of the system can therefore be written in the generic form

$$\hat{H} = \hat{H}_0(q) + \hat{H}_{env}\{\xi\} + \hat{H}_{int}(q, \{\xi\}) \tag{17}$$

where the dependences on q and ξ include possible dependence on the corresponding conjugate variables (omitted for compactness of notation).

A first remark is that on rather general grounds we would expect the effects of interaction with the environment to be qualitatively different in the context of the metastable-state decay ("tunneling") and two-well ("coherence") problems. In the former, we have simply exchanged a simple one-dimensional potential with a metastable well for a many-dimensional potential. Unless the interaction with the environment is so strong as to render the original metastable minimum unstable, the effect is simply to replace the original well by a region of metastability in the many-dimensional space (see fig. 5)so that the "escape path" no longer runs in general along the q-axis but may be a curved path in this space. Since we are still interested principally in the rate of escape from the metastable region, our problem is not qualitatively changed. Another way of putting it is that in the metastable-state decay problem we are interested only in the probability of penetration of the barrier: the phase of the wave function on the far side is irrelevant, since once it has escaped the particle goes off "to infinity" and never, as it were, returns to interfere with itself.

By contrast, in the two-well problem the very existence of the NH_3-type oscillation depends on the existence of a definite phase relation between the components of the wave function in the two wells; it is very easy to demonstrate that if e.g. at a time $\tau = \pi/2\Delta_0$, when the probabilities of finding the particle in the two wells are equal, one replaces the coherent superposition of the L and R groundstates by an incoherent mixture, this completely destroys the subsequent oscillatory behavior. Now when the system is subjected to interaction with a complex environment, one may think qualitatively of the environment as "observing" (or trying to "observe") the system, a process which will tend according to the standard quantum theory of measurement to convert superpositions to mixtures. Hence one would guess that in general such interactions would tend to destroy the pristine "two-state" behavior. In the language of many-dimensional space, what happens is that rather than tunneling along

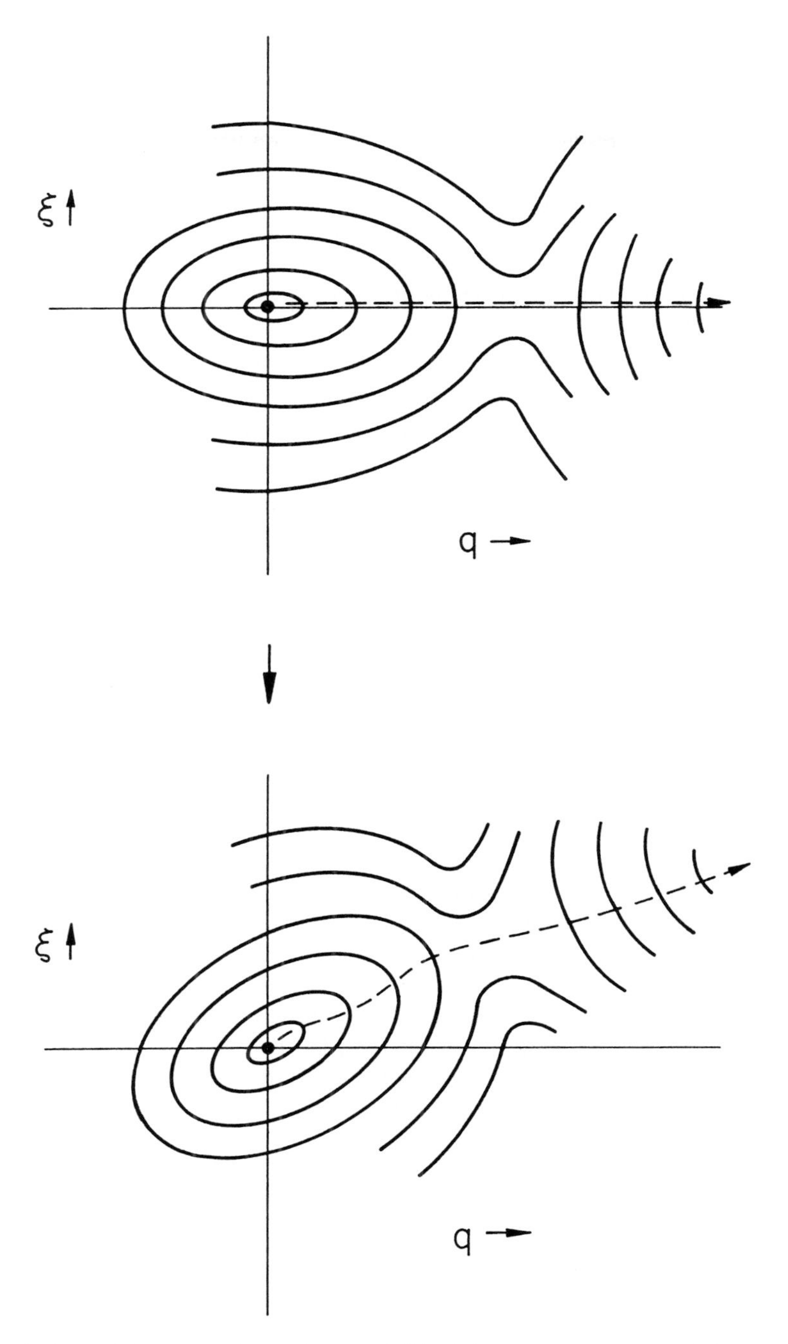

Figure 5: Potential contours (a) without and (b) with system-environment interaction

a single axis (with a single well-defined WKB tunneling amplitude) we now find ourselves tunneling between two rather extensive regions of the many-dimensional space, so that except in special cases the tunneling amplitude is not uniquely defined and the resulting oscillations tend to interfere destructively. Thus one's general prejudice is that while the effect of interaction with the environment on "tunneling" (i.e. metastable-state decay) should be only to change parameters such as $\Gamma(T)$ somewhat, without changing the qualitative aspects of the behavior, in the case of "coherence" (two-well oscillations) such interactions may change the behavior qualitatively. We should see that this prejudice is in fact correct.

The first question to decide is how to describe the interaction with the environment (this question is sometimes phrased as "how to take into account dissipation?", but, actually, the dissipative effects are only a part of the problem). Clearly there exist cases where one has enough information about the relevant "universe" (a phrase I shall use as a short-hand for "system-plus-environment") that one can confidently write down an explicit Hamiltonian of the form (17); this is the case, for example, for an α-particle tunneling out of a heavy nucleus, or a muon moving in a metal. In such cases the degree of excitation of the "environmental" degrees of freedom may be very appreciable, and we cannot necessarily expect to be able to fit the Hamiltonian into any generic model. However, there exists also a class of problems in which the "system" of interest is some macroscopic or mesoscopic degree of freedom, such as the magnetic flux in a SQUID ring, the magnetization of an (anti)ferromagnetic grain or possibly the Higgs field in the early Universe. With the possible exception of the last example, in such a case one usually does not have sufficient microscopic information to write down an explicit Hamiltonian of the form (17) in terms of the "true" microscopic variables. On the other hand, if the variable representing the "system" is indeed macroscopic or mesoscopic, there exist rather generic considerations which render plausible the assumption that any one degree of freedom of the environment will be weakly perturbed by the interaction with the system. In such a case there is a very general argument, which goes back to the classic paper of Feynman and Vernon in 1963 and is set out in detail in appendix C of ref. (8), to the effect that the environment may be modeled by a bath of (in general fictitious) harmonic oscillators, with a system-environment coupling which is linear in the oscillator coordinates and/or momenta. In the case of a general linear dissipative and/or reactive mechanism we can make the stronger statement that by appropriate canonical transformations the interaction can be cast as bilinear in the system and oscillator coordinates (see below).

This generic statement is all very well, but it is of course not much use until we have a means of determining the spectrum of the "oscillator bath" and the coupling of the system to it; and, as we have noticed, in many cases of real-life interest our information about the microscopic details is inadequate to do this. Remarkably, however, it turns out that precisely that combination of parameters which controls the effect of the environment on quantum tunneling and/or coherence also determines its effect on the motion in the "classical" regime, i.e. the (broad) regime of energies, etc., where the predictions of quantum mechanics are well approximated by those of classical mechanics. It is therefore possible to infer some or all aspects of the tunneling behavior directly from the observed behavior in the classically accessible regime, without necessarily having any knowledge of the microscopic details. In the case of a generalized linear dissipative/reactive mechanism, which I now discuss explicitly, it turns out that a knowledge of the classical equation of motion uniquely determines the behavior in the tunneling regime; for the more general case, see ref. (8). The discussion of the next few paragraphs follows ref. (9), to which I refer the reader for details.

Let's imagine, then, that we deal with a system described by a single variable q, such that the classical equation has the generic form

$$\hat{K}q(t) = -\partial V(q)/\partial q \tag{18}$$

where V(q) is an arbitrary (real) function of q (and not of the conjugate momentum p) and $\hat{K}$ is an arbitrary linear integrodifferential operator. In electrical engineering terms, we can think of V(q) as an arbitrary nonlinear inductance ("black box") and $\hat{K}$ as representing an arbitrary linear "shunt" due in part to the environment, with admittance $Y(\omega) \equiv K(\omega)/i\omega$ where $K(\omega)$ is the Fourier transform of the operator $\hat{K}(t)$. We impose three conditions:

(a) small perturbation of any one degree of freedom of the environment.

(b) "strictly linear" coupling, i.e. $\hat{K}_{int}$ linear in p and q.

(c) time-reversal invariance

Condition (a) allow us, in the way first described by Feynman and Vernon and set out in detail in appendix C of ref. (8), to model the environment as a set of linear harmonic oscillators with coordinates ξ_j, momenta π_j, masses m_j and frequencies ω_j, and to expand the interaction up to linear order in the ξ_j and π_j. Thus the most general Hamiltonian we need to consider is of the form

$$\hat{H}(q,p:\xi_j,\pi_j) \quad = \quad (\frac{p^2}{2M_o} + V_o(q)) + \sum_j \left\{ \frac{1}{2}\frac{\pi_j^2}{m_j} + \frac{1}{2}m_j\omega_j^2\xi_j^2 \right\}$$

$$- \sum_j (F_j(pq)\xi_j + G_j(pq)\pi_j) + \Phi(p,q) \quad (19)$$

where the quantity $\Phi(pq)$ (which it may be convenient to include for certain kinds of interaction)may depend on the oscillator parameters m_j, ω_j but not on the ξ_j or π_j. Here the quantities M_o and $V_o(q)$ may not have any direct physical significance. We now apply conditions (b) and (c); the former is not a priori obvious in general, but can be justified from physical considerations in may cases of real-life interest. The effect is to limit $F_j(pq)$ to the form $\kappa_j q$ and G_j to the form $\lambda_j p$, (or vice versa depending on the time-reversal properties of the ξ_j relative to q). One can then demonstrate by using appropriate canonical transformations (see ref.(9)) that the Hamiltonian can be cast in terms of new oscillator variables x_j, p_j to read

$$\begin{aligned} \hat{H}(p,q:x_j,p_j) &= \left(\frac{p^2}{2\tilde{M}} + V(q)\right) + \frac{1}{2}\sum_j \left(\frac{p_j^2}{2m_j} + \frac{1}{2}m_j\omega_j^2 x_j^2\right) \\ &+ q\sum_j C_j x_j + q^2 \sum_j C_j^2/2m_j\omega_j^2 \end{aligned} \quad (20)$$

Here the first two terms describe the "isolated" system, the second two the bath of (redefined) oscillators and the fifth term the system-bath interaction. The last term is the so-called "counter-term", whose function is to ensure that the potential $V(q)$ is the actual physical potential observed in the classical equation of motion (see below).[a] The Hamiltonian (20) is adequate to describe both the classical motion and the quantum tunneling; and the crunch is that there is a single quantity which encapsulates the effect of the environment (oscillators) on both phenomena, namely the spectral density

$$J(\omega) \equiv \frac{\pi}{2}\sum_j (C_j^2/m_j\omega_j)\delta(\omega - \omega_j) \quad (21)$$

In particular, it may be verified by directly calculating the classical equation of motion and eliminating the oscillator coordinates that $J(\omega)$ is related to the (assumed classically known) quantity $K(\omega)$ by

$$J(\omega) = \lim_{\epsilon \to 0+} Im K(\omega - i\epsilon) \quad (22)$$

[a] As eqn (20) stands, the quantity $\tilde{M}$ is not necessarily the "physically observed" mass; however, it would obviously be possible to write an appropriate counterterm to ensure this. The reason this has not usually been done in the literature is historical, in that in the early days the emphasis was mainly on "ohmic" dissipation, for which $\tilde{M} \equiv M$.

Thus, a knowledge of the classical equation of motion permits in principle a complete prediction of the tunneling behavior for any situation of interest.

The general philosophy of the approach outlined above is very similar to that which, in retrospect, justifies nineteenth-century models of the atom as a collection of independent oscillators. Nowadays we of course know very well that the atom is no such thing; nevertheless, provided any one atomic transition is only weakly excited, the results derived from the oscillator model are still valid, and this conclusion is in no way affected by the fact that the absorption of the light wave in question by the gas as a whole may be nearly 100%. In the same way, the above "oscillator-bath" model retains its validity even when the collective effect of the oscillators is to strongly overdamp the classical motion. It should be strongly emphasized that the "oscillators" represented in eqn. (20), like those of the nineteenth-century atomic model, are completely abstract and may have nothing to do with any real phonons or anything similar, even if such exist in the system considered.[b]

Once we have the description (20) of the system-environment interaction, it is a purely technical matter to derive the predictions for any given form of tunneling-related behavior. In the case of metastable-well decay, it turns out that a particularly convenient technique is the instanton one, which is as easy to apply in many dimensions as in one. Omitting the details (see. ref. (9)) we find that the WKB decay rate at T=0 is given by the expression

$$\Gamma_{QM} \;=\; A\, exp - B/\hbar \tag{23}$$

where the WKB exponent B is given by the saddlepoint value of the "imaginary-time" action $S \equiv \int \mathcal{L}(q, \dot{q} : \{x_j, \dot{x}_j\}) d\tau$, and A is a (generally rather unpleasant) fluctuational determinant. On integrating out the oscillator degrees of freedom, one gets an expression for S which involves only the quantity $q(\tau)$ evaluated over the instanton trajectory and its Fourier transform $q(\omega)$:

$$S[q(\tau)] \;=\; \int_{-\infty}^{\infty} V[q(\tau)] d\tau \;+\; \frac{1}{2\pi} \int_{-\infty}^{\infty} \frac{1}{2} K(-i \mid \omega \mid) \mid q(\omega) \mid^2 d\omega \tag{24}$$

where $K(-i \mid \omega \mid)$ is simply the analytic continuation of the linear response $K(\omega)$ into the complex plane. The result (24) may at first seem trivially obvious, given that the instanton technique relies on the substitution $t \to i\tau$, but I know of no convincing demonstration appreciably simpler than the above

[b]Unfortunately, misunderstanding of this crucial point still occasionally leads to claims that the Hamiltonian (20) cannot be a generic description of a linearly responding environment. Subject to the proviso made explicitly above, I believe these claims to be wholly unfounded.

one, and it should be emphasized that (24) in general fails when the conditions specified explicitly above are not met.

Let's consider some simple examples.

(a) "Normal" ohmic dissipation (energy dissipated $\propto \eta \dot{q}^2$). This is a very common case: the tunneling variable is subject to simple linear friction, and the appropriate form of $K(\omega)$ is simply

$$K(\omega) = -M\omega^2 + i\eta\omega. \tag{25}$$

On making the substitution $\omega \longrightarrow -i \mid \omega \mid$, we see that relative to the case $\eta = 0$ we have added to $S[q(\tau)]$ a term

$$\Delta S = \int_{-\infty}^{\infty} \eta \mid q(\omega) \mid^2 d\omega > 0 \tag{26}$$

and thus the saddlepoint value of S (i.e. B) for finite η must be greater than the value for $\eta = 0$. We thus reach the conclusion that "normal" ohmic dissipation always decreases the tunneling rate. In this particular case there is a simple way of understanding this result directly from the Hamiltonian (20), since it is a (special) characteristic of "normal" ohmic dissipation that $\tilde{M}$ is not renormalized and thus is the physical mass: The effect of the interaction terms (including the "counterterm") is to keep both the height and the q-value of the saddlepoint (now in coordinate, not function space!) fixed while shifting it out to finite values of $\{x_j\}$ (see fig. (5)). Thus the many-dimensional WKB integral

$$B \equiv 2 \int \sqrt{2m(V(q), \{x_i\})} ds/\hbar \tag{27}$$

where ds is an element of path length in the many-dimensional space, is increased because the length of path to the saddlepoint (and a fortiori to the outer edge of the barrier) is increased and the T=0 WKB rate correspondingly suppressed. By contrast, the exponent of the classical (AK) thermal activation rate, which is simply the barrier height divided by kT, is not affected by the dissipation–of course a long-known result (and not specific to the simple "normal" ohmic case).

(b) "Anomalous" dissipation. The simplest case of this is when the energy dissipated is proportional to the squared rate of change, not of the tunneling variable q itself but of the conjugate momentum p. Such cases do exist but are rather hard to find: perhaps the simplest is the tunneling of a charged particle coupled to the electromagnetic field. In that case the relevant form of $K(\omega)$ is

$$K(\omega) = -M\omega^2 + i\lambda\omega^3 \tag{28}$$

and the correction to the action S correspondingly has the form

$$\Delta S = - \int_{-\infty}^{\infty} \lambda \mid \omega \mid^3 \mid q(\omega) \mid^2 d\omega < 0 \tag{29}$$

so that the effect of this type of dissipation is to enhance the tunneling rate. Unfortunately, the discussion of this result in terms of the many-dimensional potential is more complicated than in (a), because of the finite mass renormalization.

It is clear that a similar discussion can be given for an arbitrarily complicated form of $K(\omega)$: for examples, see ref.(9).

Let's now turn to the "coherence" (symmetric double-well) problem, again describing it by the generic Hamiltonian (20). We assume that the potential $V(q)$ and the other parameters, are such that the following set of inequalities is satisfied (cf. fig. 3)($\hbar = 1$)

$$\Delta_{oo}, kT \ll \omega_o, \omega_{env} \ll V_o \tag{30}$$

where V_o is the barrier height, ω_o the frequency of small oscillations around the minimum in each well separately, Δ_{oo} the tunneling splitting calculated for the one-dimensional potential $V(q)$ (with the "physical" mass) and ω_{env} the characteristic cutoff frequency of the environment. For simplicity I will assume that for $\omega \ll \omega_{env}$ the spectral density $J(\omega)$ is proportional to $\omega^s, s > 0$, and will refer to the cases $s > 1, s = 1$, and $s < 1$ respectively as "superohmic", "ohmic" and "subohmic". The well minima are taken to lie at $\pm q_o/2$. The ensuing discussion mainly follows ref. (10), to which the reader is referred for details.

The basic physics is that the main effect of the oscillators with frequencies in the range $\omega \gtrsim \omega_o$ is to affect the actual process of transmission through the barrier, while those in the regime $\omega \lesssim \Delta_{oo}$ mostly "detune" the states in the two wells (i.e. shift the groundstate energies in a fluctuating way). Thus we choose an effective cutoff frequency ω_c such that $\Delta_{oo} \ll \omega_c \ll (\omega_{env}, \omega_o)$ and integrate out of the action S the oscillators with frequencies $> \omega_c$. In this way we find for the action of a single instanton, in which the system is regarded as moving from one well to another in imaginary time, the expression

$$S_{inst} = \int_{-\infty}^{\infty} (\frac{1}{2} M\dot{q}^2 + V(q))d\tau + \frac{1}{2} \int_{-\infty}^{\infty} d\tau \int_{-\infty}^{\infty} d\tau' \alpha(\tau - \tau') \left(q(\tau) - q(\tau')\right)^2 \tag{31}$$

where $\alpha(\tau) \equiv \frac{1}{2\pi} \int_{\omega_c}^{\infty} J'(\omega) exp - \omega \mid \tau \mid d\tau$, and the tunneling splitting is given by an expression of the form

$$\Delta_o = \text{const.}\omega_o exp - S_{inst}/\hbar \tag{32}$$

Now an"instanton" trajectory (as distinct from a "bounce"!) by definition starts at $q = +q_o/2$ at $\tau = -\infty$, and finishes at $q = -q_o/2$ at $\tau = +\infty$; thus, the quantity $q(\tau) - q(\tau')$ is finite as $\mid \tau - \tau' \mid \rightarrow \infty$ (in fact equal to q_o), and thus, if $\alpha(\tau)$ falls off as τ^{-2} or more slowly for large τ, then the second term in (31) diverges. Thus, we can afford to take $\omega_c \rightarrow 0$ only in the "superohmic" case ($s > 1$) ; for the ohmic ($s = 1$) or subohmic cases we must keep it finite. In the ohmic case we then get

$$\frac{S}{\hbar} = \frac{S_o}{\hbar} + \frac{\eta q_o^2}{2\pi\hbar} \, ln \, \left(\frac{\omega_o}{\omega_c}\right) \tag{33}$$

from which

$$\Delta_o \sim \Delta_{oo} \left(\frac{\omega_c}{\omega_o}\right)^{\eta q_o^2/2\pi\hbar} \tag{34}$$

After carrying out the above renormalization procedure, we can approximate the problem by a so-called "spin-boson" model, in which in effect the Hilbert space of the system is truncated to the 2-dimensional "groundstate" manifold. The resulting Hamiltonian is

$$\hat{H} = -\frac{1}{2}\hbar\Delta_o\hat{\sigma}_x + \frac{1}{2}q_o\sigma_z \sum_j C_j x_j + \sum_j \left(\frac{p_j^2}{2m_j} + \frac{1}{2}m_j\omega_j^2 x_j^2\right) \tag{35}$$

with a spectral density $J_{SB}(\omega)$ (different by constant factors from the $J(\omega)$ defined above) which now comes only from the low-frequency ($\omega < \omega_c$) oscillators, since the high-frequency cases have already been taken care of:

$$J_{SB}(\omega) \equiv \frac{\pi}{2} \sum_j \left(C_j^2/m_j\omega_j\right) \delta(\omega - \omega_j) \sim A\omega^s \theta(\omega_c - \omega) \tag{36}$$

(Actually, for technical reasons it is usually more convenient to implement the cutoff procedure by an exponential smoothing factor rather than a θ-function so that the $\theta(\omega_c - \omega)$ in (36) is replaced by exp-ω/ω_c).

In the special case of "ohmic" dissipation (s=1) it turns out that $J_{SB}(\omega)$ is equal to $(\eta q_o^2/2\pi\hbar)\omega$, so it is convenient to define the dimensionless factor

$$\alpha \equiv \eta q_o^2/2\pi\hbar \tag{37}$$

and then we have (see (34))

$$\Delta_o = const.\Delta_{oo} \left(\frac{\omega_c}{\omega_o}\right)^\alpha \equiv \Delta(\omega_c) \tag{38}$$

It is convenient use the notation Δ_o, (not to be confused with Δ_{oo});we must remember that $\Delta_o = f(\omega_c)$.

The general behavior of a system described by the spin-boson Hamiltonian (35) is determined by the competition between the term in Δ_o, which tends to diagonalize $\hat{\sigma}_x$, and the interaction with the environment (second term) which tends to diagonalize the incompatible variable $\hat{\sigma}_z$. To see which wins for a particular value of s we can adopt a "poor man's renormalization group" approach: we progressively reduce ω_c to $\omega'(\Delta)$, where $\Delta \ll \omega_c' \ll \omega_c$, and eliminate the oscillations in the range $\omega_c' < \omega < \omega_c$ by an adiabatic technique. This reduces Δ by a Franck-Condon factor:

$$\Delta(\omega_c) \rightarrow \Delta(\omega_c') \equiv \Delta(\omega_c) exp - \int_{\omega_c'(\Delta)}^{\omega_c} \frac{J(\omega)}{\omega^2} d\omega \tag{39}$$

We consider the result of iterating this process. For $s > 1$ the correction is finite, so Δ iterates to a finite value and we cannot continue the process indefinitely. This is consistent with the hypothesis that the interaction with the bath renormalizes the tunneling splitting but does not depress coherent tunneling qualitatively. In the subohmic case $(s < 1)\Delta$ iterates to zero, implying that there is no tunneling at zero temperature. What about the ohmic case (s=1)? In that case it is tempting to seek a self-consistant solution to the problem by putting $\omega_c'(\Delta) \sim \Delta$. If we do so, we find for $\alpha > 1$ no solution other than the trivial one $\Delta = 0$, while for $\alpha < 1$ we get

$$\Delta = \Delta_o \left(\frac{\Delta_o}{\omega_c}\right)^{\frac{\alpha}{1-\alpha}} \tag{40}$$

Further insight into the competition between tunneling and localization is given by a variational ansatz due to Silbey and Harris (11). Quite generally, we can write the many-body groundstate wave function in the $(\sigma_z, \{x_i\})$ basis in the form

$$\phi_o \equiv \frac{1}{\sqrt{2}} \{|\uparrow>| \chi_\uparrow\{x_i\} > + |\downarrow>| \chi \downarrow \{x_i\} >\} \tag{41}$$

where $|\uparrow>, |\downarrow>$ represent the eigenstates of $\hat{\sigma}_z$ with eigenvalue $+1, -1$ respectively. For the (generic) state (41), the expectation value of the last three terms in (35) (call them collectively $\hat{H}_{solv}$) is

$$< \hat{H}_{solv} >= \frac{1}{2} < \chi_{\uparrow} \mid \hat{H}_{osc} \mid \chi_{\uparrow} > \; + (|\uparrow>\rightarrow|\downarrow>) \tag{42a}$$

while the expectation value of the tunneling (first) term is

$$< \hat{H}_{tun} >= -\frac{1}{2}\hbar\Delta_o < \chi_{\uparrow} \mid \chi_{\downarrow} > \tag{42b}$$

the term $< \chi_{\uparrow} \mid \chi_{\downarrow} >$ being the standard Franck-Condon factor. Eqns. (41-42) are generically true. Now, following (11), we make the specific ansatz

$$\chi_{\uparrow,\downarrow} = exp \pm i \sum_i \lambda_i \hat{p}_i \mid 0 > \tag{43}$$

where $\mid 0 >$ represents the oscillator groundstate, and treat the coefficients λ_i (which represent the amount by which the i-th oscillator wave function is displaced) as variational parameters. Substituting (43) into (42)and varying with respect to λ_i, we find

$$\lambda_i = \frac{\frac{1}{2} q_o C_i}{m_i \omega_i (\omega_i + \Delta)} \tag{44a}$$

where Δ is defined by

$$\Delta \equiv \Delta_o exp - \sum_i m_i \omega_i \lambda_i^2 \tag{44b}$$

Combination of eqns. (44a-b)yields a self-consistent equation for Δ:

$$\Delta = \Delta_o exp - \int_o^{\omega_c} \frac{J_{SB}(\omega)}{(\omega + \Delta)^2} d\omega \tag{45}$$

Let's discuss the solution of this equation for the three cases $s > 1$, $s < 1$, and $s = 1$.

(a) $s > 1$ ("superohmic" case): The integral in the exponent of RHS of eqn. (45) is finite and approximately independent of Δ. Thus we find

$$\Delta \cong \Delta_o exp - F, \quad F \equiv \int_o^{\omega_c} \frac{J(\omega)}{\omega^2} dw \tag{46}$$

It may be seen that $< \sigma_x >=< \chi_{\uparrow} \mid \chi_{\downarrow} >= e^{-F}$ is finite in the groundstate, indicating the coherence of the noninteracting groundstate is not completely lost. The low-frequency oscillators ($\omega \lesssim \Delta$) are not much displaced. (b) $s < 1$ ("subohmic" case): The integral is (nearly) independent of ω_c but proportional to $\Delta^{-(1-s)}$. Hence the self-consistant equation (45) becomes

$$ln(\Delta/\Delta_o) = -b\Delta^{-(1-s)} \tag{47}$$

where b is the coefficient of ω^s in $J_{SB}(\omega)$. A sufficient condition for no finite-Δ solution to exist is

$$(1-s)b\Delta_o^{-(1-s)} > 1 \tag{48}$$

so that for fixed b there is no solution in the limit $\Delta_o \to 0$. Thus in this case $\Delta = 0$, the quantity $< \sigma_x >$ is zero in the groundstate and all oscillators are fully "solvated" (that is, they minimize $< H_{osc} >$, ignoring the tunneling energy).
(c) s=1 ("ohmic" case): We recover the result (40) obtained above by a heuristic argument, namely

$$\Delta = \Delta_o \left(\frac{\Delta_o}{\omega_c}\right)^{\frac{\alpha}{1-\alpha}}, \alpha < 1: \ = 0, \alpha > 1. \tag{49}$$

Thus for $\alpha > 1$ the situation is qualitatively similar to the subohmic case, while for $\alpha < 1$ it is qualitatively similar to the superohmic case; in this latter case the low-frequency oscillators are <u>not</u> fully solvated.

The qualitative conclusion, then, is that for $s < 1$, and also for $s = 1, \alpha > 1$, the groundstate in the presence of finite Δ_o is identical to the groundstate with no tunneling at all, i.e. it is doubly degenerate with the two states corresponding to localization within one or the other well. For $s > 1$ and for $s = 1, \alpha < 1$, on the other hand, the groundstate is qualitatively more like that of the uncoupled system with finite Δ_o, i.e. it is nondegenerate and possesses a finite value of $< \sigma_x >$. Thus, (in the limit $\omega_c/\Delta_o \to \infty$) the point $\alpha = 1, s = 1$ is a discriminant between two qualitatively different types of behavior. This result can be shown to be rigorous (see eg. ref. (12)).

One would think that these two different types of behavior would be reflected also in the dynamics of the SB problem, and indeed this turns out to be true. However, there are considerable extra subtleties in this case and a detailed calculation is necessary: for reasons of space I refer the reader to ref. (10), and simply summarize the results for the quantity P(t),[c] (which, recall, is the value of $< \sigma_z(t) >$ conditional on $\sigma_z(t < 0)$ being $+1$) as follows:

---

[c]Contrary to a conjecture made in ref. (10), the behavior of the correlation function C(t), though in general similar to that of P(t), shows qualitative differences over a small region of the ohmic parameter space: see e.g. ref. (13).

(1) For $s > 1$ the system performs coherent oscillations at $T = 0$, and at for at least a finite range of T (for $1 < s < 2$ there is a crossover to overdamped behavior at finite $T \lesssim \omega_c$)

(2) For $s < 1$, and for $s = 1, \alpha > 1$, the system is completely localized at $T = 0$, and at finite T undergoes incoherent transitions as a rate $\propto$ const. exp$-(T_o/T)^{1-s}$ (or $T^{2\alpha-1}$ in the case $s = 1$).

(3) For $s = 1, \alpha < 1$ the situation depends critically on the dimensionless parameters α and $\alpha kT/\Delta(\Delta \equiv \Delta_o \left(\frac{\Delta_o}{\omega_c}\right)^{\frac{\alpha}{1-\alpha}}$. For $\alpha > 1/2$, the behavior is an incoherent hopping which remains finite as $T \to 0$. For $\alpha < 1/2$, the behavior at $T = 0$ is a coherent oscillation which becomes overdamped as $\alpha \to 1/2$: for fixed α a transition to overdamping occurs when $\alpha kT \sim \Delta$. For $\alpha kT \gg \Delta$ and any value of α the behavior is incoherent hopping at a rate $\propto T^{2\alpha-1}$: note that this gives an increase in the hopping rate with T for $\alpha > 1/2$ but a decrease for $\alpha < 1/2$!

The second part of these lectures deals with some aspects of a special manifestation of quantum tunneling, the Josephson effect, as it is likely to be realized in the new system of Bose-condensed (BEC) atomic alkali gases. Originally predicted theoretically for superconductors in 1962, the Josephson effect was observed there soon afterwards by Anderson and Rowell (1963). An "internal" Josephson effect was observed in superfluid ^3He by Osheroff (1973) and the "standard" Josephson effect both there and in $^4He - II$ by Avenel and Varoquaux in the mid-eighties. There has been no report to date of any observation in the BEC alkali gases, but this is mainly due to difficulties in stabilizing the barrier which are likely to be overcome in the next few years.

In the original system, superconducting metals, the effect is usually observed by monitoring the d.c. or a.c. current-voltage characteristics of a single junction subjected to a controlled external current. However, it is conceptually simpler to consider a geometry which is in fact close to that likely to be realized in the BEC alkali gases, namely two bulk regions connected by a small barrier or pinhole, the whole being isolated from the external world (see fig. 6).

Crudely and schematically, the wave position of a Bose-centered system (or the "pseudo-Bose-condensed" system of Cooper pairs in the Fermi case) may be written

$$\Psi \sim [\psi(r)]^N \tag{50}$$

where $\underline{r}$ denotes the coordinate of a single atom (in the Bose case) or the COM coordinate of a Cooper pair (in the Fermi case). Suppose that in the above

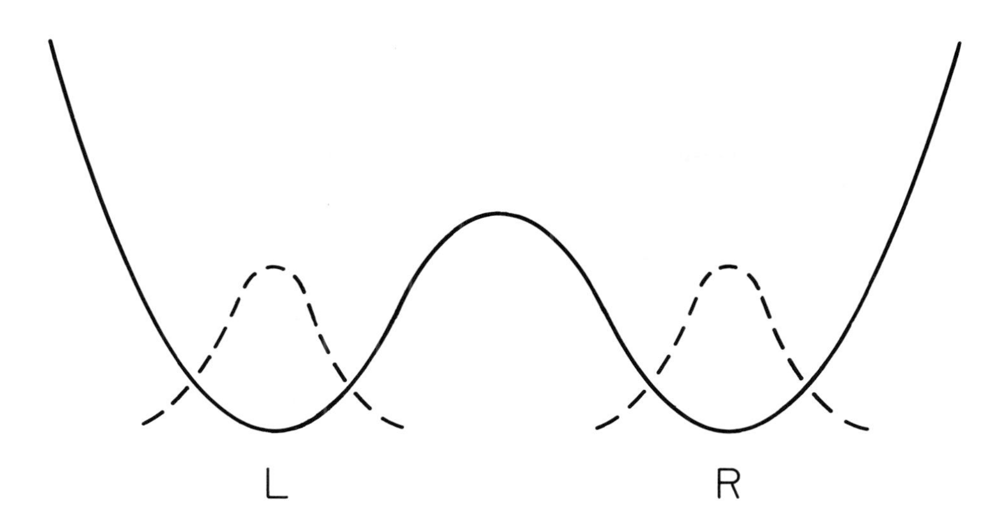

Figure 6: Schematic set-up for observation of the Josephson effect in a BEC alkali gas

geometry we put

$$\Psi(r) \sim \mid a \mid \psi_L(r) + \mid b \mid \psi_R(r) exp\ (i\Delta\phi) \tag{51}$$

where $\psi_L(r)$ and $\psi_R(r)$ are localized on the L and R bulk regions respectively, and ask how the wave function inside the junction (pinhole) behaves as we crank up the phase difference $\Delta\phi$ from zero. If the junction is "strong", then even for $\Delta\phi \gg 2\pi$ the wave function will maintain a finite value throughout the barrier region and develop a "winding" of the phase to give the necessary total $\Delta\phi$. If on the other hand it is sufficiently "weak", it will be advantageous, as $\Delta\phi$ is increased from zero, to decrease the amplitude of the wave function in the barrier region so as to decrease the kinetic energy, and eventually, when $\Delta\phi = \pi$, a node will occur in the middle of the junction region; as $\Delta\phi$ is then further raised from π to 2π, the wave function will recover to its $\Delta\phi = 0$ form (so that the cases $\Delta\phi = 2\pi$ and $\Delta\phi = 0$ are experimentally indistinguishable). I shall take the Josephson effect to be defined by the occurrence of a phase slip of 2π on every cycle, as above: it then follows that the energy associated with the junction must be a single-valued function of ϕ with periodicity 2π, and time-reversal invariance then implies that it must be a single-valued function of $\cos\Delta\phi$. In most cases of practical interest this function is well approximated by its lowest-order term, i.e. we have (though see below)

$$E_{junct}(cos\Delta\phi) = -E_J cos\Delta\phi \tag{52}$$

wehre E_J is by definition the Josephson coupling energy.

From now on we neglect the explicit variation of the wave function within the junction itself and write the many-body state schematically as in (50) but with a slightly revised notation:

$$\Psi = (ae^{+i\Delta\phi/2}\psi_L + be^{-i\Delta\phi/2}\psi_R)^N$$
$$(a, b, \text{real and positive}, \ | a |^2 + | b |^2 = 1) \tag{53}$$

If we define the operator

$$\Delta\hat{N} = \frac{1}{2}(\hat{N}_L - \hat{N}_R) \tag{54}$$

(the factor of 1/2 is a matter of convenience) there it is straightforward to show that within the manifold spanned by (53) we have

$$i\frac{\partial}{\partial\Delta\hat{\phi}} = \Delta\hat{N}, \tag{55}$$

i.e. $\Delta\hat{\phi}$ and $\Delta\hat{N}$ are canonically conjugate variables:

$$[\Delta\hat{\phi}, \Delta\hat{N}] = -i \tag{56}$$

Let's now consider the relevant energies associated with $\Delta\phi$ and ΔN respectively. First, let's reconsider the Josephson coupling energy, which is essentially the kinetic energy arising from the need to "bend" the wave function in the junction region. We can express this in terms of the splitting $\Delta E \equiv E_+ - E_-$ of the even- and odd-parity single-particle states: if we approximate $\psi_L(\psi_R)$by $(\psi_+ + \psi_-)/\sqrt{2}$ $((\psi_+ - \psi_-)/\sqrt{2})$, then we can write the general single-particle wave function in the form

$$\Psi = \frac{1}{\sqrt{2}}\left(\left(ae^{+i\Delta\phi/2} + be^{-i\Delta\phi/2}\right)\psi_+ + \left(ae^{i\Delta\phi/2} - be^{-i\Delta\phi/2}\right)\psi_-\right) \tag{57}$$

and then the expectation value of the tunneling energy is clearly

$$< E_{tun}(\Delta\phi) >= const. - ab \cdot \Delta E \cdot cos\Delta\phi \tag{58}$$

which apart from a constant can be written (since $N_R \sim | a |^2 .N_L \sim | b |^2$)

$$< E_{tun}(\Delta\phi) >= -E_J(\Delta N) cos\Delta\phi \tag{59}$$

$$E_J = \Delta E \left(1 - \left(\frac{\Delta N}{N}\right)^2\right)^{1/2} \tag{60}$$

If this were all, the application to (59) of the commutation relation (56) would lead to the standard equation of motion of the simple 2-state problem.

However, a qualitatively new aspect of the many-particle problem by comparision with the single-particle one is the fact that in general there will be another term depending on ΔN, because if one starts with $< N_L >=< N_R >$ and displaces particles from (say) L to R it costs a finite energy due to the non-infinite compressibility of the bulk systems L and R. Thus we can write a "polarization" energy in the form

$$E(\Delta\hat{N}) = \frac{1}{2}\frac{(\Delta\hat{N})^2}{\chi}, \quad \chi \equiv \left[\frac{\partial(\Delta\mu)}{\partial(\Delta N)}\right]^{-1} \tag{61}$$

Which dependence on $\Delta\hat{N}$ dominates, that from (60) or that from (61)? Expanding (60) for small ΔN, we see that the ratio of the coefficients of $\Delta\hat{N}^2$ in (61) so that in (60) is $N^2/(\chi \cdot \Delta E)$. Now, typically, because of the normalization of $\psi(r)$, the quantity ΔE is proportional to V^{-1}, while χ is proportional to V. Thus the ratio is proportional to N^2 and tends to ∞ in the thermodynamic limit. More quantitatively, e.g. for ^{4}He with χ due to gravitational effects, we have $\chi \sim V\rho/mgH$, $\Delta E \sim (\hbar/md^2)(da^2/V)$, where d and a are respectively the length and width of the aperture and H is the height of the free surface of the He; thus the term from (60) dominates as soon as $N_{bulk}/N_{junc} >> [(\hbar^2/md^2)/(mgH)]^{1/2} \lesssim 10$. We conclude that except in the very special and unusual case that $\Delta\hat{N}$ is very close to $\hat{N}$,[d] the dependence of the energy though (60) is negligible compared to that arising from the polarization term (61). Thus we may simply put, in (60), $E_J(\Delta N) = \Delta E \equiv E_J$, and arrive at the Hamiltonian

$$\hat{H} = -E_J \cos\Delta\hat{\phi} + \frac{1}{2}\frac{\Delta\hat{N}^2}{\chi} \tag{62}$$

with the commutation relation

$$[\Delta\hat{\phi}, \Delta\hat{N}] = -i \tag{63}$$

[d]We are actually interested in $\partial E/\partial\Delta N$ rather than $E(\Delta N)$ itself (cf. below), and the contribution to the former from (60) diverges as $\Delta N \to N$.

But the problem defined by eqns. (62) and (63) is simply that of the simple quantum pendulum! ($\Delta\hat{N} \to \hat{L}, \chi \to I'$ (moment of inertia), $E_J \to mgl$). If an "external" potential $\Delta\mu$ is applied, the effect is simply to add to $\hat{H}$ a term of the form $\Delta\hat{N} \cdot \Delta\mu$, which is the analog of an external "vector potential" on the pendulum.

By using the equivalence to a pendulum we can rapidly obtain a number of interesting results:
(a) semiclassical dynamics: the equations of motion are

$$\Delta\dot{N} \equiv I = \frac{\partial H}{\partial(\Delta\phi)} = E_J \sin\Delta\phi \tag{64}$$

$$\Delta\dot{\phi} = -\frac{\partial E}{\partial(\Delta N)} = -\left(\frac{\Delta N}{\chi} + \Delta\mu\right) \tag{65}$$

where I is the current through the junction. Eqns. (64-5) are known as the Josephson equations: note that $I \leq I_c \equiv E_J$ (in our units). The small oscillations around the stable equilibrium position ($\Delta\phi = 0$) have the frequency

$$\omega_J^2 \equiv \frac{E_J}{\chi} \tag{66}$$

which is usually called the Josephson plasma frequency.
(b) zero-point spread: assuming that $\Delta\phi << 2\pi$ we have approximately

$$\frac{1}{2}E_J\overline{\Delta\phi^2} = \frac{1}{2\chi}\overline{\Delta N^2} = \frac{1}{4}\hbar\omega_J \tag{67}$$

so

$$\overline{\Delta\phi^2} \sim \frac{1}{\overline{\Delta N^2}} \sim (E_J\chi)^{-1/2} \tag{68}$$

Typically, $E_J\chi$ is large compared to 1 (but small compared to N^2). Thus we have

$$\overline{\Delta\phi^2} << 1,\ 1 << \overline{\Delta N^2} << N^2 \tag{69}$$

However, as we shall see below, it may be possible to violate these inequalities for a BEC alkali gas.
(c) effect of leads with external current I: the usual belief is that this can be handled by adding a term to the Hamiltonian of the form

$$\delta\hat{H} = -I\Delta\hat{\phi} \tag{70}$$

corresponding to a finite torque in the pendulum analogy. This certainly gives the right equations of motion in the classical limit, but in the fully quantum case it gives rise to some subtle conceptual problems which there is no space to discuss here (it is unlikely to be directly relevant to the BEC alkali-gas problem). If indeed the prescription (70) is valid, then for $I < I_c$ we get the possibility of tunnelling in the "washboard potential" from one metastable potential well to another. This genuinely "macroscopic" quantum tunnelling should be carefully distinguished from the Josephson effect itself (cf. ref. (14); it has been the subject of extensive experimental investigation in the superconducting case.

I close this introduction by noting that in superfluid ^3He one encounters a phenomenon which can be regarded as an "internal" Josephson effect, namely the longitudinal nuclear magnetic resonance. If, for simplicity, we consider the A phase, the usual description of the state of the Cooper pairs is

$$\Psi \sim \sqrt{2}(e^{i\Delta\phi/2}\,|\uparrow\uparrow\rangle + e^{-i\Delta\phi/2}\,|\downarrow\downarrow\rangle)^{N/2} \tag{71}$$

where $(\uparrow\uparrow\rangle)$ indicates that both spins are up, etc. The relevant energies are the nuclear dipole interaction energy, which turns out to have the form

$$E_{dip} \sim -g_D \cos\Delta\phi \tag{72}$$

and the "polarization" energy

$$E_{pol} \sim \frac{1}{2}\frac{(\Delta\hat{N})^2}{\chi} \tag{73}$$

where ΔN is (half) the excess of up spins over down spins and χ is the usual Pauli susceptibility. Since g_D and χ both scale as N in the thermodynamic limit, there is an almost exact analogy to the usual "geometrical" Josephson effect. For normal sample sizes we find that $g_D\chi$ is large compared to 1, so that the inequalities (69) are well fulfilled; however, in the future it may be possible to devise experimental situations in which this is no longer so, in which case we should expect the NMR behavior to be interestingly different.

I now turn to the specific system of BEC alkali gases. It is by now well known that, despite the fact that two atoms of any alkali element other than H, even when polarized with parallel spins, are unstable against recombination into a molecule, the kinetics of this process at low densities is so slow that the monatomic gas is effectively stable over the timescale (typically $\sim$ 1 min) of the relevant experiments. If the nuclear spin of the isotope in question is half-odd-integral (as in the case for most stable alkali isotopes) then the atoms as wholes behave as bosons, and in free space would be expected to undergo Bose condensation when the product of the density n and the cube of the de Broglie wavelength λ_{DB} satisfy the condition

$$n\lambda_{DB}^3 = 2.612. \qquad (74)$$

It is, of course, generally believed that the experimentally observed λ-transition of liquid ^4He is an example of Bose condensation, and that the superfluid properties of the liquid in the He-II phase below the transition are a consequence of this. However, <u>*direct*</u> and unambiguous verification of the existence of Bose condensation in this system has some proved elusive. The BEC alkali gases differ from liquid ^4He in a number of respects, in particular in that they are extremely dilute and hence compressible, so that the occurrence of BEC has spectacular and easily detectable consequences.

In practice the systems in question are confined by a combination of magnetic fields and (sometimes) laser-induced potentials; although the direction of the magnetic field is not strictly constant in space, the spin direction of the atoms usually follows it sufficiently adiabatically that it is a good approximation to treat them as moving in a <u>*scalar*</u> potential which, moreover, usually has the form appropriate to an anisotropic harmonic oscillator. In such a potential it turns out that the appropriate criterion for Bose condensation is that the condition (74) is met at the center of the well (where the density is greatest); for currently attained densities (typically 10^{12}-10^{15} cm^{-3}) the required temperatures lie in the range of 100 nK to a few μK, and over the last two years these temperatures have been achieved, with the help of spectacular advances in laser cooling, by a number of groups.

When BEC occurs for a system in a harmonic trap, a spectacular change in the density distribution is predicted. For a totally noninteracting gas, all the condensed atoms would pile up in the quantum harmonic-oscillator ground state, which is typically a factor $\sim 10-100$ narrower than the width of the thermal distribution at T_c, and thus would produce a very sharp spike in the density profile on the background of the much wider thermal distribution. In a gas with repulsive inter-particle interactions[e] this effect is modified but not totally suppressed: in fact, under the conditions usually realized to date ("Thomas-Fermi limit") it turns out that the condensate profile is determined at T=0 by the condition that the chemical potential is constant in space, i.e. by the condition

$$\frac{1}{2}m\omega_0^2r^2 + U_0\rho(r) = \mu = const., \quad U_0 \equiv \frac{4\pi\hbar^2 a_s}{m} \qquad (75)$$

[e]The case of an attractive interparticle interaction is more complicated and needs separate discussion.

so that $\rho(r) = U_0^{-1}(\mu - (1/2)m\omega^2 r^2)$ (a_s is the s-wave scattering length). The value of μ, and hence the radius R_c at which the density of the condensate tends to zero, is determined by the condition $\int \rho(r)dr = N$, the total number of particles; in the "Thomas-Fermi limit" where (75) is justified we find

$$R_c \sim \zeta a_\perp, \quad \zeta \equiv (8\pi N a_s/a_\perp)^{1/5} \tag{76}$$

where $a_\perp$ is the extent of the harmonic-oscillator ground state (for simplicity I assume the isotropic case). Typically we have $N \sim 10^3 - 10^7$, $a_s/a_\perp \sim 10^{-3} - 10^{-2}$, so $\zeta \sim 2 - 10$. Thus the width of the condensate cloud, though larger than the width of the quantum ground state, is usually still small compared to that of the thermal distribution at T_c.

A point which needs to be borne in mind in the following is that the density of excited states, even in the presence of interactions, is not very different from that of the noninteracting gas in the harmonic potential, and in particular in three dimensions is proportional to $E^2/(\hbar\omega_0)^3$.

In order to observe the Josephson effect in a BEC alkali gas, one would need to create a double-well potential with a barrier height low enough to permit appreciable tunnelling through it of single atoms. Such double-well potentials have in fact already been realized with the help of intense laser beams, but the barriers used to date have been too high to permit appreciable Josephson effects. However, there seems no objection in principle to realizing an appropriate geometry.

The most obvious way to treat the Josephson effect in a BEC alkali gas is simply to take over the general scheme sketched above. That is, we define the self-consistent ground states $\psi_L(r), \psi_R(r)$ in the L and R wells respectively, write (at $T = 0$) the general many-body wave function in the general form

$$\begin{gathered}\Psi = (ae^{i\Delta\phi/2}\psi_L(r) + be^{-i\Delta\phi/2}\psi_R(r))^N \\ (a, b \text{ real and positive}, |a|^2 + |b|^2 = 1)\end{gathered} \tag{77}$$

and proceed as above. Unfortunately there are some snags, mostly connected with the fact that the self-consistently determined ground state wave functions $\psi_{L,R}(r)$ are in general functions of the <u>number</u> of particles N_L, N_R in the respective wells (a complication which we implicitly assumed was absent in the simple cases discussed above). As a result, for finite ΔN there is no symmetry principle which guarantees that ψ_L and ψ_R will be precisely orthogonal (or can easily be made so). I believe that the corrections due to this are of order e^{-B} relative to the ones kept (where B is the WKB exponent, cf. below). The fact that $\psi_{L,R}$ is itself a function of $N_{L,R}$ means that our argument to the effect that ΔN and $\Delta\phi$ are conjugate variables does not go through in its simple form, but I believe that the corrections due to this are likely to be small provided the relevant values of ΔN are small compared to N itself. Thus, with appropriate caution we write the MBWF in the form (77) and assume the standard commutation relation

$$[\Delta N, \Delta\phi] = -i. \tag{78}$$

We need now to calculate the two principal energies which depend on ΔN and $\Delta\phi$, namely the "polarization" and "tunnelling" energies. As to the former, the general expression for it is

$$E(\Delta N) = E_L\left(\frac{N+\Delta N}{2}\right) + E_R\left(\frac{N-\Delta N}{2}\right) \tag{79}$$

Since for an individual well containing N particles the energy is proportional to $N^{7/5}$, the resulting expression is quite awkward. However, for small ΔN we have a simple expansion:

$$E \cong \frac{1}{2}\frac{(\Delta N)^2}{\chi} + 0(\Delta N)^4, \quad \chi^{-1} \sim \frac{\zeta^2}{N}\hbar\omega_0 \tag{80}$$

where ζ is the dimensionless quantity defined in (76).

The calculation of the Josephson energy for a realistic potential shape is a bit more complicated. For the moment suppose that ΔN is close to zero, i.e. the condensate wave function is of the simple form $2^{-1/2}(e^{i\Delta\phi/2}\psi_L(r) + e^{-i\Delta\phi/2}\psi_R(r))$. The demonstration that the Josephson energy is proportional to $\cos\Delta\phi$ is not totally trivial,(15) but if we <u>*assume*</u> that it is we can find or bound the coefficient as follows: In the 1D case we write in the junction region for <u>*small*</u> $\Delta\phi$

$$\psi(x) = \psi_0(x) exp i\phi(x) \tag{81}$$

where $\psi_0(x)$ is the wave function corresponding to $\Delta\phi = 0$, and $\phi(x)$ satisfies the condition $\phi_A - \phi_B = \Delta\phi$ with A and B points well into the "accessible" regions L and R respectively. It is clear that the multiplication by the phase factor does not change either the external or the interparticle potential energy, so to find the total energy charge due to a finite $\Delta\phi$ we need only minimize the kinetic energy subject to the above boundary condition; it is clear that to lowest order in $\Delta\phi^2$ it does not help to modify the amplitude $\psi_0(x)$, so the ansatz (81) is justified. This gives the result that for $\Delta\phi \to 0$ the <u>*single − particle*</u> energy depends on $\Delta\phi$ as

$$E(\Delta\phi) - E(\Delta\phi = 0) \cong \frac{1}{2}(\Delta\phi)^2 \left[\int \frac{dx}{\rho_1(x)}\right]^{-1}, \rho_1(x) \equiv |\psi_0(x)|^2 \qquad (82)$$

It is clear, given our assumption, that this is just the leading term in the expansion of the cosine, so we find for general $\Delta\phi$, apart from a constant,

$$E(\Delta\phi) = -E_J cos\Delta\phi, \quad E_J = \left[\int \frac{dx}{\rho(x)}\right]^{-1} \qquad (83)$$

where $\rho(x)$ is the <u>total</u> density. The formula (82) is consistent, for a noninteracting system, with the usual WKB formula for the splitting of the even-and-odd-parity ground state combinations; however, it would make little sense to use an argument based on this idea here since (unlike in the simple cases discussed above) the interactions in general severely distort the shape of the self-consistent ground state.

In the more realistic 3D case we need to minimize the quantity

$$E = \int \rho(r)(\nabla\phi)^2(\mathbf{r})d\mathbf{r} \qquad (84)$$

subject to the condition

$$\int_A^B \nabla\phi \cdot d\mathbf{r} = \Delta\phi \qquad (85)$$

where A and B now represent <u>*surfaces*</u> deep within the accessible L and R regions respectively, and thereby find the quantity

$$E_J \equiv (\partial^2 E/\partial(\Delta\phi^2)_{\Delta\phi\to 0} \qquad (86)$$

If we make the mapping $\phi \to V$ (voltage), $\rho(r) \to \epsilon(r)$ (dielectric constant), $E_J \to C$ (capacitance), then it is clear that the problem is completely isomorphic to the process of finding the capacitance of a capacitor stuffed with a material of inhomogenous dielectric constant. Using this analogy (or directly) it is straightforward to obtain upper and lower limits on E_J: in the following let x represent the direction linking the centers of the two wells and y,z the other two dimensions.

To find an <u>*upper*</u> limit we use a variational ansatz:

$$\phi(x, y, z) = \phi(x) \qquad (87)$$

and minimize the gradient energy (84) within this class. This leads to the result

$$E_J \leq \left[\int \frac{dx}{\int \rho(xyz)dydz} \right]^{-1} \tag{88}$$

To obtain a lower limit we neglect the positive terms $(\partial\phi/\partial y)^2, (\partial\phi/\partial z)^2$ and find the exact minimum of the rest (now not assuming that ϕ will be independent of y and z!); this yields

$$E_J \geq \int \frac{dydz}{\int \frac{dx}{\rho(xyz)}} \tag{89}$$

For a factorizable form of $\rho(xyz)$ ($\rho = f(x)g(yz)$) the limits converge and thus yield the exact result. (I have a strong suspicion that the electrostatic analogs of the inequalities (88) and (89) are proved in some nineteenth-century treatise on the manufacture of condensers, but have so far failed to find it!)

The upshot of the above argument is that rather generally we expect the order of magnitude of the Josephson energy to be given correctly by a WKB-type expression of the form

$$E_J \sim N\hbar\omega_0 exp - B \tag{90}$$

where the order of magnitude of the WKB experiment B is $V_0/\hbar\omega_0$, with V_0 the height of the barrier separating the two wells. (For a more detailed discussion, see ref. (15)[f].) If now we combine this with the expression (80) for χ, we find

$$E_J\chi \sim \frac{N^2}{\zeta^2} e^{-B} \tag{91}$$

Although N is typically of order $10^3 - 10^7$, the exponential sensitivity to B means that by varying the barrier height appropriately we can "tune" $E_J\chi$ from very large to very small values. The interest of this is that according to (69) above, in the limit $E_J\chi >> 1$ we have the "standard" Josephson situation where $\Delta\phi << 1$, whereas it is clear that in the opposite limit $\Delta\phi \sim 1$ and it is ΔN which is well-defined. (This is almost certainly the limit realized in the initial stage of the interference experiment of Andrews et al. (16).) Such a crossover is much more difficult to explore in the case of the traditional superfluids (though to some extent it can be realized in mesoscopic superconducting grains).

[f] The reason formula (12) of ref. (15) looks different from (90) relates to the precise definition of V_o used there.

To the extent that the above considerations are valid, the problem of the Josephson effect in the BEC alkali gases reduces, as in the traditional superfluids, to the problem of a simple quantum pendulum, with the difference that it may not be that difficult to realize the "delocalized" regime ($\Delta\phi \sim 1$). However, this leaves open at least two (related) problems:

(1) What is the mechanism of dissipation in the motion of the "pendulum"?

(2) If ΔN is at all appreciable, is it really legitimate to consider only the *groundstate* of each potential well?

A provisional answer to question (1) is given in ref. (15): at any finite temperature the most effective mechanism of dissipation is the one which is perhaps most obvious, namely the shunting of the Josephson current by purely activated (hence incoherent) transitions of the normal component. In fact, for a small deviation $\delta\mu$ of the relative chemical potential of the two wells from equilibrium it is estimated in (15) that the ratio of the normal shunting current I_n to the supercurrent I_s is of order

$$I_n/I_s \sim (2\pi)^{-1}(k_BT/\hbar\omega_o)^2\, N_o^{-7/15}\, exp(B/2) \tag{92}$$

where B is the WKB exponent. A standard result for a single well is that $k_BT_c \sim N^{1/3}\hbar\omega_o$, so we see that for $T \sim T_c$ the ratio I_n/I_s is always large compared to 1, which implies that the Josephson effect will be overdamped. Coherent oscillations should be observable only below a temperature of order

$$T_o \sim T_c \cdot N^{-1/5} \cdot exp - B/2. \ll T_c. \tag{93}$$

Question (2) is plausibly, not a worry so long as the relative chemical potential $\Delta\mu$ excited in the oscillation does not exceed the energy of the first excited state within the individual wells, which turns out to be of order $\hbar\omega_o$ even in the presence of interactions. For disturbances larger than this (which are not necessarily particularly difficult to excite, e.g. by simply changing the level of the well floor suddenly) it is a real problem, and to my knowledge has not so far received a satisfactory resolution. It seems probable that the analysis of such large-amplitude disturbances in the BEC gases will need a new approach.

Finially, I just mention briefly that the extreme "manipulability" of the potentials which control the dynamics of BEC alkali gases may make feasible a whole set of experiments, in particular some connected with "phase memory", which has been analyzed theoretically in the context of the traditional superfluids (see e.g. ref. (17)), but never actually performed. The remarkable experiment of Andrews et. al. is a first step in this direction, but many more are in principle possible. Since I have discussed these possibilities explicitly elsewhere (18), I do not attempt to review them here.

References

1. R. Rouse, S. Han and J. Lukens, Phys. Rev. Lett. 75, 1614 (1995)
2. J. Friedman et al., Phys. Rev. Lett. 76, 3830 (1996)
3. L. Thomas et al., Nature 383, 145 (1996)
4. M. L. Goldberger and K. M. Watson, Collision Theory, (John Wiley and Sons, New York 1964), section 8.2
5. H. Kramers, Physica 7, 284 (1990)
6. A. Schmid, Ann. Phys. (NY) 170, 333 (1986)
7. P. Hänggi, P. Talkner and M. Borkovec, Revs. Mod. Phys. 62, 251 (1990)
8. A. O. Caldeira and A. J. Leggett. Ann. Phys. (NY) 148, 374 (1983): erratum, ibid 153, 445 (1984)
9. A. J. Leggett, Phys. Rev. B30, 1208 (1984)
10. A. J. Leggett, et al., Revs. Mod. Phys. 59, 1 (1987)
11. R Silbey and R. A. Harris, J. Chem. Phys. 80, 2615 (1984)
12. H. Spohn and R. Dümcke, J. Stat. Phys. 41, 389 (1985)
13. U. Weiss, Quantum Dissipative Systems (World Scientific, Singapore 1993), sections 15.5-6
14. K. K. Likharev, Usp. Fiz. Nauk 139, 169 (1983); translation, Soviet Physics Uspekhi 26, 87 (1983)
15. I. Zapata, F. Sols and A. J. Leggett, Phys. Rev. A(RC), in press.
16. M. R. Andrews, et al., Science 275, 637 (1997)
17. A. J. Leggett and F. Sols, Found. Phys. 21, 353 (1991)
18. A. J. Leggett, Physica Scripta, in press.

TUNNELLING IN TWO DIMENSIONS

S. C. CREAGH
Division de Physique Théorique[a]*, Institut de Physique Nucléaire,*
91406 Orsay Cedex, France
and
CEA-Saclay, Service de Physique Théorique
F91191 Gif-Sur-Yvette Cedex France
E-mail: creagh@spht.saclay.cea.fr

An exploration is made of the behaviour of tunnelling phenomena in low-dimensional systems taking into account classical limits with varying degrees of nonintegrability, spanning the spectrum from completely integrable to completely chaotic motion. There are three broad categories: (a) integrable and KAM systems, for which tunnelling is intimately related to the complexification of the invariant tori that characterise classical mechanics; (b) mixed systems in which coupling to a chaotic sea is important and tunnelling rates are irregular and (c) completely chaotic problems in which tunnelling rate fluctuations can be analysed in terms of complex classical orbits. Between each of these there is a certain amount of overlap in the methods used and in the structures that are important, but there are also some very stark differences. Such connections are emphasised in the discussion.

1 Introduction

The title of this tutorial is chosen to suggest two aspects of the subject covered. First, and most important, the discussion is directed at tunnelling in more than one dimension. The range and degree of complexity of classical dynamics is qualitatively greater in several dimensions than in one, and this has profound implications for the behaviour and calculation of tunnelling effects. This refers not just to analysis of the usual barrier-penetration problems, but also reflects the emergence of completely new tunnelling scenarios which have no analog in one dimension. Second, dimension two is specified because it is not a large number — the discussion here is restricted to isolated systems, and dissipation or coupling to the environment are not treated explicitly. Such effects could undoubtedly be accounted for eventually, but for the moment treatment of the isolated system is in itself sufficiently interesting to hold our attention. The discussion here actually holds for any number of degrees of freedom, as long as they are all observable, but two dimensions are sufficient to uncover the main behaviour patterns and correspond to almost all the problems treated in detail to date (if nontrivial time-dependence is counted separately as a dimension).

The clearest example of a fundamentally multidimensional effect is the

[a]Unité de recherche des Universités Paris 11 et Paris 6, Associée au CNRS

dynamical tunnelling scenario,[1,2] which will end up playing a prominent role here. Illustrated in Fig. 1 is a simple two-dimensional potential used by Davis and Heller [2] to demonstrate the existence of pairs of near-degenerate states which are just like the doublets of symmetric double wells except that in this potential there is no energetic barrier to explain them. The explanation comes instead from a fundamentally multidimensional feature of the underlying classical dynamics — the restriction of certain trajectories in phase space to two-dimensional invariant tori even though, from purely energetic considerations, a three-dimensional constant-energy surface is open to exploration. Approximate localised states can be associated with certain of these tori [3] and, when they come in congruent pairs, approximately degenerate pairs are formed analogous to states localised on one side or other of an energy barrier. The small energy level splitting between exact states is related to the tunnelling rate between the tori in the usual way.

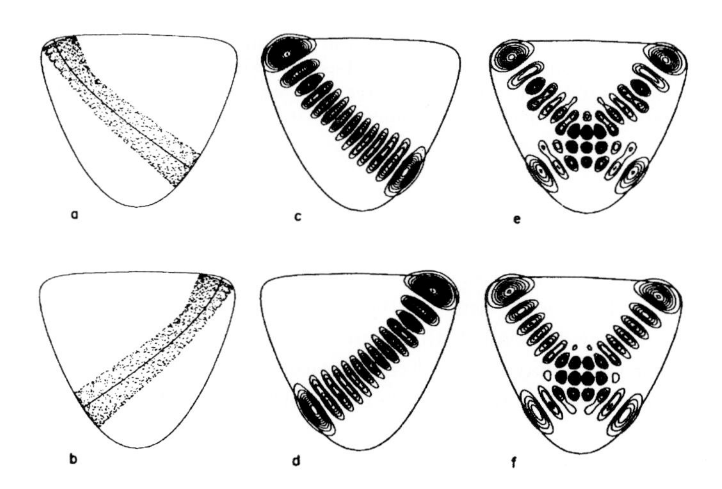

Figure 1: An illustration of dynamical tunnelling from [2]. In (a) and (b) are shown classical trajectories confined to two mutually-congruent invariant tori. Shown in (c) and (d) are corresponding approximate quantum states, or quasimodes. The exact states, respectively even and odd in x, are shown in (e) and (f) — they are approximated by even and odd combinations of the quasimodes. It is interesting to note that the allowed regions of the two quasimodes overlap in configuration space. In other words, localisation is in phase space.

In this and in other classes of multidimensional tunnelling, full account must be taken of the underlying phase space dynamics if a quantitative, or sometimes even qualitative, analysis of tunnelling rates is to be pursued. This is because, even in an initial treatment of wavefunctions and energy levels in which tunnelling effects are ignored, the structure of the solutions depends very heavily on the nature of dynamics in the classical limit. Only after this is accounted for can we begin to incorporate tunnelling effects. This is fundamentally unlike one-dimensional problems, where all wavefunctions are essentially alike. The problem of classifying quantum-mechanical solutions according to classical structure has long been the object of intense study and quite a lot can now be said — a summary of the main ideas will be given in Sec. 2, in the hope that the ensuing discussion might be more accessible to nonspecialists (detailed reviews can be found in [3]). The fundamental object here will be to see how tunnelling changes across the spectrum of classical behaviour as a result of this morphology.

Even in the specific situation described by Davis and Heller, explicitly calculated tunnelling rates exhibit a considerable range of behaviour, depending on the details of the dynamics, and the methods of analysis vary accordingly. Their discussion takes up a significant part of this review. In the integrable and nearintegrable regimes, discussed in Sec. 3, explicit semiclassical formulas of the form $\Delta E \sim Ae^{-K/\hbar}$ can be given for the energy-level splitting, and dependence on parameters is smooth. Within this class, there is a further division into exactly integrable and perturbed, KAM systems and the form of the amplitude A differs accordingly. For integrable systems, the calculation is relatively straightforward and similar to in one dimension,[4–6] with $A \propto \hbar$ (unlike one dimension, however, potential problems still have the possibility of dynamical tunnelling). The KAM regime is considerably more complex because the underlying classical structure (the complexification of an invariant torus) is difficult to compute. There does exist a semiclassical formula,[7] however, predicting $A \propto \hbar^{3/2}$, whose behaviour has been confirmed numerically in certain cases.[8–11]

States associated with invariant tori can exist even in mixed systems, where significant regions of phase space are filled with chaotic trajectories. Then tunnelling rates between tori behave very differently.[12–27] They are larger and display a much more complicated dependence on parameters than predicted by a semiclassical formula $\Delta E \sim Ae^{-K/\hbar}$. This is the chaos-assisted tunnelling regime, discussed in Sec. 4. The intuitive picture is that, whereas tunnelling in integrable and nearintegrable systems proceeds by direct coupling between regular states, in the chaos-assisted regime intermediate coupling to states associated with chaotic parts of phase space dominates. Large fluctuations in the tunnelling rate are then associated with avoided crossings between regular and chaotic states.[14,16] The involvement of chaotic states precludes for now a completely semiclassical calculation, but statistical analyses are possible.[16,23] In addition, in certain billiard problems [15] a more complete investigation is possible using scattering matrix techniques,[21,26,27] described in Sec. 4.2.

A common feature of the dynamical tunnelling scenarios mentioned up to this point is that the starting point of calculation is EBK quantisation, associating the basic wavefunctions with invariant tori. As chaotic dynamics becomes more prominent, different approaches begin to play a role. The statistical analysis of Sec. 4.1 and the scattering matrix approach of Sec. 4.2 are each representative of one of the two basic approaches in treating the quantum mechanics of classically chaotic systems — the first is a statistical treatment in which a chaotic Hamiltonian is modelled by a representative of one of the standard random matrix ensembles and the second is a collective approach in which objects like propagators and Green's functions are approximated by sums over trajectories. These are pursued more fully in Sec. 5, where attention is turned to the chaotic parts of phase space.

The root of many semiclassical calculations for chaotic systems is the approximation of a unitary operator, such as a propagator or scattering matrix, in terms of a sum over trajectories of the corresponding classical dynamical system,[3] as suggested, for example, by saddle point-integration of the path-integral.[28] Tunnelling effects are incorporated by including complex trajectories.[29–42] An extensive study of the use of complex trajectories in approximating the propagator of a particular chaotic map has been performed by Shudo and Ikeda[38,39] and this is described in Sec. 5.1. Even though centered on chaotic aspects, their analysis is related to the scattering-matrix approach to chaos-assisted tunnelling between tori. In time-independent problems, the energy-dependent Green's function, and its trace, is the natural object of study since it encodes within it a complete specification of quantum-mechanical solutions, as discussed in Sec. 5.2. This is also where completely chaotic tunnelling, in which regular parts of phase space are negligible, is introduced. The specific method used there is the trace formula, which, following the inclusion of complex periodic orbits and some trickery to eliminate the real ones, is capable of directly predicting collective fluctuations in the tunnelling rate.[40–42]

Before moving on to the discussion proper, a couple of disclaimers are in order. First, this is not a balanced review of the work that has been done on the topic. It focusses with a certain amount of detail on a selection of calculations which are hoped to be representative of what is possible in the field as a whole. However, much of the literature is given undeservedly short shrift. In particular, with the exception of Sec. 5.1, attention is directed primarily at spectral problems, and problems with splittings at that. There has been a considerable amount of calculation on the dynamical evolution of wavefunctions which is hardly addressed. In spectral problems, the tendency to look at splittings, rather than, for example, resonance widths of metastable systems, is simply because such calculations are often a little cleaner in practice — the general ideas should be easily translated. Second, while semiclassical calculation of tunnelling is now possible with a certain degree of rigour in one dimension [43–47] (even exact calculation using WKB is sometimes attainable [44,47]), the discussion here is often decidedly heuristic. It is probable that at some point exactly integrable systems could be treated with the same degree of rectitude as in one dimension, but in more generic systems the heuristic nature of the discussion seems unavoidable — the dynamics, especially complex dynamics, of generic Hamiltonian systems does not lend itself to a rigourous treatment where semiclassical applications are concerned. With this in mind, we now go on to an overview of the connections between dynamics and the solutions of quantum mechanics.

2 Dynamics and the zoology of wavefunctions

To understand the influence of dynamics on tunnelling, it helps first to have a clear idea of how dynamics can be used to classify the more basic building blocks of quantum mechanics, such as wavefunctions and energy level spectra. To this end, we begin in the present section with a summary of the principal regimes and the standard techniques that are used in each case.[3] Most of the section could probably be skipped safely by readers already familiar with the field of quantum chaos. The most important technical content is an outline of the basic theory behind quantisation of invariant tori in integrable and near-integrable systems. This EBK, or torus quantisation is a starting point for much of the analysis of tunnelling that follows and in order to understand the calculations involved there, it is important to begin with a firm grounding basic structures and concepts used.

The division of this section reflects the division of the article as a whole. The three principle regimes are (a) integrable and near-integrable dynamics, (b) mixed dynamics and (c) complete chaos. The principle methods used are divided accordingly. Torus quantisation is the mainstay of integrable and near-integrable dynamics and collective spectral methods, such as the trace formula and random matrix theory (RMT from now on), are the principle approaches for the chaotic case. The mixed case simply used a mixture of these methods and will not receive much attention in this section. It should be stressed, however, when it comes to tunnelling characteristics, the mixed regime will play a much more distinctive role than it does here.

2.1 Integrable and near-integrable dynamics

We illustrate the construction and quantisation of invariant tori with the following two-dimensional potential,

$$V(x, y) = (x^2 - 1)^2 + y^2 + \lambda x^2 y^2, \tag{1}$$

We have deliberately chosen this to be in the form of a symmetric double well so that when we come to consider quantum mechanics later, tunnelling effects will be relevant, but in this section we will be interested mainly in dynamics restricted to one side.

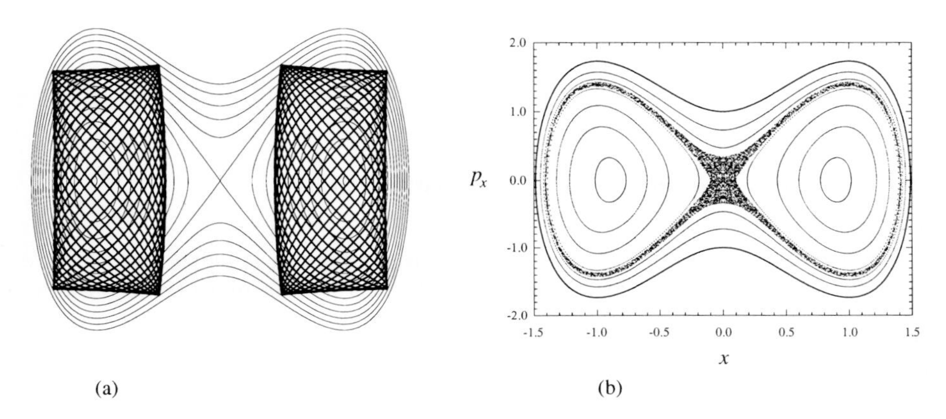

Figure 2: Typical trajectories are shown in (a). There are two of them, gradually filling two congruent tori. Notice that the trajectories do not cross from one well to the another even though at $E = 3/2$ their energies are greater than the barrier energy $E_{\rm sad} = 1$. In (b) is shown a Poincaré section at the same energy recording trajectories as the pass through $y = 0$. Most trajectories lie on invariant circles, but there is a band of chaos where the separatrix used to be. The outer bounding curve corresponds to the limiting case of a one-dimensional trajectory which always stays on the x-axis.

This potential is separable when $\lambda = 0$ and mixed when the coupling parameter is nonzero. Typical dynamics is illustrated in Fig. 2 for $\lambda = 1/20$. In Fig. 2(a) are shown two trajectories in configuration space, each lying on one of two congruent tori. Localisation on one side occurs even though their energy $E = 3/2$ is higher than the saddle-point energy $E_{\rm sad} = 1$. In Fig. 2(b) is shown a Poincaré section in which (x, p_x) is recorded every time a trajectory passes upwards through $y = 0$. Most orbits remain confined to invariant circles and there is a band of chaos which shrinks to a separatrix if we let $\lambda \to 0$. The third invariant circle from each center is a phase-space cross-section through one of the invariant tori in Fig. 2(a).

Confinement to invariant tori, or invariant circles in a Poincaré section, occurs when, in addition to energy, there is another conserved quantity I. In exactly integrable problems such as the separable limit of the potential above, we are lucky enough to find a globally-defined function $I(\mathbf{q}, \mathbf{p})$ that is analytic in its arguments $\mathbf{q}$ and $\mathbf{p}$. The separable limit of the current example has $I = p_y^2/2 + y^2$. Let us develop our notation under this assumption and consider the more generic, perturbed situation later.

A remarkable fact is that it can be shown, under quite general conditions, that the confining surface must be topologically equivalent to a torus. This is more than an academic point for us because the toral structure will play a fundamental role in constructing semiclassical approximations to wavefunctions. A second essential ingredient will be the fact that for any curve C lying on the torus, the action integral,

$$S_C = \int_C \mathbf{p} \cdot \mathrm{d}\mathbf{q} \tag{2}$$

remains unchanged when C is continuously deformed while keeping the end-points fixed. In two degrees of freedom, any two-dimensional surface with this property will be called a *Lagrangian manifold* — this terminology will recur because the phases of multidimensional WKB approximations are always associated with Lagrangian manifolds.[48] Both of these properties are direct consequences the fact that a torus is the simultaneous level set of two phase space functions, H and I, which Poisson-commute. A good demonstration from a geometrical point of view can be found in Chapter 10 of Arnold.[49]

For future reference, we take the opportunity here to state how this structure relates to the construction of action-angle variables. The Lagrangian property guarantees that $S_C = 0$ for any closed curve C that can be shrunk continuously to a point. There are, however, closed curves which cannot be trivialised in this way and for which the action is nonzero. We denote by C_1 and C_2 two topologically independent closed curves for which this is the case — the topology of the torus is such that they can be chosen so that any other closed curve can be continuously deformed onto multiple iterations of them. To each of the curves we associate an action variable, $I_a = S_{C_a}/2\pi$, which are together Poisson-commuting functions on phase space taking fixed values on any given torus. They can be extended to a set of canonical variables $(\mathbf{I}, \boldsymbol{\varphi}) = (I_1, I_2, \varphi_1, \varphi_2)$ such that the angles (φ_1, φ_2) are 2π-periodic. The Hamiltonian $H(\mathbf{q}, \mathbf{p}) = H(\mathbf{I})$ is a function of the actions alone and motion reduces to uniform translation in the angles $\dot{\boldsymbol{\varphi}} = \partial H(\mathbf{I})/\partial \mathbf{I} = \boldsymbol{\omega}(\mathbf{I})$.

We are now in a position to construct an approximation to quantum-mechanical wavefunctions and their energies in what is known as EBK (for Einstein-Brillouin-Keller) or torus quantisation. A complete description can be found in [3]. Substitute an ansatz of the form,

$$\psi(\mathbf{q}) = \sqrt{\rho(\mathbf{q})}\, e^{iS(\mathbf{q})/\hbar} \tag{3}$$

into the Schrödinger equation. This works at leading order in $\hbar$ if $S(\mathbf{q})$ satisfies the Hamilton Jacobi equation,

$$H(\mathbf{q}, \nabla S) = E, \tag{4}$$

which is solved by choosing $S(\mathbf{q})$ to be the action, defined according to Eq. (2), of a curve on the torus ending with position $\mathbf{q}$. At next to leading order, $\rho(\mathbf{q})$ should be a density in the coordinate variables which is invariant under the dynamics. In angle variables, the invariant density is a simple constant $\tilde{\rho}(\boldsymbol{\varphi}) = 1/(2\pi)^2$ (the normalisation is chosen so that we will finally end up with $\int d\mathbf{q}\,|\psi(\mathbf{q})|^2 = 1$). This projects on to the variables $\mathbf{q}$ using the Jacobian for the variable change,

$$\rho(\mathbf{q}) = \tilde{\rho}(\boldsymbol{\varphi}) \det \frac{\partial \boldsymbol{\varphi}}{\partial \mathbf{q}} = \frac{1}{(2\pi)^2} \left[\det \frac{\partial \mathbf{q}(\boldsymbol{\varphi}, \mathbf{I})}{\partial \boldsymbol{\varphi}}\right]^{-1}. \tag{5}$$

We have thus far constructed a function which satisfies the Schrödinger equation within semiclassical approximation but which does so only locally. In general there will be several sheets on the torus above a given neighbourhood of $\mathbf{q}$, each of them offering a distinct local wavefunction $\psi_\alpha(\mathbf{q})$. Whenever two of these sheets coalesce (in a fold generically) it will be the case that $\det\,[\partial \mathbf{q}(\boldsymbol{\varphi}, \mathbf{I})/\partial \boldsymbol{\varphi}] = 0$ and the amplitudes in the two local wavefunctions will diverge. To treat this caustic structure consistently, it is sufficient to patch both local wavefunctions together by matching the actions and carefully choosing the branch on the square root in Eq. (3) (remember that the density changes sign on passing from one branch of the torus to the other). Explicit rules for this are given in Maslov theory,[48] but in this article we will tend to brush over the issue.

By systematically covering the whole torus we can hope to associate a quantum state with a single-valued function on it (albeit singular at caustics), from which the wavefunction is obtained simply by summing over all points above a given $\mathbf{q}$. For this to work we need the wavefunction to return to its initial value after following any closed loop on the torus. It is sufficient to check that, for the two basis loops C_1 and C_2, the total accumulated phase change is a multiple of 2π. This results in the two quantisation conditions,

$$I_a = \left(n_a + \frac{\mu_a}{2}\right)\hbar, \tag{6}$$

where the Maslov index μ_a is an even integer expressing the accumulated effect of phase changes coming from the changing sign in $\rho(\mathbf{q})$ every time a caustic is crossed. This condition specifies a discrete set of tori with the discrete energy levels $E_{n_1,n_2} = H(\mathbf{I}(n_1, n_2))$ labelled by the quantum numbers n_1 and n_2. We will see in the following section that this procedure of quantisation by single-valuedness will extend to a discussion of tunnelling effects in exactly integrable systems, though the situation after perturbation is more complicated.

For exactly integrable systems, analytical formulas are available for all the classical ingredients used to construct the approximate quantum states. So we have a rather complete understanding of the energy level spectrum and, using the wavefunction, other quantum properties such matrix elements can be computed.[50]

A wider and more important class, however, is that of near-integrable systems, of which the system in Fig. 2 is an example. In this case, motion is still predominantly confined to tori, but globally-defined constants of motion do not exist. Instead, although each individual torus defines a pair of actions $\mathbf{I}$ which serve locally as integrals of motion, the dependence of various classical properties on $\mathbf{I}$ as we pass from one torus to another is very complicated. The main tool for understanding this regime is the KAM theorem,[51] which examines the success of perturbation expansions in reconstructing integrals of motion when an integrable Hamiltonian is perturbed. The result is that the expansions converge and tori survive if their frequencies are far enough from resonance, that is, if ω_1/ω_2 is not well-approximated by a rational number, but that they are destroyed otherwise. The generic situation is that almost all tori satisfy the survival criterion and phase space is filled with them. However, between every pair of surviving tori there are resonant tori which are destroyed, just as between every two real numbers there are rational numbers. The resulting structure on all scales is responsible for a pathological dependence of any phase space variable on $\mathbf{I}$.

For the simplest semiclassical calculations of interest in this section, such complications do not form much of a barrier to practical calculation, as we will see now. It is in tunnelling effects that such structure will turn out to be of prime importance. For the moment however, we are content to know that in near-integrable systems it is possible to parametrise motion on tori by expressing phase space coordinates as 2π-periodic functions $\mathbf{q}(\varphi, \mathbf{I})$ and $\mathbf{p}(\varphi, \mathbf{I})$

of the angles $\boldsymbol{\varphi}$. A good way of representing them is in the Fourier expansions,

$$\mathbf{q}(\boldsymbol{\varphi}, \mathbf{I}) = \sum_{\mathbf{m}} \mathbf{q}_{\mathbf{m}}(\mathbf{I})\, e^{2\pi i \mathbf{m} \cdot \boldsymbol{\varphi}}$$

$$\mathbf{p}(\boldsymbol{\varphi}, \mathbf{I}) = \sum_{\mathbf{m}} \mathbf{p}_{\mathbf{m}}(\mathbf{I})\, e^{2\pi i \mathbf{m} \cdot \boldsymbol{\varphi}} \tag{7}$$

In practice, for example, even if little is known analytically about the system, the coefficients $\mathbf{q}_{\mathbf{m}}(\mathbf{I})$ and $\mathbf{p}_{\mathbf{m}}(\mathbf{I})$ can be calculated very accurately for a given torus by Fourier analysis of a numerically integrated trajectory. From this, all the ingredients for the quantisation condition and wavefunction are easily constructed.

2.2 Mixed dynamics

In mixed systems, a large part of phase space is occupied by stochastic dynamics, but there remain regular islands filled with invariant tori. This may happen in a problem such as illustrated in Fig. 2 where a perturbation parameter is increased enough that separatrix chaos expands and engulfs the regular regions or it may simply be that, without reference to an unperturbed problem, some generic Hamiltonian produces islands which are surrounded by chaos. For such problems, a first working hypothesis for understanding quantum-mechanical structure is given by Percival's conjecture.[53] This says simply that we can separate quantum states into those supported in the integrable regions ("integrable states" for short) and those supported in the stochastic sea ("chaotic states"). By "supported" it is meant here that some representation of the state in phase space, such as a Wigner function or Husimi distribution,[3] is concentrated in such a region.

For the integrable regions, or islands, calculation of wavefunctions can proceed just as before. Even if there is no nearby integrable Hamiltonian with which to compare, a numerical evaluation of Fourier series such as in Eq. (7) is possible. So, for the basic quantum-mechanical questions, little more needs to be said. It will turn out, however, that tunnelling rates will often behave qualitatively differently depending on whether the islands are surrounded by stochastic seas or not. So the distinction will become very important later.

Similarly, the primary direction of attack for chaotic states is the same as that for completely chaotic systems, which we will discuss in the next section.

On the whole, therefore, mixed systems are treated simply by mixing methods appropriate for completely integrable and completely chaotic systems. It should be pointed out that while this is effective for the dominant features, it sweeps under the rug quite a few subtleties that are not well understood.

These have to do with the dynamics and quantum mechanics at the interface between integrable and chaotic regions. These regions of marginal dynamics are very complex with many small islands and classical structures that are not easily incorporated into semiclassical calculations. Thus, even though some minority of states can be supported there, it will be very difficult to come up with a completely semiclassical theory for them. They can nonetheless be important. Frischat and Doron,[27] for example, discover that these states play a leading role in tunnelling from one island to another in mixed systems.

2.3 Chaotic dynamics

Our canonical example of a completely chaotic problem will be the double well defined by,

$$V(x, y) = (x^2 - 1)^4 + x^2y^2 + \mu y^2. \tag{8}$$

Notice the fourth power on the first term which, for reasons uninteresting to us here, has the effect of amplifying the tendency to chaos. A typical trajectory in this potential, such as illustrated in Fig. 3(a), fills out the region of configuration space enclosed by a potential contour. Unlike the example in Fig. 2, localisation of a trajectory only occurs if the energy is below the saddle energy $E_{\rm sad} = 1$. A more complete picture is given by the Poincaré section in Fig. 3(b). A typical trajectory fills out almost all of the region of the (x, p_x)-plane enclosed by the boundary curves defined by $p_x^2/2 + (x^2 - 1)^4 = E$ (which corresponds to putting all the energy into the x degree of freedom so that motion is restricted to the symmetry axis $y = 0$). The only flaw is that two small islands exist near the outer extremities of the x-axis. The occurence of islands like these is generic. They impede the more sophisticated semiclassical approaches,[54] but can be ignored for the purposes of understanding dominant features.

When wavefunctions are computed in a problem like this, they fill out the regions defined by a typical trajectory. This is the case whether we examine the state in a configuration or a phase space representation. This is in contrast to integrable states which are localised around the quantised tori associated with them. Loosely speaking, when the classical dynamics is "ergodic," so are the wavefunctions (there are qualifications to this that we will not go into). This change in the structure of wavefunctions is accompanied by a change in the semiclassical methods used to analyse them. For such delocalised states there is no longer a theory which tells us how to construct individual cases. Instead there are collective methods which deal with the whole spectrum at once. Broadly speaking, these are either statistical, using random matrix theory

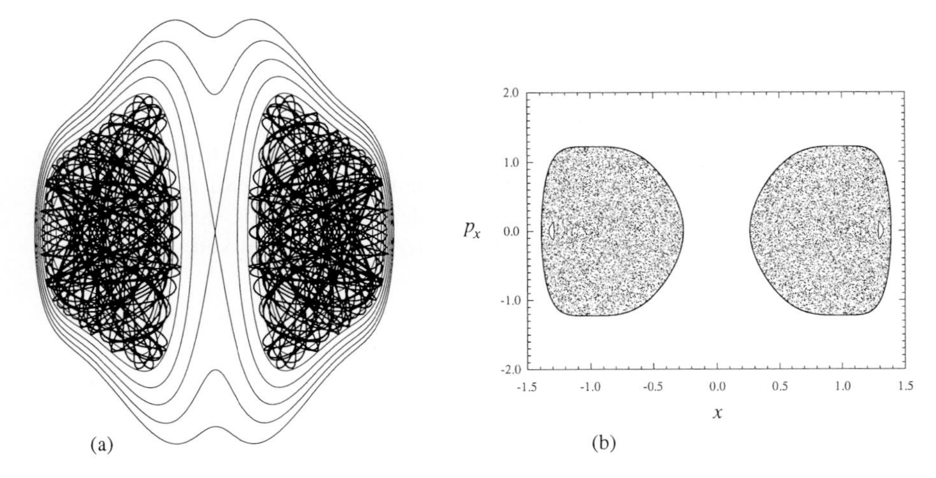

Figure 3: In a chaotic potential, most trajectories fill out the Poincaré section defined by $y = 0$, as shown in (b). Since the energy $E = 1/2$ is less than the saddle energy $E_{\text{sad}} = 1$, trajectories remain localised on one side of the barrier. In configuration space, a typical trajectory, shown in (a), fills the area enclosed by the corresponding potential contour.

(or RMT), or semiclassical, approximating objects such as propagators and Green's functions using classical trajectories.

The RMT approach says simply that a chaotic Hamiltonian can be modelled for statistical purposes by taking a representative of various ensembles of random matrices, the particular ensemble depending on the symmetries imposed on the Hamiltonian. We will confine our attention to the Gaussian orthogonal ensemble (GOE), which models Hamiltonans with a time reversal symmetry. In the standard references explicit formulas are available for various correlation functions and spacing distributions.[52]

Of the semiclassical methods, the most important is the trace formula.[3] This is a generalisation of the Poisson summation formula,

$$\sum_{m=-\infty}^{\infty} \delta(x-m) = \sum_{n=-\infty}^{\infty} e^{2\pi i n x} = 1 + 2\sum_{r=1}^{\infty} \cos(2\pi r x). \tag{9}$$

To see the connection, let us identify the set of integers m with the spectrum of the operator $-i\partial/\partial\theta$, acting on the space of functions periodic in θ. Then the sum of delta-functions at left is the density of eigenvalues on the real line. The Poisson summation formula tells us that this can be constructed by adding up oscillatory functions with successively increasing frequencies in the argument x. To complete the preparation, let us note that the phases $2\pi r x$ can be considered to be the actions taken from the periods $2\pi r$ of the classical motion when we use p_θ as a Hamiltonian on the phase space (θ, p_θ).

The Gutzwiller trace formula is a generalisation of this to the spectrum of a general Hamiltonian operator H. It is simpler if we restrict our attention to one-dimensional systems or chaotic systems in higher dimensions. Then all the periodic solutions γ of the classical limit at a given energy E are isolated in phase space, forming a discrete set. The trace formula expresses the density of quantum levels E_n as the following sum over the orbits γ,

$$\sum_n \delta(E - E_n) = \rho_0(E) + \mathrm{Im} \sum_\gamma A_\gamma(E)\, e^{iS_\gamma(E)/\hbar}. \tag{10}$$

The Thomas Fermi density of states $\rho_0(E)$ is a slowly varying function of E analogous to the 1 at right in Eq. (9) above. It will not interest us very much. Each orbit γ is responsible for a rapidly varying oscillation to be superimposed on $\rho_0(E)$. The phase is determined by the action $S_\gamma(E) = \oint_\gamma \mathbf{p} \cdot d\mathbf{q}$ of γ and the amplitude $A_\gamma(E)$ is computed from a linearisation of dynamics in its immediate neighbourhood.

In a naive approach, we expect that if we keep adding longer and longer orbits, we determine structure in the density of states at shorter and shorter energy scales, until eventually the delta-function singularities are revealed. This certainly works in one dimension, where the orbits are all repetitions of some primitive oscillation and the series can be summed geometrically, giving a density of states with singularities at the EBK levels.[55] For a chaotic system in more dimensions, however, the series diverges before the singularities are revealed and the summation must be treated with more respect. The divergence arises because in a chaotic system the orbits proliferate very rapidly with length and this is enough to overpower the decaying amplitudes. The resolution of this problem is to reorganise the sum in so-called zeta-functions, which are capable of converging well enough to give individual states.[54] We will not use them, but will rely on their existence to reassure ourselves that the summation can ultimately be made to make sense.

A less ambitious approach is to convolve both sides of the equation with a function that damps longer orbits. The right hand side then effectively reduces to a summation over some finite, and possibly small, number of orbits. The penalty is that the right hand side no longer distinguishes individual states, but does reveal long range coherent fluctuations. The fluctuations can be pushed to smaller energy scales by including longer orbits. In a sense this is the natural use of the trace formula because relatively complicated quantum structure can be found with few orbits. As structure on smaller energy scales is required, an inordinately larger amount of work is required to compute the periodic orbits, due to their exponential proliferation with length.

If, besides the energy level, we were interested in some other properties of the states, the approach adopted is similar. If w_n is some some matrix element or other property of the state the procedure is always the same. We construct a sum

$$\sum_n w_n \delta(E - E_n) \tag{11}$$

and express it as a sum over trajectories such as in Eq. (10). The main difference is generally that the orbits participating are no longer necessarily periodic but have some other conditions placed on them, depending on w_n. The Green's function, for example, corresponds to the case $w_n = \psi_n^*(\mathbf{q})\psi_n(\mathbf{q}')$ and is approximated by a sum over orbits starting at $\mathbf{q}'$ and finishing at $\mathbf{q}$. Another example will appear in our discussion of tunnelling in chaotic systems where we weight the delta functions with energy level splittings and the orbits contributing are periodic, but in a generalised sense.

In all of these calculations, if it is desired to pass beyond the shortest orbits and longest fluctuations, and certainly to distinguish individual states, it is vital to find some organising principle for the classical orbits so that they might be found systematically. In completely chaotic systems this is achieved by finding a symbolic dynamics. This encodes periodic orbits efficiently as sequences of symbols. It should be noted that this works effectively only when phase space is dominated by chaos. When dynamics is mixed and a significant degree of marginal behaviour exists, organisation of periodic orbits is problematic in practice. For certain classes of nonperiodic orbits, so-called homoclinic structures have been found to be an effective organising principle.[56,41]

3 Tunnelling in integrable and near-integrable systems

We saw in Sec. 2.1 that the calculation of basic quantum mechanical properties was very similar for integrable and for near-integrable systems. The main difference is one of practical implementation, where analytical formulas are available for exactly integrable problems but perturbative or numerical methods are used in near-integrable systems. This similarity disappears when we consider tunnelling effects. The difference arises because, to compute tunnelling effects, we must examine the behaviour of the wavefunction in the classically forbidden regions, which is constructed semiclassically from an analytical continuation of a torus to complex values of the angle variables $\boldsymbol{\varphi}$. The way in which this analytical continuation behaves seems to be strikingly different for exactly integrable and KAM systems.

Consider the energy level splitting ΔE between states supported on two congruent tori with actions $\mathbf{I}$. In both the exactly integrable and the near-integrable cases, the dependence of the splitting on parameters is regular and given by semiclassical formulas of the form

$$\Delta E = A(\mathbf{I}, \hbar)\, e^{-K(\mathbf{I})/\hbar}. \tag{12}$$

Here $S = iK$ is an imaginary action computed from complexified dynamics and $A(\mathbf{I}, \hbar)$ is a smooth function of its arguments. However, the structure underlying the computation and the form of A are very different depending on whether the system is exactly integrable or not. The most noticable consequence is that, whereas for exactly integrable systems A is linear in $\hbar$ (as in one dimension), for large enough perturbations it scales differently, ($A \sim \hbar^{3/2}$ in two dimensions, for example). The underlying reason is that, when analytically continued into the region between the tori, exactly integrable tori join smoothly to form a single manifold, whereas KAM tori intersect at an angle.

In the next section, yet another regime of chaos-assisted tunnelling in mixed systems is examined in which understanding of the complexified tori is not sufficient because coupling to third-party chaotic states is important. When this arises, dependence on parameters is irregular. In this section we concentrate on cases where tunnelling is confined to a direct route between tori. We begin with exactly integrable systems and then discuss the changes that occur when perturbations are introduced.

3.1 Exactly integrable case

In a one-dimensional symmetric double well, such as defined by the potential,

$$V(q) = (q^2 - 1)^2, \tag{13}$$

the semiclassical splitting between energy levels under the barrier is;[57]

$$\Delta E = \frac{\hbar\omega}{\pi}\, e^{-K/\hbar}, \tag{14}$$

where $S = iK$ is an imaginary tunnelling action and ω is the frequency of oscillation in one well. The basic features of this result are unchanged in several dimensions if we confine our attention to exactly integrable systems (it may be necessary to sum over different tunnelling actions, possibly with nonzero real parts). Analyses that work in this limit can be found in [4-6]. This observation is evident for the separable limit $\lambda \to 0$ of the potential in Eq. (1), but holds also in problems where integrability does not derive from a separation of variables. An account is given here of how the splitting might be calculated in an intrinsically multidimensional manner, without reference to reduction to one-dimensional problems. The underlying procedure, continuation of the

EBK wavefunction into the forbidden regions by complexification of invariant tori, is important for a discussion of tunnelling in the KAM regime also.

Real EBK quantisation gives doubly degenerate energy levels when applied to a pair of congruent tori such as illustrated in Fig. 2(a). This is the case because the real tori, and the wavefunctions constructed from them, are disconnected and effectively independent. When complexified, however, the tori are seen to be different parts of the same manifold and, when this is taken into account in the quantisation conditions, the degeneracy in energy is lifted.

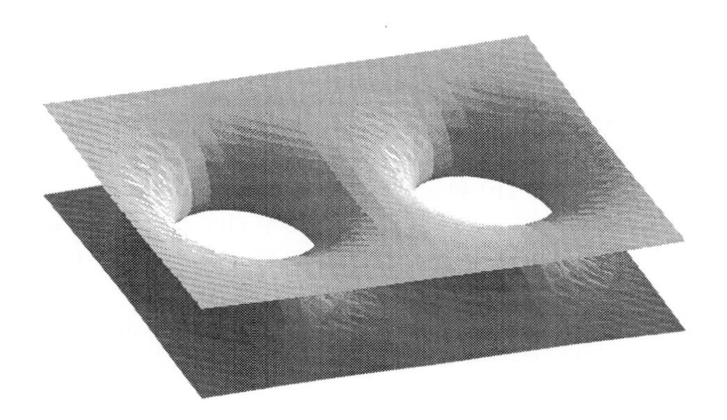

Figure 4: For a one-dimensional quartic oscillator $V(q) = (q^2 - 1)^2$, analytic continuation of the two real invariant circles produces a single Riemann sheet structure with the topology illustrated here. The two holes are encircled by the two real orbits. Complexification introduces a topologically distinct loop, which encircles the bridge in the middle and which in one dimension can be associated with evolution in imaginary time. It is assumed later that, in the KAM regime, this structure splits into two distinct surfaces meeting at isolated points (either as a time-dependent perturbation of one-dimensional tori or as a restriction of two-dimensional tori to a surface of section).

While in real EBK the wavefunction is simply set to zero in the forbidden region between two tori, we know that it is nonzero, though small and decaying rapidly as we move away from the allowed region. This evanescent part is captured by allowing complex actions in the WKB ansatz — the complex Hamilton Jacobi equation is then satisfied if we define the action by integration over an analytic continuation of the torus into the forbidden region. This analytic continuation is easy to implement for exactly integrable systems (and very difficult in the KAM regime). To be explicit, consider the separable limit $\lambda \rightarrow 0$ of the potential in Eq. (1), which has the two integrals of motion $H_x = p_x^2/2 + (x^2 - 1)^2$ and $H_y = p_y^2/2 + y^2$. If we fix the specific values $H_x = E_x$ and $H_y = E_y$, then for every $\mathbf{q} = (x, y)$ we can define four distinct momenta according to the choice of signs in

$$
\begin{aligned}
p_x &= \pm\sqrt{2(E_x - (x^2-1)^2)} \\
p_y &= \pm\sqrt{2(E_y - y^2)}.
\end{aligned} \tag{15}
$$

The two real tori on either side of the barrier project onto the two rectangular regions in configuration space where p_x and p_y are real [Fig. 1(a)]. The topology of the torus is recovered if we glue the corresponding four sheets of $\mathbf{p}(\mathbf{q})$ together at the boundaries, matching them pairwise as appropriate. Solutions can be defined, however, even if $\mathbf{q}$ is not in one of the rectangles covered by a torus, in which case one or both of the momenta is complex. Of most interest for tunnelling is that between the two allowed regions is a rectangle where p_y is still real but p_x is imaginary — when viewed in phase space this also has the topology of a torus.

So the two real conguent tori, when continued under the barrier along real position coordinates, are connected in the present case by a third torus with complex momenta (other possibilities for connection are discussed in [9]). Branches in other sectors extend to infinity and end up not playing a role in quantisation, so we ignore them. In complex EBK we associate a wave function with the whole structure, extending underneath the barrier as well as into both allowed regions. Quantisation, including estimation of the splitting, is achieved by demanding that the wavefunction be single valued when looping around the tunnelling torus as well as around the two real tori.

The calculation proceeds similarly for any exactly integrable Hamiltonian. The construction of the complexified torus above relied not on separability but on the existence of two independent, globally-defined first integrals. More generally, we can suppose a two-degree-of-freedom Hamiltonian $H(\mathbf{q},\mathbf{p})$ with a first integral $I(\mathbf{q},\mathbf{p})$, both of which are assumed to be analytic functions of their arguments. We can therefore solve for $\mathbf{p}(\mathbf{q},E,I)$ and, gluing branches together at caustics, form a multidimensional Riemann sheet structure. This surface is actually a coordinate-invariant construction in complex phase space, defined simply as a simultaneous level set level of H and I. The discussion above singled out $\mathbf{q}$ by restricting it to be real, but for full invariance both $\mathbf{q}$ and $\mathbf{p}$ should be allowed to take complex values, defining a manifold of two complex or four real dimensions, generalising Fig. 4.[6] The real-$\mathbf{q}$ cross-section discussed above is necessary in practice since the full four-dimensional manifold is impossible to visualise — the essential topological features were nonetheless evident even in this restricted view.

As with real WKB, we can associate a local wavefunction $\psi(\mathbf{q})$ with any point on the manifold and we get quantisation by demanding that the local solutions combine globally to define a single-valued function. The primary complication of complex WKB is the appearance of the Stokes phenomenon, where the coefficient of an exponentially small solution changes on crossing certain surfaces of codimension 1 which emerge from each caustic surface (called Stokes surfaces, or lines in one dimension). The Stokes phenomenon plays the role that Maslov theory does in real WKB. It is more difficult because it is nonlocal — one can never tell from purely local considerations whether or not a Stokes surface from a distant caustic passes through a given point. A basic review in one dimension is given in [57], though the reader might also want to take account of the significant advances made since then.[43–47] In the present context of exactly-integrable systems, the multidimensional version is not much more complicated.[6] We will not go into it further except to give the main features of the resulting prediction.

The result is essentially that the real EBK quantisation conditions of Eq. (6) are modified in the following way. The loops C_a on the real torus still define nontrivial loops in the complexified case and singlevaluedness around them quantises their actions according to,

$$I_a = \left(n_a + \frac{1}{2}\right)\hbar - \frac{\hbar T_a}{2\pi}. \tag{16}$$

The quantity T_a represents an effective phase shift that has arisen due to tunnelling effects. Its presence is related to the existence of small components of an exponentially growing solution on the forbidden side of a caustic. It is computed in simple cases by conditions of the form,

$$T^2 = e^{iS_C} = e^{-2K/\hbar}, \tag{17}$$

obtained after demanding single-valuedness around a closed loop C which appears in the complexified manifold.[6] The two solutions of this result in an action splitting of the form $\Delta I_a = (\hbar/\pi)e^{-K/\hbar}$ or an energy splitting $\Delta E = \partial H/\partial I_a \cdot \Delta I_a = (\hbar\omega_a/\pi)e^{-K/\hbar}$ (when several tunnelling loops C are present, a summation is necessary). The one-dimensional result in Eq. (14) is a special case of this.

This has been a very complicated way to rederive an obvious result, but it was worth doing for two reasons. First, the general method of analytically continuing tori works for all integrable problems even if the dynamics is not separable (and therefore not reducible to a one-dimensional problem). Second, an understanding of the complex geometry of the integrable case is a useful starting point for the investigation of KAM systems, whose analysis is nothing like the one-dimensional case.

3.2 Analytic continuation of KAM tori

If we now allow λ to be nonzero in Eq. (1), what happens to the complexified torus? The response to this simple question seems at present to be very poorly understood. A calculation of tunnelling rates by Wilkinson [7] is based on the premise that, when analytically continued into the forbidden region from each well, the left and right complexified tori meet not as a single surface but at an angle, defining two distinct manifolds. This is described in the next subsection, but first let us explore the classical complex structure, because this is a surprisingly difficult issue. We will give a simplified account which is hoped to convey the main features. For a more detailed technical account of various strategies for continuing the wavefunction into the forbidden region, see [11].

Under perturbation, there is no simple analog of the explicit solutions in Eq. (15) because there are no globally-defined constants of motion. Therefore we seek other methods. In practical calculation, perturbative or numerical, a real KAM torus is usually presented to us as a Fourier expansion, as in Eq. (7). Initial analytic continuation is straightforward because we simply let φ take complex values. However, the series will diverge at imaginary parts of φ determined by the singularity closest to the real axis, and it turns out that this is guaranteed to occur before the center of the barrier is reached. In order to get to the center, the series has to be resummed, which turns out to be problematic in the KAM case.

Let us restrain our attention to a horizontal cut through the torus, along $y = 0$, choosing the branches $p_y > 0$. In other words, we are looking at the intersection of the torus with a Poincaré section of the type shown in Fig. 2(b). With this restriction, we can confine our attention to functions $x(\phi)$ and $p_x(\phi)$ of a single angle variable ϕ. In the unperturbed case, $x(\phi)$ is just the solution of a one-dimensional quartic oscillator. It has the form of an elliptic function with two periods in the complex ϕ-plane — a real one 2π and an imaginary one $i\alpha_0$ corresponding to a tunnelling loop. In half the imaginary period, a trajectory crosses from one well to the other, $x(\phi + i\alpha_0/2) = -x(\phi)$. As the angle sweeps over the unit cell defined by the periods $(2\pi, i\alpha_0)$, the solution covers a surface in complex phase space with the topology of Fig. 4. This includes two infinities corresponding to extension of the upper and lower sheets. (If the Riemann surface is compactified by including these infinities, it attains the topology of a torus, naturally parametrised by $[\mathrm{Re}(\phi), \mathrm{Im}(\phi)]$. However, it is important to note that this toral structure is not related to that of a two-degree-of-freedom real torus, and $[\mathrm{Re}(\phi), \mathrm{Im}(\phi)]$ cannot be treated as independent angle variables since all physical quantities must be analytic in $\mathrm{Re}(\phi) + i\mathrm{Im}(\phi)$).

The two infinities, reached in finite time, correspond to two poles of the function $x(\phi)$ at $\pi + i\alpha_0/4$ and $\pi + 3i\alpha_0/4$. The existence of singularities is guaranteed once two periods are found, and is not at all restricted to the quartic oscillator — since any nonconstant function on the complex plane must have singularities, $x(\phi)$ must have a singularity in the unit cell if it is biperiodic. Therefore, the Fourier expansion *must* diverge at the center of the barrier, where $\mathrm{Im}(\phi) = \pm\alpha_0/4$, and we can never hope to get there simply by inserting complex angles into a Fourier expansion. In the integrable case, the expansion is still useful, however, because techniques such as Padé approximation can use the information in the Fourier coefficients to yield expressions with larger radii of convergence.

To see how this works in a practical case, we will use an algebraic model for a Poincaré section to avoid having to integrate differential equations. We use the following map on the plane,

$$\begin{aligned} \bar{q} &= q_n + \frac{\delta}{2} p_n \\ p_{n+1} &= p_n + \delta F(\bar{q}) \\ q_{n+1} &= \bar{q} \ + \frac{\delta}{2} p_{n+1}, \end{aligned} \tag{18}$$

where $F(q) = 2q - 3q^3$ is a force function for the quartic potential. This is actually a rudimentary symplectic integrator for the one-dimensional quartic oscillator with time step δ. In the limit $\delta \to 0$ it approximates the differential equation but even for finite δ it defines a canonical transformation. It is a version of the standard map [51] reorganised to make time-reversal symmetry more apparent. It turns out that perturbation expansions are even in δ so $\epsilon = \delta^2$ is more properly thought of as the perturbation parameter. If a trajectory of the perturbed system lies on a torus, we expect a smooth function $q(t)$ to be defined so that $q_n = q(t_0 + n\delta)$. We prefer to rescale t by the period T of $q(t)$ to define a 2π-periodic function $q(\phi)$ of $\phi = 2\pi t/T$. A trajectory is then $q_n = q(\phi_0 + n\omega)$ where $\omega = 2\pi\delta/T$.

In Fig. 5 we show the Fourier coefficients of $q(\phi)$ for the two values $\delta = 1/10$ and $\delta = 1/2$ (in each case the initial condition has energy $E = 3/4$ in the limiting Hamiltonian). They were computed simply by iterating a trajectory and Fourier-analysing the result. For the smaller time step, the coefficients decay rapidly and smoothly (until the limit of numerical precision is reached). This is similar to the integrable limit where a simple pole is responsible eventually for an exponential decay. In this case it is not a problem to sum the exponential tail using Padé techniques and approximate the torus at the center of the barrier.

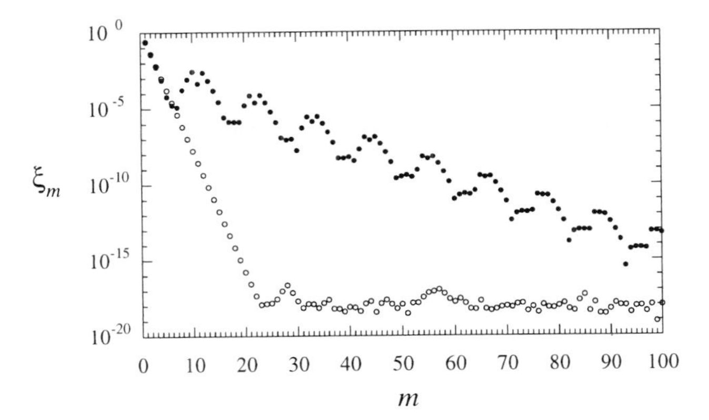

Figure 5: Fourier coefficients of $q(\phi) = \sum_m \xi_m e^{im\phi}$ are shown for the two time steps $\delta = 1/10$ (open circles) and $\delta = 1/2$ (filled circles). For the smaller perturbation, the coefficients decay essentially exponentially until the limits of numerical precision are reached. For the larger perturbation, the decay is slower, signalling that singularities are closer to the real ϕ-axis, but also less regular, indicating that the singularity is not as simple.

The larger time step presents a very different picture, however. The first few coefficients follow the integrable case but then peel off. Not only do the coefficients decay at a slower rate, signifying that singularities are closer to the real axis, but they fluctuate strongly. In this case Padé resummation, which tries to fit the decay to an exponential (or the singularity to a pole) will not work. We note, however, that the fluctuations are quite regular, becoming almost periodic in the later terms. The periodicity coincides with that of the resonance condition, which arises every time $m\omega$ is an integer multiple of 2π. (To check against the figure, note that $\omega = 2\pi/10.94577\cdots$ for the case $\delta = 1/2$, predicting a periodicity $\Delta m \approx 11$.)

To see what properties the singularity might have, consider the following caricature for $q(\phi)$ in which the apparent periodicity is real. We set

$$f(\phi) = \sum_{m=1}^{\infty} e^{-m\alpha} P(m\omega)\, e^{-im\phi}, \tag{19}$$

where P is some periodic function,

$$P(\theta + 2\pi) = P(\theta) = \sum_k P_k\, e^{ik\theta} \tag{20}$$

modulating the otherwise exponential decay of the coefficients. If we insert the Fourier series for $P(m\omega)$ and exchange the order of summation, we can sum geometrically over m as follows,

$$f(\phi) = \sum_k P_k \sum_{m=1}^{\infty} e^{-im(\phi - i\alpha - k\omega)} = \sum_k \frac{P_k}{e^{i(\phi - i\alpha - k\omega)} - 1}. \qquad (21)$$

Thus there is a pole, with strength proportional to P_k, at every point where $\phi - i\alpha - k\omega$ is an integer multiple of 2π. If ω is irrational, the line $\mathrm{Im}(\phi) = \alpha$ is densely filled by such poles! Even this picture is optimistic, for if we include a factor m^β in the Fourier coefficients of f, which is probably more realistic, the result is a dense set of branch points of the type $(\phi - i\alpha - k\omega)^{\beta-1}$. To model a real function, we should add the conjugated series to Eq (19) and this would lead to a similar set of singularities on $\mathrm{Im}(\phi) = -\alpha$. Analytic continuation beyond these frontiers is meaningless, at least for physical purposes. Since Eq. (19) is the simplest possible model consistent with the numerically calculated coefficients, the continuation of the KAM torus must be at least as complicated.

The boundary formed by $\mathrm{Im}(\phi) = \pm\alpha$ is known as the natural boundary of the torus. A detailed investigation was made in [58] for a model very like Eq. (18). In general terms, one can understand that if a singularity is found at some complex position in phase space, it is likely to be mapped to another singularity under the action of a Poincaré map. In angle coordinates a singularity is thus mapped from ϕ_0 to $\phi_0 + \omega$ to $\phi_0 + 2\omega$ and so on, eventually filling up a horizontal line in the complex ϕ-plane when ω is irrational (remember that ϕ is 2π-periodic). In the original coordinates (q, p), the natural boundary is quite complicated. Explicit calculations are shown in [58], where it is seen to have a fractal structure. Any numerical attempt to continue the torus must falter when the natural boundary is reached, even if Fourier series are avoided. For example if the Hamilton-Jacobi is integrated numerically, small uncertainties in the initial data must eventually dominate the propagated surface.

When δ is small the problem in Eq. (18) has a separation of time scales which allows a straightforward implementation of perturbation theory on the whole complex plane. Because the kicking frequency is much faster than the frequency of unperturbed motion, an expansion is initially free of the resonance structure leading to natural boundaries. It is therefore tempting to try to estimate the complexified tori in this regime. This calculation holds its own surprises, however. It turns out that, to all orders in perturbation theory, the prediction for the complexified tori is that they form a single smooth manifold as in the integrable case. Abstractly, the mapping can be thought of as an operator

$$\mathcal{U}_\delta = e^{\frac{\delta}{2}L_T} e^{\delta L_V} e^{\frac{\delta}{2}L_T} \qquad (22)$$

where, for any function F on phase space, L_F is the Lie operator [51,59] $L_F G = \{G, F\}$ and e^{tL_F} is the flow on phase space defined by using F as a Hamiltonian (think of it as a Taylor expansion). Formally, we can resum $\mathcal{U}_\delta$ as follows,[59]

$$\mathcal{U}_\delta = e^{\delta(L_{H_0}+\delta^2 L_{H_2}+\delta^4 L_{H_4}+\cdots)}. \tag{23}$$

where the H_n's are themselves functions on phase space. For example, the first couple of terms are,

$$\begin{aligned} H_0 &= \frac{1}{2}p^2 + V(q) \\ H_2 &= \frac{1}{24}p^2 F'(q) + \frac{1}{12}F(q)^2. \end{aligned} \tag{24}$$

This expresses the mapping as the end result of a Hamiltonian flow with the Hamiltonian $H_\delta = H_0 + \delta^2 H_2 + \delta^4 H_4 + \cdots$, which is then an invariant whose level curves are the tori.

Because it is an analytic function of q and p, the level surface of any truncation of H_δ defines a single complex manifold for which the two real tori are its intersection with real space. This does *not* imply that the tori actually join smoothly however, just that any transversality is smaller than every power of δ. Extensive analysis of a similar effect exists in the case of separatrix splittings.[60] A separatrix connecting two maxima of a one-dimensional potential is replaced under perturbation by two distinct manifolds, an unstable manifold approaching the first maximum in negative time and a stable manifold approaching the second maximum in positive time. In nonintegrable systems they intersect transversally in phase space, at so-called heteroclinic points, with a nonzero angle of intersection. An analysis of the way in which the perturbation series diverges has allowed an estimate "beyond all orders" — the result is proportional to $\delta^{-3}e^{-A/\delta}$ where the constant A is determined by the position in the complex t-plane of the first singularity of the integrable limit.[60] Similar methods might very well allow an investigation of the splitting of complexified tori in the case above. Such an analysis would be useful in explicitly illustrating the structure of intersection assumed in the next subsection, though in practice the result is likely to be too small to play a role in quantum mechanics.

The most significant problem remains that in moderately to strongly perturbed problems, the formation of natural boundaries due to resonant effects offers a severe impediment, practical and theoretical, to the explicit construction of the complex intersection of tori.

Nevertheless we are going to describe a semiclassical calculation of energy-level splittings based on the assumption that complexified manifolds can be defined and calculated at the center. We do so because, even though the classical ingredients going into it are poorly understood, there have been verifications of the main features of its prediction [7–11] and above all because, at present, it is the only means available to us by which we can attempt to understand tunnelling in the KAM regime. Ultimately, justification might come from the fact that quantum mechanics should not "see" structure in phase space below a length scale set by $\hbar$. Since the pathology of the continued torus is a reflection of miniscule oscillations of the real torus it is easy to believe that it should have nothing to do with quantum mechanics. Perhaps it is the case that some microscopically smoothed version of the torus is responsible for the wavefunction and that a well-defined analytic continuation can be defined with the properties we demand. An observation which helps in this regard is that, for the purposes of semiclassical approximation, it is not necessary that the action $S(\mathbf{q})$ solve the Hamilton-Jacobi equations exactly because errors of $O(\hbar^2)$ are admissable. In order to proceed, we will assume without justification that some procedure of this type is available to us.

3.3 Wilkinson's formula

The basic result for tunnelling in near-integrable systems is a formula derived by Wilkinson in [7]. Even though the discussion there focussed on calculation of the avoided crossing arising from noncongruent tori, we will concentrate on the case of energy-level splitting between congruent tori. We note also that a version which works for the life-times of metastable states can be found in [11].

The starting point is a formula variously attributed to Landau,[61] Bardeen[62] or Herring.[63] Following Wilkinson, we will refer to it simply as Herring's formula. Let $|L\rangle$ and $|R\rangle$ be approximate states localised around two congruent tori. We will refer to them as the left and right states even though the symmetry need not be a spatial reflection (time-reversal symmetry is common). Exact states are odd and even combinations of them. Using the localised states as a basis, the eigenvalue equation can be written in the form,

$$\begin{aligned} \hat{H}|L\rangle &= \bar{E}|L\rangle + (\Delta E/2)|R\rangle \\ \hat{H}|R\rangle &= \bar{E}|R\rangle + (\Delta E/2)|L\rangle. \end{aligned} \tag{25}$$

Now let $\hat{\Theta}$ be any Hermitean operator such that $\hat{\Theta}|L\rangle \approx 0$ and $\hat{\Theta}|R\rangle \approx |R\rangle$. Applying $\hat{\Theta}$ to each of the equations above, respectively taking matrix elements with $|R\rangle$ and $|L\rangle$ and subtracting the complex conjugate of the first from the second, we get,

$$\langle R|\left[\hat{\Theta}, \hat{H}\right]|L\rangle = \frac{\Delta E}{2}\left(\langle R|\hat{\Theta}|R\rangle - \langle L|\hat{\Theta}|L\rangle\right) \approx \frac{\Delta E}{2}. \tag{26}$$

however, some numerical confirmation does exist. The first was in [8], where the dependence of avoided crossings on $\hbar$ with fixed classical dynamics was shown for a spin model to be consistent with Eq. (12). In [10], a calculation of band gaps for motion along a periodic channel resulted in a structurally similar formula, and in that case classical calculation was also possible, giving agreement provided coupling to third states played no role. This calculation is strongly related to the scattering matrix approach to chaos-assisted tunnelling which will be described in detail in the next section. Successful calculation of the ground state splitting of a double well was reported in [11]. Application to wider classes of problem, however, waits for some WKB method for extending the wavefunction beyond the natural boundaries of KAM tori.

4 Chaos-assisted Tunnelling

It was pointed out by Bohigas, Tomsovic and Ullmo [14,16] that the tunnelling mechanism described in the previous section often breaks down when a perturbation parameter is increased until dynamics is strongly mixed, so that large regions of phase space become filled with chaotic trajectories. Roughly speaking, direct tunnelling between tori becomes less favoured than a route with intermediate coupling to states supported in the chaotic sea.

It should be noted that, in addition to the spectral questions focussed upon here, there have been dynamical studies, initiated by Lin and Ballentine,[12] which examine the evolution of a wave packet initially localised within an island. There it is found that tunnelling of the wavepacket into a congruent island is greatly accelerated by the presence of surrounding chaos, an effect which is obviously strongly related to enhancement of splittings discussed in this section. Note that it is also possible in certain circumstances that suppression occurs.[66] In addition to the calculations discussed in detail below, there exists a considerable body of literature on spectral [24] and dynamical questions,[13] and on connections between the two.[17,18,19,22,25]

4.1 Overview and statistics

The first hint that something different is going on comes if splittings are examined as a function of a parameter. Whereas Eq. (12) predicts a smooth variation of splittings, the real tunnelling rate becomes irregular. We illustrate this with an example taken from [14,16]. In the potential,

$$V(x, y) = x^4/b + by^4 + 2\lambda x^2 y^2, \tag{34}$$

with $b = \pi/4$ chosen to break exchange symmetry, a strongly mixed regime is found when the coupling constant is $\lambda = -1/4$. The Poincaré section shown in

Fig. 7(b) exhibits two islands on the momentum axis and these are surrounded by a single chaotic sea. The islands correspond to two distinct tori with trajectories of the type shown in Fig. 7(a), related to each other by reflection in one of the coordinates. They can be quantised with a transverse quantum number n_1 and a larger longitudinal quantum number n_2.

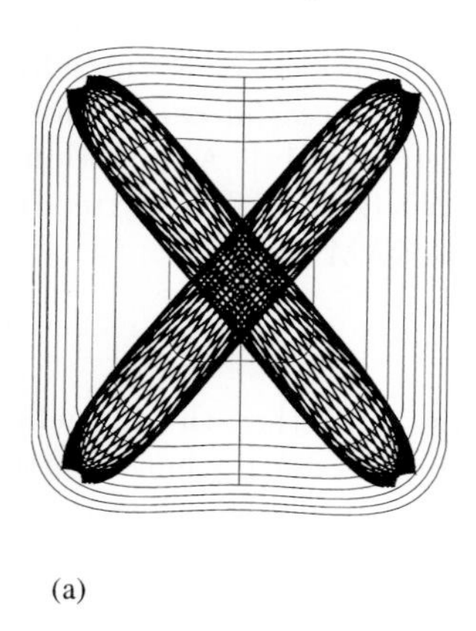

(a)

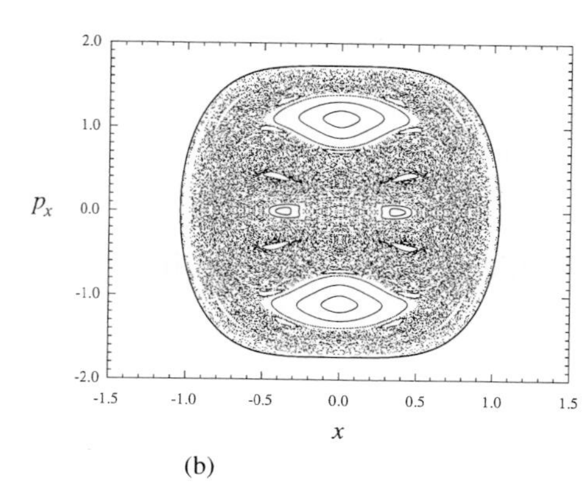

(b)

Figure 7: Dynamics of the potential in Eq. (34) is illustrated with the same conventions as in Figs. 2 and 3. In this case the energy is irrelevant because all energies are equivalent, up to a recaling of variables. Notice that tunnelling occurs in this problem even though there is nothing like a potential barrier.

Being congruent, the tori give rise to a doublet with splitting $\Delta E(\lambda)$. It is is plotted as a function of λ in Fig. 8 for three fixed pairs of tori at three different values of $\hbar$, decreasing by a factor of 3 each time. When $\lambda \to 0$ these tori cease to exist, so care needs to be taken in interpretation of this limit. However it is clear that for smaller values of λ, where the chaotic component of phase space is less important, the variation is smooth, consistent with our previous discussion. For larger values however, there are dramatic fluctuations, with peaks covering several orders of magnitude. A hint as to the origin of this structure comes from noting the values of λ at which states supported in the chaotic sea appear to cross the mean energy of the doublet in an avoided crossing. These are indicated by arrows in Fig. 8. Every peak coincides with an avoided crossing, though some crossings occur where there is no peak. Therefore we should associate the fluctuations with coupling to the chaotic sea.

An important point to note here is that there is no absolute dynamical partition of the chaotic region into mutually exclusive symmetric parts as there there is for integrable regions, so there is no *a priori* reason to expect chaotic states to appear as doublets. Therefore each chaotic state has fixed parities

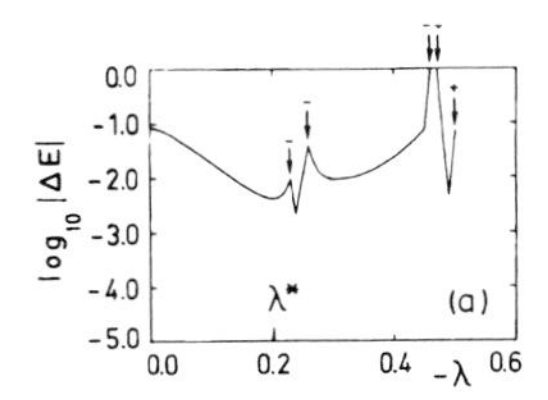

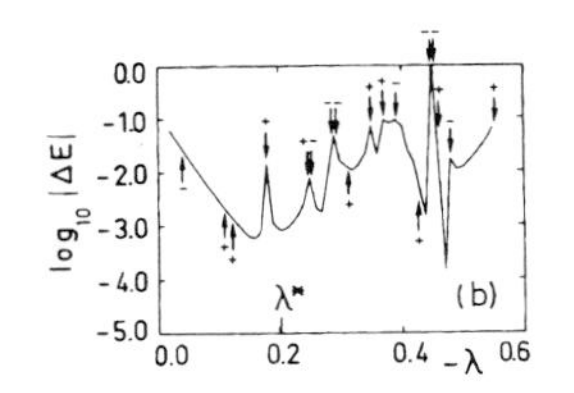

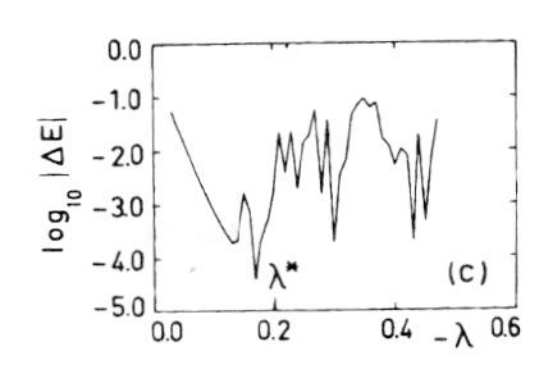

Figure 8: From [14,16], the dependence of a splitting on λ for three different values of $\hbar$, the first and last chosen so that the same classical torus is responsible in each case. The arrows indicate points where a chaotic state, whose parity is indicated by the sign above, undergoes an avoided crossing with one of the two toral states.

with respect to the reflection operations. This statement will be weakened later on, but for now we assume it to be the case. If $|L\rangle$ and $|R\rangle$ are as usual the localised integrable states, we will discuss cases where the exact states $|\pm\rangle = (|L\rangle \pm |R\rangle)/\sqrt{2}$ are either totally symmetric or totally antisymmetric, so we will only consider chaotic states $|C\rangle$ in those symmetry classes. The total symmetry of each state $|C\rangle$ at an avoided crossing is noted next to the arrow on the figure.

The following three-state Hamiltonian was invoked in [14,16] to model this situation,

$$H_3 = \begin{pmatrix} E+\epsilon & 0 & 0 \\ 0 & E-\epsilon & v \\ 0 & v & E^c \end{pmatrix}. \tag{35}$$

Here we assume that an even-parity chaotic state $|C\rangle$ with energy E^c couples to $|+\rangle$ with coupling strength v. We might suppose in addition that there is a direct coupling of strength ϵ between $|L\rangle$ and $|R\rangle$. In practice one of v or ϵ is usually dominant however. If it is ϵ then we are back in the situation described by Eq. (25), which has already been dealt with, so we will assume from now on that coupling to $|C\rangle$ is dominant and set $\epsilon = 0$. The splitting is therefore determined by the shift imposed on E_+ by v. It is easily calculated and seen to observe the limits,

$$\Delta E \longrightarrow \begin{cases} \dfrac{v^2}{E-E^c} & E - E^c \gg v, \\ |v| & E - E^c \ll v. \end{cases} \tag{36}$$

As we vary λ, we therefore get a peak of height $|v|$ as $E^c(\lambda)$ crosses E^+.

In an ideal world, we might like to have a semiclassical model for the states taking part in this event and arrive at explicit expressions for the splittings in terms of classical quantities. But, since even an analysis of the two-state coupling parameter ϵ has proved nontrivial, this is out of the question for now. It is still possible, however, to make an analysis of the statistics of the splittings.

More globally, the complexity of the parameter-dependence of ΔE is now seen to arise from the complexity of the chaotic spectrum and of the coupling to that spectrum. In [16] and [23], explicit models for the coupling to the chaotic spectrum were introduced to investigate the statistics, which we will now describe.

The simplest possible model is to assume that the integrable states are simultaneously weakly coupled to many states taken from an ensemble which models the chaotic spectrum. For time-reversal symmetric problems the ensemble is the GOE. To be more precise, since the even and odd chaotic spectra are independent, $|+\rangle$ and $|-\rangle$ are each coupled to an independent GOE. The most common situation is that the states are in the regime $E - E^c \gg v$, so the splitting is,

$$\Delta E = \delta_+ - \delta_- = \sum_i \frac{v_{i+}^2}{E - E_{i+}^c} - \sum_j \frac{v_{j-}^2}{E - E_{j-}^c} \tag{37}$$

where $\delta_\pm$ are the accumulated energy shifts of each symmetry class and E_{i+}^c and E_{j+}^c are taken from independent GOE's. It should be noted, however, that the tail of the distribution will be dominated by the exceptionally large splittings that arise from occasional coupling to single states with $E - E^c \sim v$ and the exact form of Eq. (36) has to be used in an analysis of that case.

It remains to construct a model for the v's. Each v represents an overlap between a localised integrable state and a state in the chaotic sea. We will imagine that a sequence of splittings are examined in which the decay of a localised integrable wavefunction into the chaotic sea is held fixed, which sets the scale of v at some small number. Fluctuations about the mean of the overlap are then naturally modelled in RMT by assuming the v's to be Gaussian-distributed (which is how the components of $|C\rangle$ are distributed in any generic basis). It is natural to rescale the splittings to the universal dimensionless variable,

$$y = \frac{\Delta E \cdot \Delta}{\langle v^2 \rangle}, \tag{38}$$

where Δ is the mean spacing between chaotic states.

Then, given the assumptions above it is shown in [23] that at small and moderate values the y's are distributed according to the Cauchy distribution,

$$p(y) = \frac{1}{\pi^2 + y^2}. \tag{39}$$

At larger values, there is a cutoff due to the fact that the behaviour $\Delta E \sim \sum v^2/(E - E^c)$ is replaced by $\Delta E \leq v$. This distribution is very extended. In fact, were it not for the large y cutoff, all its moments would be infinite. Remember also that this is not due particularly to a distribution in tunnelling rates from the torus to chaotic region, whose scale v is held fixed, so much as to the interaction of the integrable wavefunction with the chaotic states once the chaotic region is reached. It is also very different from the distribution of splittings between chaotic states separated by an energetic barrier, which is narrower (typically Porter-Thomas,[42] see Sec. 5.2). The dominant feature in producing a Cauchy-distribution is the nature of the singularity in Eq. (36) around avoided crossings. As stressed in [16,23], it is not strongly dependent on how the E^c's are distributed. It holds equally well in the extreme cases that the chaotic energy levels are completely uncorrelated (Poisson distribution) or completely correlated (rigid spectrum).[23,27]

In practice, it is often necessary to modify the model above to take into account the fact that partial dynamical barriers might exist in the chaotic region.[16,23,27] If a chaotic trajectory is followed in a Poincaré section, it often does not fill the stochastic region uniformly but instead gets stuck in certain subsets for long periods before breaking through to neighbouring regions, where again it gets stuck, and so on. Such barriers are often associated with KAM tori that have broken up but left behind partial barriers that allow flux to penetrate only weakly. When this happens, chaotic states tend to be localised within partially decoupled parts of phase space. It is analogous to the localisation of states near tori, but is weaker, since the classical localisation is not absolute. This introduces a correlation between the even and odd chaotic spectra that can strongly alter the distribution of splittings. Roughly speaking, even though δ_+ and δ_- in Eq. (37) remain Cauchy-distributed, their difference gives something else.

Many different scenarios are discussed in [16,23], but here we will just discuss the simplest case. Suppose that the two tori are embedded within two chaotic regions separated by a partial dynamical barrier which is invariant under symmetry. Suppose, in addition, that the (classical) time scale t_c for a trajectory to remain trapped in one region is longer than the Heisenberg time $t_H \sim \hbar/\Delta$. Then, chaotic states will themselves come in doublets whose separation is smaller than the mean spacing Δ. In this case the E^c_{i+} and E^c_{j+} should no longer be considered independent in Eq. (37), but to satisfy

$$E^c_{i+} - E^c_{i-} = \epsilon_i < \Delta. \tag{40}$$

An explicit formula is given in [16] for the variance of ϵ_i in terms of the classical flux crossing the partial barrier. In general the chaotic splittings ϵ_i are much bigger than the splittings between integrable states because the barriers are not absolute. If we assume $v_i \ll \epsilon_i \ll E - E^c_{i\pm} \sim \Delta$, we should continue as before but replace Eq. (37) by

$$\Delta E = \sum_i \frac{v_i^2}{E - E^c_{i+}} - \frac{v_i^2}{E - E^c_{i-}} \approx \sum_i \frac{v_i^2 \epsilon_i}{(E - E^c_i)^2} \tag{41}$$

The important point is that the nature of the tail of the divergence at avoided crossings is changed, which means that the Cauchy distribution for $p(y)$ must be replaced by something else. Explicit formulas are given in [23] but here we just give the behaviour of the tail for moderately large splittings,

$$p(y) \sim \frac{1}{y^{3/2}}. \tag{42}$$

For very large splittings, $\epsilon_i \ll E - E^c_{i\pm}$ no longer holds and instead the discussion leading to the Cauchy distribution is once again valid. Therefore there is a transition to the $1/y^2$ decay encountered earlier, which itself is cutoff at the very largest splittings as we have already discussed. From the stronger divergence near avoided crossings an even wider distribution has emerged.

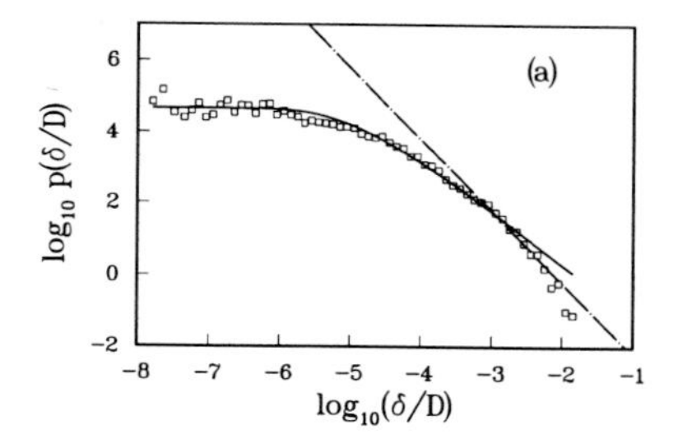

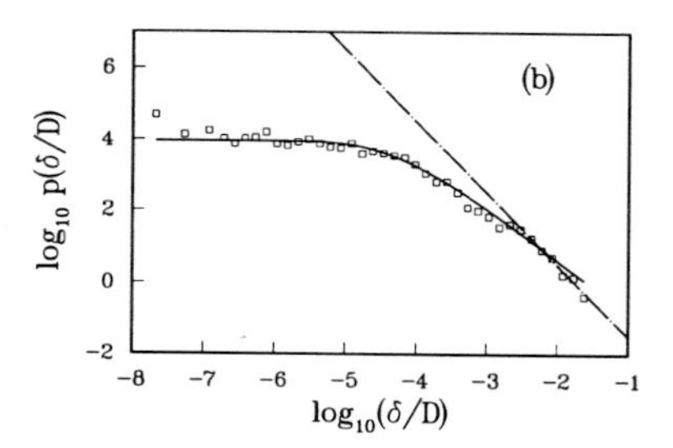

Figure 9: From [16,23], splitting distributions for two ensembles produced by letting λ vary over a small interval. In our notation, $\delta = \Delta E$ and $D = \Delta$. Both distributions are calculated for states localised in the islands in Fig. 7, (a) and (b) corresponding respectively to states with the lowest ($n_1 = 0$) and the first excited ($n_1 = 1$) transverse quantum numbers. Transport barriers in the chaotic region are identical in each case, but the variance of v is smaller in (a) than in (b). The solid curve is the full calculation of [23] (not given here) and the dashed line is the Cauchy tail.

In Fig. 9 a histogram, taken from [16,23], is shown of splittings calculated in the quartic oscillator of Eq. (34). The actual distribution agrees well with the full calculation of the distribution as given in [23]. The transition to a Cauchy tail is also seen. The various regimes here have also been seen in the annular disk problem.[27] The splittings are of eigenphases rather than of energy levels (next section), but the idea is the same. Finally, we note that it has been emphasised in [21,26,27] that an important role can be played by coupling to "beach states" supported in the marginal regions at the interface between the island and the stochastic sea.

4.2 Scattering matrix approach

The picture up to now of chaos-assisted tunnelling has been a largely quantum-mechanical one of interaction between the toral states and chaotic states supported in the stochastic sea. It is tempting to invoke a "classical" picture complementary to this, which is as follows. In order for a particle to tunnel from one torus to the other, it first has to reach the chaotic region, which it can cross by the comparitively rapid process of classical diffusion, and then tunnel from there to the other torus. Tunnelling rates are then related to the structure of classical transport across the chaotic sea.

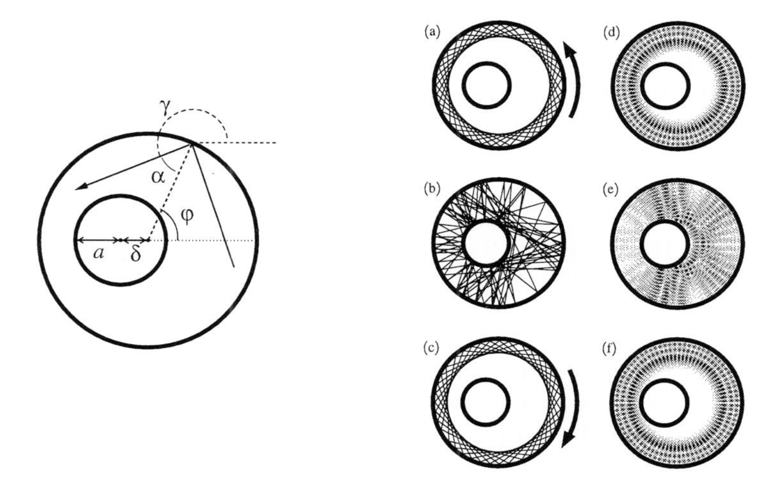

Figure 10: An illustration of the annular disc billiard, taken from [27]. At right are shown some typical trajectories and wavefunctions. The top and bottom wavefunctions are the real parts of the counterclockwise and clockwise states, respectively, associated with the tori shown in (a) and (c). The middle shows a typical chaotic state, along with a typical chaotic trajectory.

This picture finds its natural expression in an analysis of the annular disk, introduced by Bohigas et al.[15] The intuitive classical picture has been quantified for that system in the scattering matrix approach of Doron and Frischat,[21,26,27] which we describe here. Note that systems similar to the one dealt with here may have direct practical relevance for optical traps.[20]

The annular disk billiard is a two-dimensional circular cavity of radius R with a circular obstacle of radius a inside, as shown at left in Fig. 10. If the circles are concentric, the system is separable and wavefunctions are of the form

$$\psi_{mn}(\mathbf{x}) = B_m(k_{mn}r)\, e^{im\theta} \tag{43}$$

where $B_m(x)$ are linear combinations of Bessel functions of order m. Since the radial solution is independent of the sign of m, the states are doubly degenerate in this limit. The states are associated semiclassically with tori which cover the same annulus in configuration space, but whose trajectories encircle the obstacle in opposite senses and which are mapped into each other by time reversal symmetry. For an individual pair of states we will still use the notation $|L\rangle$ (clockwise) and $|R\rangle$ (anticlockwise). An important feature is that if the angular momentum is high enough that the impact parameter is greater than the inner radius, as in the top and bottom of Fig. 10, the classical solution does not change of the inner obstacle is deformed or, in particular, moved off-center by a distance δ.

Quantum-mechanically, there is a change, but only at the level of complex WKB. The real WKB solutions remain exactly the same after perturbation provided $m > k(a + \delta)$, but the boundary conditions placed on their analytic continuation into the center are different — the part of the continued torus following reflection from the obstacle is therefore affected by the perturbation. This results in the degeneracy being lifted to be replaced by a doublet whose splitting is entirely analogous to those we have examined before. In this case an explicit calculation of Wilkinson's formula is possible, since the torus is known analytically. However, the limit of zero perturbation [67] is not of the kind discussed earlier because the tunnelling action diverges (it corresponds more to separable limit discussed in [7]).

We now give a brief description of the scattering matrix approach in which many technical issues are brushed aside.[68,69] The wavefunction is first expanded in a cylindrical basis as follows,

$$\psi(\mathbf{x}) = \sum_m \left[\alpha_m H_m^-(kr) + \beta_m H_m^+(kr)\right] e^{im\theta} \tag{44}$$

where $H_m^-(x)$ and $H_m^+(x)$ are Hankel functions. The center of coordinates will always be taken to coincide with the center of the outer boundary. The coefficients α_m and β_m measure respectively the incoming and outgoing components of the wavefunction. We then consider separately reflections from the inner and outer boundaries to define scattering matrices as follows,

$$\alpha_m = \sum_l S^O_{ml}(k)\,\beta_l \qquad \text{and} \qquad \beta_m = \sum_l S^I_{ml}(k)\,\alpha_l. \tag{45}$$

The inner and outer scattering matrices are respectively $S^I_{ml}(k)$ and $S^O_{ml}(k)$ and explicit formulas can be written for them (S^O is trivial and S^I is obtained using addition formulas for cylindrical waves). Quantisation occurs if we can find a value of k such that, as a vector equation,

$$\boldsymbol{\alpha} = S^O \boldsymbol{\beta} = S^O S^I \boldsymbol{\alpha}. \tag{46}$$

So $S = S^O S^I$ should have an eigenvector $\boldsymbol{\alpha}$ with eigenvalue 1. The infinite matrix S is unitary and its spectrum lies on the unit circle. We expect quantisation to occur when an eigenphase $e^{i\theta_j}$ crosses 1. This simple picture is marred by the fact that there is generically an accumulation of eigenphases on one side of 1 (cylindrical waves with very high angular momentum are hardly affected by the scattering matrix). Therefore what is here described as a crossing of an isolated eigenphase is actually a sequence of avoided crossings, which, on a larger scale looks as if one eigenphase comes from below the accumulation, crosses all the eigenphases there, and then emerges on the other side of 1. We will keep to the simplistic description, which appears to work for practical numerical and semiclassical calculations, and defer elsewhere[68,69] for a resolution of the technicalites.

The spectrum of eigenphases (at fixed k) behaves much like an energy-level spectrum, as long as the accumulation near unity is discarded, so lessons learned from it can be translated without difficulty to more physical spectra. In particular, the eigenphases themselves come in doublets with splittings $\delta\theta_j(k)$, which will be taken from now on to be entirely analogous to energy-level splittings. If physical quantities are required, they can always be calculated later from the crossings of unity by eigenphases. In particular, one can in practice associate an effective derivative $\partial\theta_j/\partial k$ with each crossing by a doublet, and the physical splitting arising from this is related to the eigenphase splitting by $\delta k = (\partial\theta_j/\partial k)^{-1}\delta\theta_j$.

Before going ahead with a discussion of splittings, it is important for later discussion to stress an interpretation of $S(k)$ as the quantisation of a classical surface of section mapping, a portrait of which is shown in Fig. 11. This mapping is defined by recording the coordinates (γ, L) of a classical trajectory after each bounce off the outer boundary — γ is the angle the trajectory makes with the x-axis and L is the orbital angular momentum. This mapping is symplectic. That is, area in the γ-L plane is conserved from one bounce to the next.

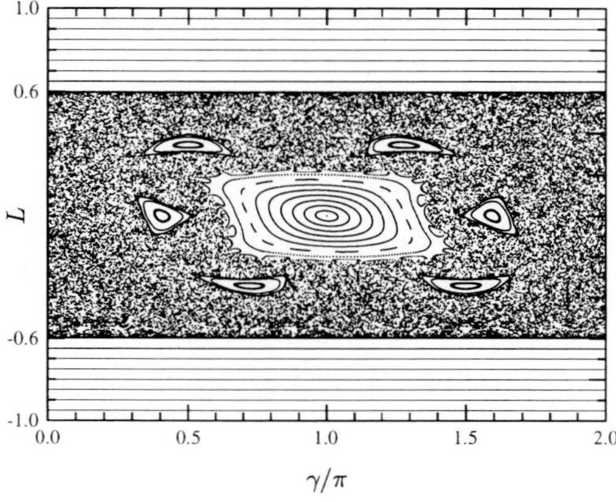

Figure 11: Also taken from [27], the surface of section mapping in the coordinates (γ, L). Trajectories with an angular momentum L greater in magnitude than 0.6 never hit the obstacle, as with (a) and (c) of Fig. 10, and fill out the horizontal lines at top and bottom. In the middle are found chaotic trajectories of the type (b) of Fig. 10.

A general feature of the semiclassical limit is that unitary transformations are related to symplectic transformations on the corresponding classical phase space.[70] For example, the classical transformation in Eq. (22) is the classical limit of the unitary operator,

$$\hat{\mathcal{U}}_\delta = e^{-i\delta\hat{T}/2\hbar} e^{-i\delta\hat{V}/\hbar} e^{-i\delta\hat{T}/2\hbar}, \tag{47}$$

where $\hat{T}$ and $\hat{V}$ are the kinetic and potential energy operators. In the same way, the mapping on (γ, L) is the classical limit of $S(k)$. More generally, we will find it helpful to make an analogy between $S(k)$ and the time evolution operator $\hat{\mathcal{U}}(T)$ integrating over a period a time-periodic Hamiltonian $\hat{H}(t+T) = \hat{H}(t)$. The eigenphases θ_j are then analogous to the Floquet energies defined by $\hat{\mathcal{U}}(T)|n\rangle = e^{-iE_nT/\hbar}|n\rangle$.

For the scattering matrix, and for kicked problems such as in Eq. (47), the dynamics is simple enough that exact and explicit formulas are available for their matrix elements (in the present case, $S(k)$ can be expressed as an infinite sum over products of Bessel functions [21]). More generally, there is a semiclassical approximation for matrix elements of a unitary operator as a sum over classical trajectories of the classical mapping. For a time evolution operator $\hat{\mathcal{U}}(t)$ it is called the Van Vleck approximation[3] and when the operator can be interpreted as a scattering matrix, as here, the semiclassical formula is often refered to as the Miller formula.[30,70]

For large numbers of iterations, approximation as sums over (long) classical trajectories is possible, but implementation is difficult enough that it is in itself a topic of discussion, to be described in the next section. The strategy of [21,27] is to calculate with the quantum-mechanical matrix elements but to use this calculation to build up an interpretation in terms of paths in the m basis, in what one might call a semiquantal approach. Using the exact forms for the matrix elements is analogous to using the path-integral in problems with continuous time evolution except that, in place of small time-steps, a single iteration of the mapping forms the atom from which quantisation is built up.

To motivate use of the scattering matrix to calculate splittings, consider first a time-independent problem, and follow the time evolution of a state initially given by $|\psi(t=0)\rangle = |L\rangle$. We suppose for the moment that $|L\rangle$ is exact, in the sense that Eq. (25) holds exactly. The amplitude in the state $|R\rangle$ after a time t is then easily calculated to be $\langle R|\psi(t)\rangle = ie^{-iEt/\hbar}\sin\Delta Et/2\hbar$, whose magnitude is proprtional to ΔE if t is small enough. A similar formula holds for Floquet problems or for the scattering matrix except that time is restricted to discrete values. In the annular disk billiard, an important simplification is that each pair of localised states $|L/R\rangle$ is approximated by a pair of basis states, $|\pm n\rangle$. As a result, the corresponding eigenphase splitting can be approximated by,

$$\delta\theta_n \approx \frac{2}{N}\left|\langle R|S^N|L\rangle\right| \approx \frac{2}{N}\left|\left[S^N\right]_{n,-n}\right|. \tag{48}$$

The first approximation holds provided $N\delta\theta_n \ll 1$. The second, in which the exact localised states are replaced by their basis-state approximations, works provided $N\delta\theta \gg \kappa$, where κ is a measure of the error $|R\rangle - |n\rangle$. Since κ is small, both conditions can be met simultaneously. A more careful discussion of this can be found in [27] (note also that a slightly different version, using the imaginary part of a matrix element, is preferred in [27] but we stick here to the simplest formulation).

By explicitly expanding powers of the scattering matrix according to,

$$\left[S^N\right]_{n,-n} = \sum_{\lambda_1\lambda_2\cdots\lambda_{N-1}} S_{n\lambda_1}S_{\lambda_1\lambda_2}\cdots S_{\lambda_{N-1},-n} \tag{49}$$

each eigenphase splitting is expressed as a sum over paths in a discrete angular momentum space. By grouping paths into different types and seeing which are most important, a precise meaning can be given to such questions as which "classical" routes are dominant.

For example, a direct tunnelling route would be one for which $\lambda_1 = \lambda_2 = \cdots\lambda_i = -n$ and $\lambda_{i+1} = \lambda_{i+2}\cdots = \lambda_{N-1} = n$. An interpretation in terms of paths says that a trajectory stays at $L = -n\hbar$ for i iterations of the map and then "tunnels" directly to $L = n\hbar$, where it stays from then on. This mechanism corresponds to a calculation in the Wilkinson scenario. In fact, the calculation in [10] is of a very similar kind, except that in place of the scattering matrix a transfer matrix corresponding to a unit cell of a periodic channel is used. In the present case, we would have $\delta\theta_n \propto S_{n,-n}$, which, using the Miller approximation for S_{nm} as a sum over complex trajectories,[30] is formally very similar to Wilkinson's formula.[10]

In the chaos-assisted regime, however, direct tunnelling is negligible in comparison with the contributions of more complicated routes. Two distinct kinds are identified in [21], which together are enough to capture all the essential behaviour of tunnelling rates. The first and simplest consists of paths which, having stayed on n for some initial period, make a transition to l in the stochastic region in which it then evolves for some time before making a second jump from l' to $-n$, where it stays from then on. A sum over all such paths gives a contribution $\delta\theta_n^{\mathrm{rcr}}$ where rcr refers to the simplest chaos-assisted mechanism regular $\rightarrow$ chaotic $\rightarrow$ regular. While relevant in principle, this contribution is found to be less important than a second, more complicated one. The most "expensive" parts of an rcr path are the jumps $n \rightarrow l$ and $l' \rightarrow -n$. In order to optimise the corresponding product $S_{-n,l'}S_{ln}$, l should be as close as possible to the boundary with integrable clockwise states and l' to the anticockwise boundary. Motion in this part of phase space is only marginally chaotic, with very slow rates of escape. As a result, one finds there quantum states that are effectively disconnected from the rest of the chaotic states and behave a lot like integrable states (even though there are no underlying tori). They are called "edge" or "beach" states in [27]. In the second class of paths, the transition from the regular to the chaotic parts is a two step process, in which the path spends some time on beach states. The resulting contribution to the splitting is labelled $\delta_n^{\mathrm{recer}}$, where recer stands for regular $\rightarrow$ edge $\rightarrow$ chaotic $\rightarrow$ edge $\rightarrow$ regular.

This picture is confirmed by looking explicitly at the evolution in a phase space representation of an initially localised state.[27] It is found that a significant build-up of amplitude occurs in the marginal region after some time, which is what one would expect if the recer mechanism were dominant.

In order to make further progress, the matrix elements in the sum over paths are taken from a simplified "block-matrix" model.[27] The beach states are taken to correspond simply to basis states labelled by l in two narrow bands just inside the chaotic region. It is assumed that the block of S corresponding to the truly chaotic region is diagonalised. This leads to a basis labelled by γ in the chaotic block with eigenphases θ_γ taken from a COE (eigenphase analog to the GOE). Matrix elements $S_{\gamma n}$ coupling chaotic states to regular or beach states are taken from a Gaussian ensemble whose variance is chosen to match that of the corresponding block of the original matrix S. This model returns us to the statistical modelling of the previous subsection, but for the annular disk there is the advantage that explicit models are available for coupling matrix elements.

With the simplifications of the block-matrix-model, path-sums within a single block can be performed explicitly. For example, summing over paths in the recer class yields a formula

$$\delta\theta_n^{\mathrm{recer}} = \sum_{\gamma} \frac{v_{n\gamma}}{2\sin(\theta_\gamma - \theta_n)/2}, \tag{50}$$

which is an eigenphase version of Eq. (37). Now, however, the effective coupling constants $v_{n\gamma}$ can themselves display some structure. They can be expressed as the following sums over beach states,

$$v_{n\gamma} = -2\mathrm{Re} \sum_{l,l'} e^{-i(5\theta_n + \theta_\gamma + \theta_l + \theta_{l'})/2} \frac{S_{n,l} S_{l,\gamma} S_{\gamma,-l'} S_{-l',-n}}{[2\sin(\theta_l - \theta_n)/2]\,[2\sin(\theta_{l'} - \theta_n)/2]} \tag{51}$$

In certain cases it may be legitimate once again to assume the coupling constants $v_{n\gamma}$ to be Gaussian-distributed, independently of the eigenphases θ_γ. In that case the Cauchy distribution is recovered as before. If, however, an ensemble of splittings is examined in which many beach states are sampled, the inverse square dependence of the $v_{n\gamma}$ on phase differences leads to modifications analogous to those implied by the approximation Eq. (41). In fact, a distribution with a $\delta\theta^{-3/2}$ tail can be recovered once again — though the underlying classical explanation is somewhat different.

A very natural theoretical development of chaos-assisted tunnelling would be to pursue further semiclassical analyses of the interaction with chaotic states. As will be discussed in the following section, when this is done coherence effects are often found which are not described by statistical models. In the present context, this would amount to using semiclassical approximations for the scattering matrix. Even though the following section focusses on chaotic regions, the calculations there may very well have direct relevance to chaos-assisted tunnelling between tori.

5 Chaos and complex trajectories

We now shift focus to dwell more on the quantum mechanics associated with chaotic regions of phase space. As discussed in Sec. 2.3, the natural semiclassical approach in this case is to use summation over trajectories to approximate various collective properties of quantum states, such as propagators, Green's functions and densities of states. The discussion here is still relevant to certain aspects of mixed systems. For example, the work of Shudo and Ikeda described in the first part has a very direct relevance to chaos-assisted tunnelling. The difference is that from now on quantisation of KAM tori no longer forms the starting point for investigation.

The first part of this section is a description of dynamical studies of the propagation of states under unitary evolution operators. The second looks at the calculation of spectral properties in time-independent systems. The distinction between the two can be somewhat blurred. For example, it is shown in [65] that calculation of spectra can be achieved using a semiclassically-defined unitary operator on Poincaré sections. However, we maintain the distinction because the types of question asked in each case tend to be different. A second division is that the dynamical part discusses tunnelling from regular to chaotic regions and the spectral part chaotic to chaotic tunnelling. There is no deep reason for the first of these restrictions — the general approach of the first part, if not all the details, would work equally as well to investigate tunnelling from chaotic to chaotic regions as it would from regular to chaotic. In the second part it is more natural to focus on tunnelling from chaotic regions because it is for completely chaotic problems that spectral methods like the trace formula are most powerful and best understood. Much of the discussion in the second part is taken from a collaboration of the author with N. D. Whelan.[40,41,42]

5.1 A dynamical perspective

In the scattering matrix formalism of the previous section, a very tempting semiclassical approach would be to approximate matrix elements of the scattering matrix in terms of classical trajectories.[30] The same holds any time a unitary operator arises in quantum mechanics which has a well-defined classical limit as a symplectic (area-preserving) transformation. Evaluation of the resulting Van Vleck or Miller approximation turns out to be a decidedly nontrivial task, however, all the more so when the presence of tunnelling demands that we calculate complex classical trajectories. The difficulty lies in the calculation of long classical orbits when the dynamics is nonintegrable.

It is no exaggeration to say that every semiclassical method in use for systems with chaotic dynamics traces it roots in one way or another back to an approximation of this type, usually for the propagator. Therefore their study is of more than academic interest. If an understanding of tunnelling in chaotic systems is to be achieved, we must begin here.

To date, there have been two extensive studies, using two different representations, of the use of complex orbits in such approximations. Adachi[35] has investigated coherent-state matrix elements $\langle z|\hat{\mathcal{U}}^n|z'\rangle$ of a unitary propagator defined similarly to Eq (47). Here $z = (q,p)$ is a point in phase space and the coherent state $|z\rangle$ is a minimum-uncertainty wavepacket centered there. If z and z' do not lie on the same orbit, semiclassical approximation uses complex orbits even if z and z' label points in real phase space The second study, by Shudo and Ikeda,[38] is of matrix elements in momentum representation. Since the representation used in this second case has more in common with our discussions elsewhere, we will concentrate on that calculation here.

Shudo and Ikeda consider a kicked system, defined quantum-mechanically by the unitary operator

$$\hat{\mathcal{U}} = e^{-iT(\hat{p})/\hbar} e^{-iV(\theta)/\hbar} \tag{52}$$

acting on the Hilbert space of 2π-periodic functions of θ. The obvious classical counterpart is $\mathcal{U} = e^{L_T} e^{L_V}$ in the notation of Sec. 3.2. It is very similar to the kicked systems we have considered before except that the kinetic-energy operator $T(\hat{p})$ is chosen somewhat unconventionally:

$$T(\hat{p}) = \omega p + \frac{p^2}{2}\frac{p^6}{p^6 + D^6} \qquad \text{and} \qquad V(\theta) = K\cos\theta. \tag{53}$$

This choice is made in order to manufacture a phase portrait in which there is a band of KAM tori around $p = 0$ and chaos at larger momenta, the transition being controlled by D. The general conclusions drawn from it should be equally applicable to more natural systems. A phase portrait is shown in Fig. 12 for the parameter values $D = 5$, $\omega = 1$ and $K = 1.2$.

Matrix elements $\langle p|\hat{\mathcal{U}}|p'\rangle$ in momentum representation of an operator like this share many formal similarities with the matrix elements $S_{mm'}$ of the scattering matrix considered in the previous section. For example, the momenta should be quantised according to $p = m\hbar$ because the conjugate variable θ is periodic. (The only real difference is that, for any fixed value of k, classically allowed phase space appropriate to $S(k)$ is finite whereas in the map above it is unbounded in p.) In particular, the issues facing calculation of a matrix element $\langle p|\hat{\mathcal{U}}^n|p'\rangle$, for an initial momentum p' in the integrable band and a final momentum p in the chaotic region, are the same as would face calculation of eigenphase splittings in the chaos-assisted regime using Miller approximation of $S(k)$. This calculation is exactly the subject of [38]. Note also that a numerical study of matrix elements like these has also been performed in [71] and comparison made with classical transport (though details of the dynamics involved are different).

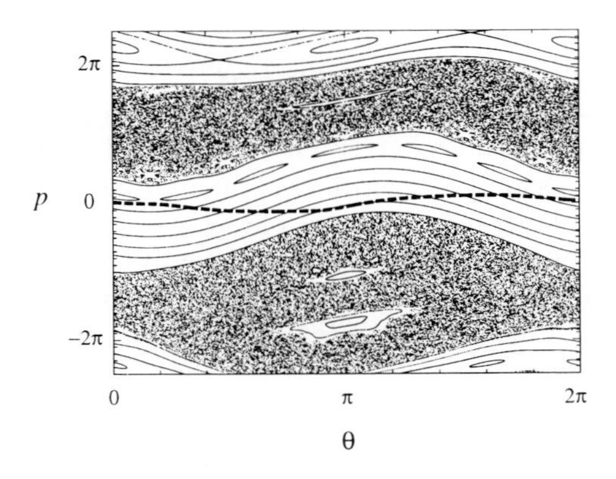

Figure 12: Phase portrait of the map used by Shudo and Ikeda. The heavy dashed curve is the image after 6 iterations of the horizontal line $p = 0$. In notation introduced below, it is the real part of $\lambda_{t=6}$. Later on, the natural branches of $\mathcal{M}_{t=6}$ will emerge from the two points on the curve where $\partial p/\partial\theta = 0$.

An advantage of dealing with a kicked system is that quantum-mechanical matrix elements can be expressed exactly as simple integrals. For example, a momentum-representation matrix element is obtained by inserting a resolution of the identity as follows,

$$\begin{aligned}\langle p|\hat{\mathcal{U}}|p'\rangle &= \oint \mathrm{d}\theta\ \langle p|e^{-iT(\hat{p})/\hbar}|\theta\rangle\langle\theta|e^{-iV(\theta)/\hbar}|p'\rangle \\ &= e^{-iT(p)/\hbar}\oint \mathrm{d}\theta\ \langle p|\theta\rangle\langle\theta|p\rangle\ e^{-iV(\theta)/\hbar} \qquad (54)\end{aligned}$$

Since the remaining matrix elements are just phase factors, this is a simple one-dimensional phase integral (corresponding to a Fourier coefficient of $e^{-iV(\theta)/\hbar}$). Therefore a careful and systematic semiclassical analysis can be made by approximating it with saddle-point integration. Not surprisingly, saddle points can be identified with classical trajectories, with the boundary conditions that the initial and final momenta should be p' and p respectively. Such trajectories can also be complex, arising when the contour of integration is pushed into the complex θ plane to pick up complex saddle points. For an example of a complete analysis of this kind, including systematic calculation of corrections at higher order in $\hbar$, see [72].

For higher iterates $\hat{\mathcal{U}}^n$ of the map, the calculation is *in principle* similar, but in practice is impossible to implement with the same degree of care. In this case, resolutions of the identity have to be inserted between exponentials in a long sequence, resulting in a many-dimensional integral (much the same as in the normal path-integral except that there is a minimum time-step). It is still possible to argue on general grounds that saddle-point integration results in a sum over classical trajectories, this time going from p' to p in n iterations (see Schulman [28] for a detailed discussion in the case of the regular path-integral or [38] for kicked systems). The problem is that it is impossible to visualise the contours so that rigorous statements might be made, for example, about which complex saddles contribute (only a subset is picked up when the contour is deformed to complex coordinates). Finding empirical rules to overcome this problem and finding systematically the very many long orbits that exist in the first place, are among the objectives of [38,39].

For the record, let us write down the general form of the resulting Van Vleck approximation,

$$\begin{aligned} K(p,p',t) &\equiv \langle p|\hat{\mathcal{U}}^t|p'\rangle \\ &= \sum_\gamma \sqrt{\frac{i\hbar}{2\pi}\frac{\partial^2 R}{\partial p \partial p'}}\; e^{iR(p,p',t)/\hbar} \end{aligned} \tag{55}$$

where the sum is over orbits γ with action

$$R(p,p',t) = \int_\gamma -\theta \mathrm{d}p - H\mathrm{d}t. \tag{56}$$

It is written it as if time were continuous, but it should be obvious how to interpret the action integral for kicked systems also. The normalisation assumes that p is quantised as $p = m\hbar$ and plane waves are normalised according to $\langle p|p'\rangle = \delta_{mm'}$. As well as from saddle point integration of path-integrals, this approximation can also be derived directly from WKB. This is sometimes a better approach in more difficult problems or when a more global view is required. It is also the dominant point of view in [38]. We will not go into it in the here, but a review can be found in [73].

As mentioned in the previous section, approximations of this type play a role in a wide class of problems. In scattering systems, the Miller approximation [30,70] looks very similar — this was used in [10] to approximate a transfer matrix, and from there to approximate splittings in the KAM regime. In multidimensional time-independent problems, there is a version for a quantisation of a Poincaré mapping [65] — in general the quantisation is only defined approximately but in billiards it can be defined exactly (as in the scattering matrix of the previous section). This latter case is essentially a reinterpretation of approximations for the energy-dependent Green's function [32] in which a time-translational symmetry along trajectories is removed by restricting dynamics

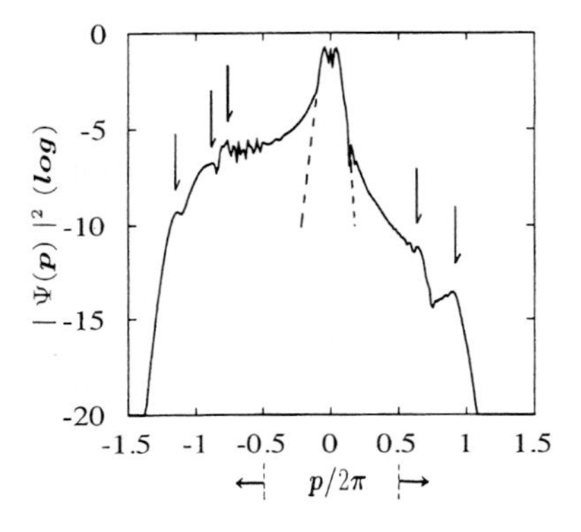

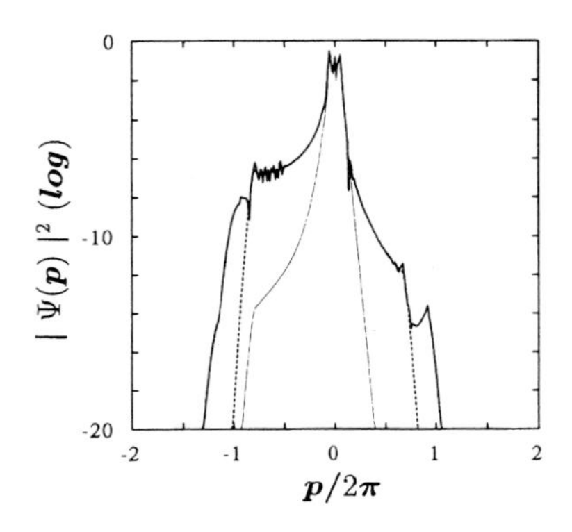

Figure 13: Plots of the transition probability $|\Psi_t(p)|^2$ for $t = 6$. The figure at left is computed quantum-mechanically. The right is semiclassical: the thin line is from the natural branches, the dashed line from the lower parts of Laputa chains and the thick line from complete Laputa chains.

to a Poincaré section. It is strongly related to the techniques used in the second part of this section. The general lessons learned from the coming discussion are therefore relevant to a much wider class of problems than the kicked systems described here.

Let us now consider what happens if this approximation is used to calculate the propagator when $p' = 0$. Since $p = 0$ lies entirely within the integrable region, real trajectories will remain forever bounded in p. Therefore the only contributions to $\Psi_t(p) \equiv K(p, p' = 0, t)$ when p is in the chaotic regions will come from complex trajectories and the propagator will be small. This is visible in Fig. 13(a), which plots $\log|\Psi_t(p)|^2$ for $t = 6$, following an exact quantum-mechanical calculation. There is a narrow band around $p = 0$ within which real trajectories can be found and the propagator is not small. Outside this band there is a rapid decay, as is to be expected when amplitude can build up only by tunnelling effects. The nonintegrability of the dynamics is responsible for the irregular nature of the decay — initially it is smooth, and monotonic on each side, but then complicated peaks and plateaus appear. This structure is explained by calculating and classifying the complex orbits entering in the Van Vleck approximation.

Let $\lambda_{t=0}$ be the horizontal line in phase space defined by $p = 0$ and let λ_t be its image after t applications of the classical map. Then a point (θ, p) on λ_t defines a classical trajectory that goes from p' to p in t iterations of the map, which are precisely the trajectories required by Eq. (55). A calculation of the

Van Vleck approximation therefore requires a calculation of the curves λ_t. If complex trajectories are required, we need the complexifications of λ_t, which topologically are two-real-dimensional planes. The initial $\lambda_{t=0}$ is naturally parametrised by the initial angle θ' of the trajectory, whose real and imaginary parts are denoted by

$$\theta' = \xi + i\eta. \tag{57}$$

This parametrisation by the *initial* angle is used even for the subsequent images λ_t. It is free of the singularities and multivaluedness that can arise if p or θ are used. If we want to calculate $\Psi_t(p)$ for real values of p, but using possibly complex trajectories, we need the intersection of λ_t with $\mathrm{Im}(p) = 0$. As one real condition placed on the two-real-dimensional λ_t, this defines a one-dimensional subset — which we will denote by the symbol $\mathcal{M}_t$. An extensive numerical investigation of the subset $\mathcal{M}_t$ was performed by Shudo and Ikeda. It is shown in Fig. 14 in the $\xi\eta$ plane, for $t = 6$ and the parameters of Fig. 12. It is clearly quite complicated, and needs some interpretation.

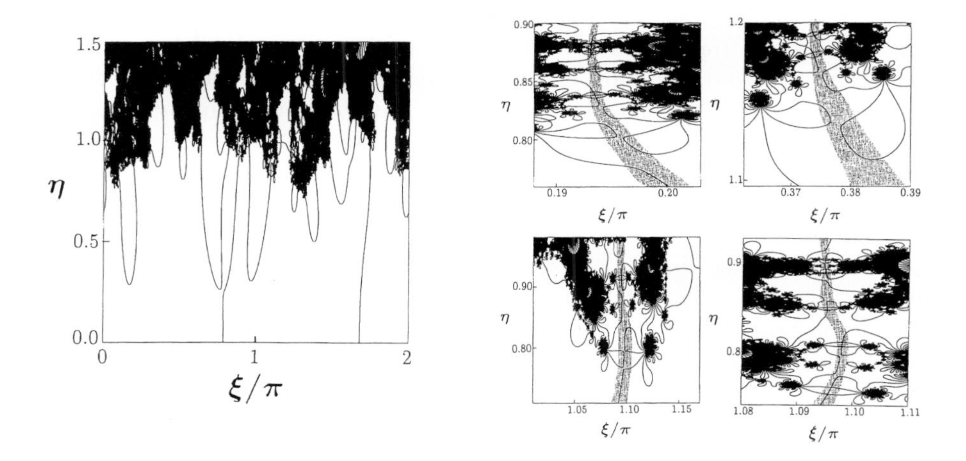

Figure 14: At left is shown the set $\mathcal{M}_{t=6}$ of complex initial conditions leading to real final momenta. Parts of $\mathcal{M}_t$ are magnified on the right hand side and Laputa chains highlighted.

The real axis $\eta = 0$ is part of the set and corresponds to the real trajectories. Two approximately vertical lines, called the "natural branches" in [38], emerge from it at points defining momentum space caustics — where λ_t folds over when projected on the momentum axis, and where $\partial p/\partial\theta' = 0$ $(= \partial p/\partial\theta)$.

These caustics are exactly analogous to the caustics that arose in torus quantisation, except that everything is rotated by $\pi/2$ in phase space. On the allowed side on the momentum axis, two real solutions can be found, which coalesce at the caustic itself, leaving two complex solutions on the forbidden side. One of the two complex solutions would give an exponentially large contribution to the Van Vleck approximation and can be discarded because deformation of an initially real contour in the steepest descents approximation could never pick up such a saddle point. The other complex solution contributes an exponentially small term and defines one of the natural branches. There are two natural branches because there are two momentum-space caustics. They determine the initial decay of $\Psi_t(p)$ on each side of the allowed region.

Had the dynamics been integrable, we might expect to be finished now. In fact, Shudo and Ikeda explicitly show that this is the case in the limit $D \to \infty$, which is integrable. In that case the decay of $\Psi_t(p)$ is smooth and monotonic and well described by the natural branches. This structure is the latest manifestation of the "direct" tunnelling route, encountered before in different guises.

In the nonintegrable case, the initial decay slows and becomes irregular. This is due to the appearance of additional branches of $\mathcal{M}_t$, which have smaller imaginary parts to their action. These branches eventually become quite dense, defining fractal-like self-similar structures in the $\xi\eta$ plane. Implementing the Van Vleck approximation is not simply a question of computing them and inserting them Eq. (55) because not all parts contribute. Note that this selection occurs even among exponentially small contributions. It arises because deformation of the integration contour in the method of steepest descents can miss even low-lying saddle points, depending on the topography of the phase function. Saddles may switch on and off as paramaters or variables are varied. This is the Stokes phenomenon. As already mentioned, an analysis for one-dimensional integrals is unproblematic, but a problem as difficult as the present one has to be treated semi-empirically for the moment. A more detailed account of this issue can be found in [39], including general rules to decide which branches of $\mathcal{M}_t$ contribute, according to position on a tree structure. Here we will be content with a minimalist treatment.

Shudo and Ikeda identify what they call "Laputa chains" in the set $\mathcal{M}_t$, and find that they dominate the propagator once p reaches the stochastic regions. Examples are shown in Fig. 14, where Laputa chains are highlighted by shading them. Each Laputa chain consists of a sequence of collisions between different branches of $\mathcal{M}_t$, in which one branch enters the chain and the other leaves. As this happens, a dance is performed in which responsibility for the dominant contribution to $\Psi_t(p)$ is exchanged among them. Starting with a part of a branch in the chain, let us follow the evolution of its contribution in either direction. As it is followed in the direction of increasing imaginary action, it collides with an entering branch and leaves, fading in importance as it goes on. Meanwhile, the entering branch takes its place in the chain. If it is followed in the other direction, its contribution increases. If this could continue forever, it would explode and yield an unphysical result. What must have happened is that its contribution switched on as it entered the chain, near the collision with the branch it replaced. The switching on and off of branches at a collision is related to the existence of a caustic point between them, where the corresponding solutions for $p(\xi + i\eta)$ coalesce. There is not enough space here for a detailed determination of exactly where this Stokes phenomenon kicks in. A detailed description can be found in [39]. We will satisfy ourselves with knowing that each branch in a Laputa chain starts contributing somewhere near its entry into the chain and continues to do so from there on, its relative importance diminishing when it leaves.

We are therefore left with a situation in which the most important contributions to the propagator arise from the parts of $\mathcal{M}_t$ within Laputa chains. This is important because it allows us to concentrate on relatively tractable parts of $\mathcal{M}_t$, while passing over the infinitely dense, fractal concentrations. The relative importance of parts within the chains is enhanced by the feature that variation of the imaginary part of the action is much slower there than on the parts outside. To understand what this means for the propagator, we need to translate the picture from the $\xi\eta$ plane to the p coordinate. If we move along a Laputa chain, switching from one branch to another at collisions, it is found that p oscillates back and forth, covering roughly the same interval between each pair of collisions, but changing direction when switching from one branch to the next.[38] The Laputa chain therefore gives rise to a plateau in $\Psi_t(p)$ over that interval, within which there is a complicated interference pattern which arises from adding up contributions from the various segments of the chain. Outside the interval there is a steep drop arising from the rapid increase of $\mathrm{Im}[R(p, p', t)]$ on branches leaving the chain. When contributions from various Laputa chains are superimposed, the picture is of a sequence of plateaus which end by dropping down to the level of a next important plateau, and so on (actually, a plateau can itself have a significant slope, it is just smaller than where the drops take place). Carrying this program out in detail leads to quantitative agreement with the quantum-mechanical propagator, as can be seen by the comparison in Fig. 13.

This analysis explains the broad features that were present in chaos-assisted tunnelling, and does so while using purely classical calculations (albeit very complicated ones). Enhancement of the tunnelling rate relative to the integrable case is explained by the appearance of nonnatural branches in $\mathcal{M}_t$, which encroach on the integrable interval of p as t increases. Erratic dependence of the tunnelling rate on parameters is unsurprising in view of the fractal nature of $\mathcal{M}_t$.

It is also indicative of the general strategy that works when chaotic behaviour is present. By definition, extended calculations in such cases are difficult, or at least numerically demanding. Therefore for practical implementation of any semiclassical approximation, some organising principle must be found for the classical trajectories that take part. In the present case, this amounted to the identification of Laputa chains. On other chaotic problems, it might correspond to an encoding of periodic orbits using a symbolic dynamics. It is true that the resulting calculations are highly sensitive to perturbation or changes of parameter, so detailed predictions using the most intricate structures are unlikely to be relevant to any concrete experiment. However, the general features they predict will be generic, and a detailed analysis of model problems such as this one will help in analyzing and classifying real-world situations.

5.2 Complete chaos in the energy domain

In time-independent problems it is more helpful to deal with energy-dependent Green's functions and related quantities than with the propagator discussed in the previous subsection. The two are related by a one-sided Fourier transform as follows,

$$G(q, q', E) = \frac{1}{i\hbar} \int_0^\infty \mathrm{d}t\, e^{iEt/\hbar} K(q, q', t), \tag{58}$$

where, by force of convention, we revert to configuration- rather than to momentum-space representations. If we have already investigated propagators and understood how to treat tunnelling by including complex orbits, it seems simple now to carry this over to Green's functions by evaluating this time integral by the method of steepest descents. The novel feature is that to carry out this program we are required to construct the propagator for complex t, which we haven't done so far. The same statement applies to other manipulations, such as taking traces, which require complexification of (q, q'). It is easy to argue in general terms that the result is a sum over orbits with energy E (usually chosen real), with possibly complex times of evolution, and to give explicit forms for their contributions.[32] It is difficult, however, at least for chaotic problems, to give a rigorous account of the nuances associated with Stokes phenomena etc. As usual we are going to resort to heurisitic arguments and educated guessing to overcome this hurdle.

Armed with the methodology of Green's functions and trace formulas, the natural problems to tackle are those with completely chaotic dynamics. So here we start looking at tunnelling out of chaotic quantum states.

While on KAM tori localisation in phase space is present by definition, it is less ubiquitous for purely dynamical effects to be responsible for localisation of a chaotic quantum state. It is still possible however. An example might be a phase portrait like the one in Fig. 12, but with an additional time reversal symmetry $p \longrightarrow -p$, so that states might exist within two congruent stochastic regions, separated by KAM tori. Splittings would then appear even though there is no energetic barrier. Dynamical studies of such quantum transport across phase space barriers can be found in [71,74]. (Note that, in higher dimensions, classical localisation within KAM barriers is no longer absolute. However, the classical mechanism for transport across KAM regions, known as Arnold diffusion,[51] is often very slow and might well be longer than a quantum-mechanically defined tunnelling rate.) The examples used in this section will be restricted to tunnelling across energetic barriers because such systems seem to be more common in practice. However, the basic principles should apply to dynamical tunnelling also.

Tunnelling from chaotic regions of phase space has recently become relevant to the interpretation of experiments involving transmission of electrons through resonant tunnelling diodes.[76–79] In these experiments, energetic barriers are formed by inserting thin layers of $(Al_{0.4}Ga_{0.6})As$ between layers of GaAs. When a quantum well is formed between two such barriers, and a magnetic field applied at some arbitrary angle, dynamics in the well can be chaotic in the presence of an electric field. The appearance of chaos has a direct and measureable imprint on current-voltage characteristics. A regular series of peaks emerges which cannot be due to simple resonance with states in the well because there are many more states than peaks. Rather, the peaks can be related to a collective spectral behaviour that is very characteristic of the quantum mechanics of classically chaotic problems. The periodicity of the peaks as a function of applied voltage can be associated with quantisation of the action of a periodic orbit, $\int p dq \sim n\hbar$. Because the orbit is unstable, this is not directly a quantisation condition for individual quantum states. It is, however, a standard feature of fluctuations in collective spectral quantities such as defined in Eq. (11). (In practice, peaks defined by such action quantisation are often highly correlated with the appearance of "scarred" states which concentrate near the periodic orbit. So in this sense it *is* possible to associate peaks with resonance with individual states, as long as they are taken from the particular subsequence of scarred states. This is the point of view in [76]).

The barriers in this problem are very thin and the tunnelling aspect is in a sense trivial — in semiclassical calculations one can simply assign a reflection and transmission coefficient to each orbit bounce, for example. So analyses of the problem have thus far not not been unduly preoccupied with the kinds of calculation of interest here. It is clear, however, that one could easily arrive at an experimental situation where account needs to be taken more explicitly of tunnelling dynamics (meaning, for example, the properties of complex trajectories crossing the barrier). The calculations outlined here go some way towards achieving this.

To discuss the general characteristics of tunnelling in chaotic systems, we turn once again to symmetric double wells. An example with predominantly chaotic dynamics was illustrated in Fig. 3. In any problem with a chaotic spectrum, the obvious semiclassical method to use is the trace formula. Tunnelling can be treated straightforwardly if the trace formula can be summed analytically. This was shown in one dimension by Balian and Bloch [32] for metastable systems and then by Miller [34] for double wells, where explicit geometric summation of orbits is possible. A uniformisation and extension to other symmetries is given in [36]. In chaotic systems, a completely analytic treatment is possible in the antisemiclassical case where $\hbar$ is large, so that only a few short orbits suffice to determine the spectrum. A calculation of band spectra, incorporating tunnelling effects through complex orbits, was performed in that case by Lebœuf and Mouchet.[37] More generally, orbits in chaotic systems cannot be summed analytically in the semiclassical regime — it is usually done numerically — and one faces the problem that complex trajectories offer very small contributions which are completely swamped by those of the large numbers of real orbits. A more direct method to calculate tunnelling effects is then required.

One way to achieve this is to use the following spectral function, introduced in [40],

$$f(E) = \sum_n \Delta E_n \, \delta(E - E_n), \tag{59}$$

where ΔE_n is the splitting and E_n the mean energy of a doublet. This function has encoded within it a complete specification of tunnelling rates in the problem. Semiclassically, it is approximated by a sum over orbits that are necessarily complex, which is seen as follows. If we construct the spectral staircase functions $N_\pm(E)$ counting odd and even energy levels, their difference gives a very good approximation to $f(E)$,

$$f(E) \approx N_+(E) - N_-(E). \tag{60}$$

The approximation consists of replacing $\Delta E_n \delta(E - E_n)$ with a step function of unit height and width ΔE_n. A trace formula exists for each of the odd and even densities of states $\rho_\pm(E)$, summing over all orbits with energy E which are periodic up to application of the symmetry operation $\mathcal{R}$.[75] We distinguish between periodic orbits, which satisfy $\phi_t(p_0) = p_0$, and pseudoperiodic orbits, which satisfy $\phi_t(p_0) = \mathcal{R}p_0$, where ϕ_t is the classical flow in phase space. When the difference $\rho_+(E) - \rho_-(E)$ is formed, orbits which are genuinely periodic cancel because they contribute identically to both symmetry classes. Only pseudoperiodic orbits, which contribute to $\rho_\pm(E)$ with alternating signs, remain. Note that pseudoperiodic orbits are necessarily complex, since real orbits cannot cross the barrier.

By integrating the difference in densities of states, we arrive at an approximation for $f(E)$ as a sum over the pseudoperiodic orbits γ of energy E which is of the following form,

$$f(E) \approx \frac{2}{\pi} \operatorname{Im} \sum_\gamma \beta_\gamma \frac{e^{iS_\gamma/\hbar}}{\sqrt{-\det(M_\gamma - I)}}. \tag{61}$$

Here, S_γ is the action and M_γ a symplectic matrix which linearises motion in degrees of freedom transverse to the orbit. The dimensionless factor β_γ is a sort of reflection coefficient which is usually equal to 1 but which is 1/2 for the most important orbit, which is the well-known "instanton" or "bounce" orbit [33] described below.

To implement this formula, we have to start searching for and classifying pseudoperiodic orbits, the order of importance being decided by the imaginary parts of their actions. The minimum imaginary action is found by the instanton orbit, computed in potential problems by turning the potential upside-down, $V(\mathbf{q}) \to -V(\mathbf{q})$. If $[\mathbf{q}(\tau), \mathbf{p}(\tau)]$ is a pseudoperiodic orbit traversing the inverted barrier in a time τ_0 with action K_0, then $[\mathbf{q}(-i\tau), i\mathbf{p}(-i\tau)]$ is a pseudoperiodic orbit of the original problem with an imaginary period $-i\tau_0$ and an imaginary action $S_0 = iK_0$. In a problem without time reversal symmetry, there can still be an instanton orbit with an imaginary action and period. However there is no longer a simple transformation to a real trajectory of a real Hamiltonian and the coordinates have both real and imaginary parts during evolution.

As an explicit example, in a problem with a magnetic field, the orbit can still be transformed into one with a real period, but the transformed dynamics is defined by making the magnetic field imaginary ($B \to -iB$) in addition to inverting the potential. Even though dynamics is for real time in the transformed problem, the resulting Hamiltonian vector field is complex and trajectories $[\mathbf{q}(\tau), \mathbf{p}(\tau)]$ have complex coordinates. Assuming $\mathbf{p}(\tau)$ to be kinetic momentum, the trajectory of the original problem is $[\mathbf{q}(-i\tau), i\mathbf{p}(-i\tau)]$ as before. This is easily checked by direct substutution. An example of such an orbit is shown in Fig. 15(a). Notice that, even though the Hamiltonian is not

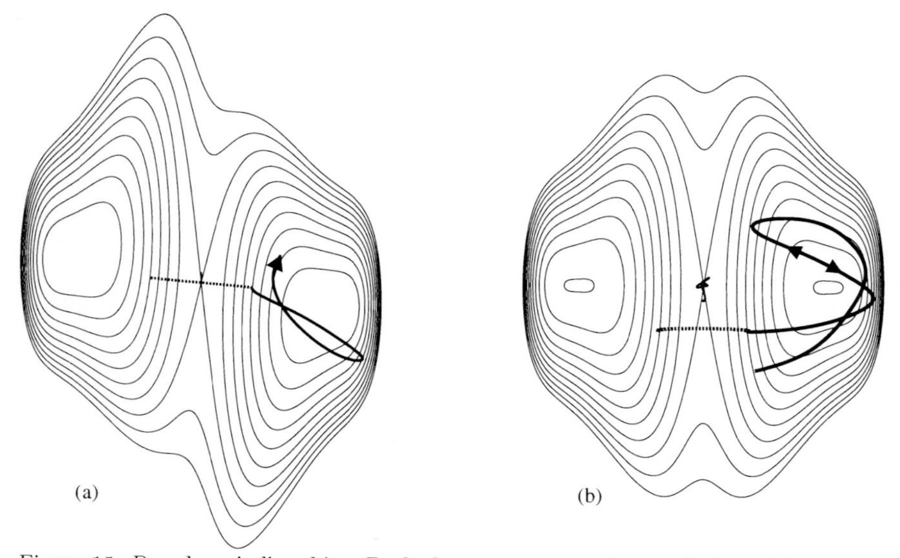

Figure 15: Pseudoperiodic orbits. Dashed parts correspond to evolution of time along the imaginary axis and solid curves to real time evolution. The system in (a) does not have time-reversal symmetry — the addition of a term $-xy/2$ to the potential in Eq. (8) breaks reflection symmetry in y and there is a magnetic field of $B = 1/2$ in natural units. Nevertheless, an instanton orbit with imaginary action and period is found for which the real part of the trajectory is self-retracing. At the turning point, all coordinates are real and the trajectory can be followed into the well under real dynamics. Unlike the case treated in detail in the text, the real continuation is not periodic. The orbit in (b) is a multidimensional pseudoperiodic orbit of the problem defined by Eq. (8). In imaginary time it crosses the barrier and in configuration space ends at the symmetry-image of the initial condition. The momentum is oriented incorrectly, however, and this is corrected by the self retracing extension in real -time dynamics. Both of these orbits also have small but nonzero imaginary parts during time-evolution, visible as the small squiggles at the center.

time-reversal symmetric, the real part is self-retracing — the origin of time can be chosen so that $[\mathbf{q}^*(i\tau), -i\mathbf{p}^*(i\tau)] = [\mathbf{q}(-i\tau), i\mathbf{p}(-i\tau)]$ (because the original Hamiltonian is real, dynamics is invariant under complex conjugation). Notice also that the turning point at $\tau = 0$ has real coordinates so we are free to continue with real time evolution once the barrier has been crossed, as shown by the solid curve in Fig. 15(a).

Hopefully, this is convincing evidence that in a wide class of barrier penetration problems, one can define an instanton orbit with imaginary action. When its contribution is included in $f(E)$, the result is a smooth monotonic function of energy $f_0(E)$. It turns out that this function gives the average behaviour of $f(E)$ — more exactly, $f_0(E)$ approximates the average value of $x_n = \Delta E_n \rho_0(E_n)$, the splitting measured in units of the mean spacing between doublets. It plays a role similar to that played by the Thomas-Fermi density of states $\rho_0(E)$ in the regular trace formula. (It is important to note, however, that the analogy is not complete because $f_0(E)$ is associated with finite-length orbits and has a nonzero action in it). In the top part of Fig. 16, an explicit comparison between $f_0(E)$ the values of x_n, computed numerically for the potential in Eq. (8), verifies the claim.

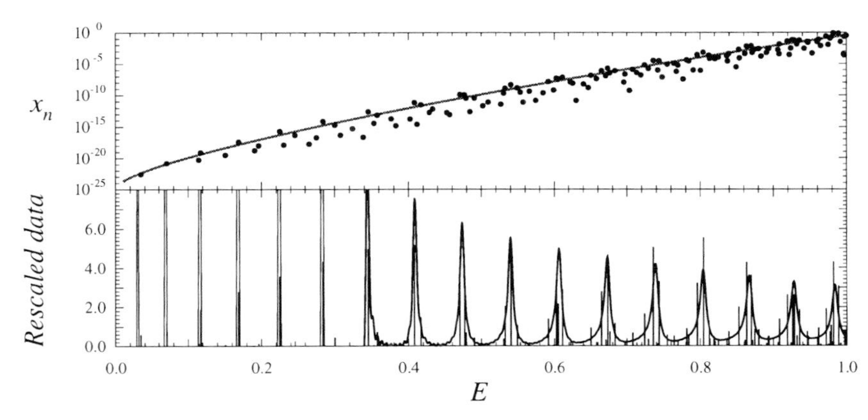

Figure 16: Results of a numerical calculation of the spectrum of the double well in Eq. (8), with $\mu = 1/10$ and $\hbar = 8^{1/2}/100$, are shown on top as filled circles. The horizontal position is the mean energy of a doublet and the vertical axis gives the splitting x_n in units of the mean level spacing. For simplicity, the data here is restricted to states even in y. The curve is $f_0(E)$ and is a good predictor of the average behaviour of the x_n's. Following rescaling by $f_0(E)$, splitting fluctuations are visible on the linear plot at the bottom, shown as a grass of spikes, each of height y_n. A periodic fluctuation in the splittings is accounted for by the prediction $y_{sc}(E)$ of the quasi-one-dimensional pseudoperiodic orbits, shown as a continuous curve.

In one dimension, $f_0(E)$ gives a complete description of splittings since it turns out that a separation $\rho_0(E)f(E) \approx f_0(E)\sum_n \delta(E - E_n)$ is possible. Another way of saying this is that in a one-dimensional version of Fig. 16, the values of x_n would fall on top of the curve $f_0(E)$. In chaotic systems, the situation is qualitatively different because there is a lot of scatter about the average. The essential problem in the chaotic case is to arrive at an understanding of these fluctuations. As with all chaotic problems there are two complementary ways of doing this. The obvious one in the present formalism, discussed in [40,41], is to pursue the periodic orbit approach to understand long-range fluctuations. The statistical approach of RMT is also possible and works at the shortest energy scales. This is discussed in [42].

Tunnelling-rate fluctuations are analysed in the periodic orbit approach by finding orbits whose action has a nonzero real part, $S_\gamma = S_{0\gamma} + iK_\gamma$. The contribution $f_\gamma(E)$ to $f(E)$ then oscillates on an energy scale $\delta E = \hbar/T_{0\gamma}$, where $T_{0\gamma} = \partial S_{0\gamma}/\partial E$ is the real part of the period. The easiest way to start finding these is if we begin with a potential that has a symmetry $V(x,-y) = V(x,y)$ in addition to the symmetry $V(-x,y) = V(x,y)$ responsible for the splittings (and has no magnetic field). In that case motion restricted to the symmetry axis is essentially one-dimensional and described by the potential $V(x,0)$. In addition to the instanton orbit, there are pseudoperiodic orbits defined by following the imaginary time evolution of the instanton with real evolution over an integer multiple rT_0 of the period T_0 of a real orbit in one of the wells. The fact that a sequence of pseudoperiodic orbits is defined in this way, whose actions $S_r = rS_0 + iK_0$ all have the same imaginary part, is specific to problems with a symmetry axis. It is expressly not the case for the Hamiltonian shown in Fig. 15(a) where the real continuation of the instanton is not periodic. We will have to keep this in mind, but a lot of intuition can be built up by first considering the more symmetric case.

In one dimension, the sum over r can be performed geometrically and we arrive at a function $f(E)$ which exhibits explicit singularities at the values of E given by Bohr-Sommerfeld quantisation. This is equivalent to the calculations in [32] and [34]. If the orbit is embedded in a higher-dimensional potential, the monodromy matrix M_r must be included in the amplitude of Eq. (61). This can be shown to be of the form,[40]

$$M_r = {M_0}^r\, W = \begin{pmatrix} a_0 & b_0 \\ c_0 & d_0 \end{pmatrix}^r \begin{pmatrix} t & -iu \\ iv & w \end{pmatrix} \tag{62}$$

where M_0 is the real monodromy matrix of the real orbit and W the monodromy matrix of the instanton. M_r has complex entries and the square root of the determinant enters Eq. (61) as a complex number. It therefore gives rise to a shift in the phase of oscillation. This shift plays a role analogous to that played by the Maslov index in the regular trace formula.[83] If the real orbit is unstable, which is what we expect in predominantly chaotic dynamics, the resulting amplitude decays exponentially with r. The constructive interference that leads to singularites in one dimension is no longer strong enough to do so and the result is a smoothly oscillating function $f_{\text{osc}}(E)$ with a local period $\delta E = \hbar/T_0(E)$.

Actual splitting fluctuations are compared with $f_{\text{osc}}(E)$ in the bottom part of Fig. 16 for the double well potential in Eq. (8), taken from [40]. The data is presented on a linear scale following division by the mean $f_0(E)$, which removes the systematic variation of scale seen in the original data. For each doublet, we can define a scaled splitting $y_n = x_n/f_0(E)$ which should average to one and whose fluctuations should be modelled by $y_{sc}(E) = [f_0(E) + f_{osc}(E)]/f_0(E)$. Certain aspects of the scaled splittings y_n are universal and they play a role similar to that played by the variables defined in Eq. (38) for the case of chaos-assisted tunnelling. Quantitative agreement can be obtained between $y_{sc}(E)$ and the fluctuations of the data y_n by averaging both of them with a Gaussian.[40] Here we will focus on a more direct comparison between the discrete data and the semiclassical prediction. To do this we identify two regimes. There is a bifurcation of the real orbit around $E \approx 1/3$, above which it is unstable and below which it is stable. We deal with the stable case first.

Below the bifurcation, the real orbit lies at the center of an island which is embedded in a stochastic sea. In this regime the amplitude from the monodromy matrix does not decay and constructive interference from repetitions is strong enough to produce singularities, as in one dimension. Thus at lower energies, $f_{osc}(E)$ explicitly gives δ-functions, which correspond to states supported in the island. There are also states outside the island for which tunnelling rates are much smaller because access to the instanton is denied to them. These are not described by singularities of $f_{osc}(E)$, though it is important to take them into account if quantitative agreement is to be achieved for averaged quantities. Geometric summation of $f_{osc}(E)$ is possible, as in one dimension, and gives rise to estimates for the island-state splittings which are of the form $(2\hbar/T_0)A(M_0, W)e^{-K_0/\hbar}$. This works quantitatively.[40] Notice that this has the same $\hbar$-dependence as in one dimension. The only difference is a factor $A(M_0, W)$ which is determined entirely by a linearisation of the transverse dynamics.

Above the bifurcation, where the instanton tunnelling route is shared more democratically among all states, the $\hbar$-dependence is different — while we cannot easily calculate splittings state-by-state, we can assign an average behaviour $\Delta E \sim \rho_0(E)^{-1}e^{-K_0/\hbar} \sim \hbar^2 e^{-K_0/\hbar}$. It is amusing to compare the integrable and chaotic dependences with the intermediate dependence $\Delta E \sim \hbar^{3/2}e^{-K/\hbar}$ predicted by Wilkinson for KAM systems. In d dimensions we would respectively have $\hbar e^{-K/\hbar}$, $\hbar^{(d+1)/2}e^{-K/\hbar}$ and $\hbar^d e^{-K/\hbar}$ for the integrable, KAM and chaotic regimes. The usual disclaimer applies that in practice the exponential is much more important in determining overall size.

In the regime where the orbit is unstable, $f_{osc}(E)$ is a smooth function. Though it does not predict with precision splittings of individual states, it can be sharply peaked and one finds that exceptionally large splittings occur under those peaks (see Fig. 16, bottom part). In other words, the fluctuations of y_n are not random but are heavily correlated with $f_{osc}(E)$. This is very reminiscent of the correlation observed between the occurence of scarred states and peaks in the conductance of resonant tunnelling diodes.[76] In fact, in the present tunnelling case, observation of phase-space representations shows that states with exceptionally large splittings are scarred by the real orbit and it is actually possible to calculate y_n explicitly as an appropriately-defined scarring weight of the wavefunction [42] [see Eq. (63)].

It is natural to ask how common it is to find a dominant periodic fluctuation of this kind in chaotic potentials. In a system without reflection-symmetry in y, such as that giving the trajectory in Fig. 15(a), there will certainly not exist such a coherent sequence of repeating orbits, all with the same imaginary part in the action. There will exist a similar sequence. For example, orbits initially on the symmetry axis could be followed as the system is deformed to one with less symmetry. The differences are that they are less easy to compute and, more importantly, they would have differing imaginary actions. Therefore one would still observe oscillations, but of smaller amplitude.

The issue of including genuinely multi-dimensional pseudoperiodic orbits is an important one, even in problems with a symmetry axis. Even though $f_{\rm osc}(E)$ successfully describes the most striking fluctuations in the systems examined, there remain strong fluctuations even about this curve. In periodic orbit theory, such fluctuations are generally described by including longer and more complicated orbits, accounting for fluctuations on shorter energy scales as their length increases. Qualitatively different pseudoperiodic orbits can be found in the present case also — an example is shown in Fig. 15(b). The question is how important they can be since they necessarily have larger imaginary actions and, formally at least, are completely dominated by the instanton and its extensions. That such orbits should be relevant is indicated rather obviously by the fact that $f_{\rm osc}(E)$ does not give a complete description of tunnelling fluctuations. Additional assurance comes from looking at a Fourier transform of the spectral function $\sum_n y_n \delta(E - E_n)$. This shows peaks at the repetitions of the quasi-one-dimensional period, as would be expected, but also shows large peaks at periods for which there is no quasi-one-dimensional explanation.[40] So multi-dimensional orbits must be important in practice even though they are formally negligible.

The resolution of this paradox is given in [40]. Any real unstable orbit has a set of homoclinic trajectories which approach it in the infinite past and the infinite future. They do so at an exponential rate $d(t) \sim e^{-\lambda |t|}$. In particular, this is true for the real continuation of an instanton, so it is possible to find real orbits which start and finish quite close to its exit point. By a slight change of initial condition, it is possible to glue them together to define a pseudoperiodic orbit. We can come arbitrarily close to the instanton by allowing the time of evolution to increase, and since the approach is exponentially fast, we can quite soon reach a situation where the differences in imaginary action are smaller than $\hbar$. In problems with a symmetry axis, a relatively complete classification of such orbits is possible and, at least in certain regimes, is enough to give a fairly complete description of the extra Fourier peaks.[40] The orbit in Fig. 15(b) is actually an early member of this family for which the real time is short enough that exponential approach to the center has yet to begin with a vengeance.

This is not the first time that homoclinic orbits have made an appearance in semiclassical calculations. They were used in [80] to organise periodic orbits in the trace formula and to deduce the existence of related chaotic states. More directly related to their occurence here was the observation in [56] that they control the recurrences of a wavepacket when semiclassical propagation using classical trajectories is implemented. The common feature in both these calculations is that the relevant semiclassical quantity is dominated by trajectories that begin and end in some region of phase space, and such trajectories are given naturally by the homoclinic structure.

In order to describe tunnelling fluctuations on the very smallest energy scales, the statistical methods of RMT are the most appropriate, or at least the most efficient. The desire to use random models might seem to be in slight contradiction with the earlier observation that strong periodicities are present, but this is a common feature of chaotic problems — there is a departure from the universal features of RMT at larger energy scales due to the system-dependent influence of short orbits, but the properties of the longest orbits conspire to restore universality at the shortest scales.[81]

A quantitative basis for statistical modelling is provided by a calculation of Wilkinson and Hannay[82] in which Herring's formula, together with continuation of wavefunctions under the barrier using the Green's function, yields an explicit formula for each individual splitting. The formula requires the wavefunction to be known, however, albeit only in the allowed region. While in chaotic problems we have no simple theory for it, we can use statistical models to make predictions. In [42] it is shown that the formula of Wilkinson and Hannay can be developed to give an expression for y_n in terms of a matrix element computed in the Poincaré-section representation of [65]. Formally, it can be written,

$$y_n \propto \langle \tilde{\psi}_n | e^{-\lambda \mathcal{O}/\hbar} | \tilde{\psi}_n \rangle. \tag{63}$$

This formula arises from a quadratic expansion in the phase of the Green's function about the contribution of a dominant tunnelling orbit, assumed here to lie on a symmetry axis y=0. The tildes indicate that the amplitude of the wavefunction ψ_n is interfered with to make it comply with certain conventions in [65]. The integration defining the matrix element is restricted to a coordinate along the Poincaré-section Σ and the operator $\mathcal{O}$ is quadratic in section coordinates (y, p_y). The parameter λ acts like a harmonic frequency and is determined from motion linearised about the tunnelling orbit.

Distributions for y_n can be constructed once we assume $|\tilde{\psi}_n\rangle$ to be defined by a GOE. If all goes well, the first eigenvalue of $\mathcal{O}$ dominates semiclassically and y_n turns out to be given by a Porter-Thomas distribution,[42]

$$p(y) = \frac{e^{-y/2}}{\sqrt{2\pi y}}. \tag{64}$$

This is found to work well in practice as long as λ is not too small. It can be compared directly with the distributions calculated in chaos-assisted tunnelling and it is seen to be very different. The fluctuations in this completely chaotic case are considerably more restrained than they were in chaos-assisted tunnelling. It should be noted that cases have been observed in practice where strong deviations away from Porter-Thomas are obtained. In such cases, however, the trend is towards even less distribution — this kind of behaviour can be explained by taking models in which $\lambda/\hbar$ is small enough that secondary eigenvalues of $\mathcal{O}$ play a role.

We note that, even though the discussion here dealt mainly with splittings in a double well, many of the potential applications would demand calculation of lifetimes in metastable systems. It is expected that most of the discussion would carry over if, in $f(E)$, ΔE_n is simply replaced by Γ_n, the resonance width of a quasi-bound state.[41] This would not only relate the discussion more directly to that of tunnelling diodes, but could be useful in interpretation of Coulomb blockade peaks in quantum dots.[84] Statistical analyses similar to the discussion above have already been brought to bear on the problem, and it would be interesting if coherent effects such as suggested by complex orbits could also be detected.

Finally, it should be acknowledged that the role of complex orbits in chaotic systems is not yet fully explored. The use of pseudoperiodic orbits must also work if tunnelling is across KAM rather than energy barriers. In that case a much more direct comparison to the structures identified by Shudo and Ikeda should be possible — and it is not yet clear how Laputa chains would relate in detail to the homoclinic structure envoked here, though a connection must exist.

6 Conclusion

The behaviour of tunnelling rates in multidimensional systems can be startlingly different, depending on the nature of dynamics in the classical limit. The smooth parameter-dependence seen in integrable and quasi-integrable systems gives way to wild fluctuations when chaotic parts of phase space become significant. Paradoxically, fluctuations were seen to be weaker in the completely chaotic systems examined, although still important. In each of these situations, the main features can be understood using basic techniques of the field of quantum chaos. However, there remain significant (and tractable!) open problems.

In KAM systems, continuation of wavefunctions deep into the forbidden regions is an important problem which still does not have a satisfactory solution, either as a matter of principle or as a matter of practice. A resolution of this issue will be important, not only for calculation in the KAM regime, but also if more detailed semiclassical approaches are to be found in the chaos-assisted case. Of the classes of problem discussed here, chaos-assisted tunnelling has probably received most attention. As a result, a comprehensive understanding of many aspects of it exists. Statistical analyses have worked very well, for example. It is far from a closed subject, however. For example, a completely semiclassical analysis of coupling to chaotic states has yet to be achieved, though it seems reasonable to hope that this should eventually be possible. The scattering matrix approach of [21,27] goes some way to achieving this, but a fully general analysis would require an understanding of the evanescent parts of KAM states on the one hand and semiclassical treatment of the chaotic states on the other. Treatment of tunnelling from chaotic states is achieved semiclassically using complex trajectories. Detailed structure is understood by organising and calculating long orbits, an issue which is well understood in the real case,[54] but which has only been partially explored in the complex case, (though there is already some detailed understanding, as discussed in Sec. 5).

Besides their theoretical interest, these problems have relevance to a wide range of practical situations. Nuclear fission, field theory and chemistry have long been fields which inspired investigation of multidimensional tunnelling. Examples are given in [9,24] of chemical processes which are explicitly in need of dynamics-sensitive analyses of the kind discussed here. More recently, mesoscopic systems have seen the emergence of tunnelling processes in which chaos can play a role. Chaos can be an important aspect in both resonant tunneling diodes [76–79] and in quantum dots.[84] Applications also exist outside of the quantum-mechanical context, as shown by the analysis of optical cavities in [20]. Therefore, there seems to be ample motivation to push hard on the problems outlined here.

Acknowledgments

The author would like to thank the INT, during whose program this tutorial was prepared. He is also grateful to the authors of [14,27,38] for making their figures available. Part of the work reviewed here comes from a collaboration of the author with Niall Whelan. Many helpful discussions were had with Eyal Doron, Stefan Frischat, Akira Shudo, Steve Tomsovic, Denis Ullmo, André Voros, Paul Walker and Niall Whelan.

References

1. R. T. Lawton and M. S. Child, Mol. Phys. **37**, 1799 (1979); *ibid.* 40, 733 (1980).
2. M. J. Davis and E. J. Heller J. Chem. Phys. **75**, 246 (1981).
3. M. C. Gutzwiller, J. Math. Phys. **12**, 343 (1971); *Chaos in Classical and Quantum Mechanics* (Springer Verlag, New York, 1990); A. Ozorio de Almeida, *Hamiltonian Systems: Chaos and Quantization* (Cambridge University Press, 1988); M. Brack and R. K. Bhaduri, *Semiclassical Physics*, Frontiers in Physics **96**, (Addison Wesley, 1997).
4. E. L. Sibert, W. P. Reinhardt and J. T. Hynes, J. Chem. Phys. **77**, 3595 (1983); T. U. Uzer, D. W. Noid and R. A. Marcus, J. Chem. Phys. **79**, 4412 (1983); A. Ozorio de Almeida, J. Chem. Phys. **88**, 6139 (1984).
5. R. E. Meyer, SIAM J. Appl. Math. **51**, 1585 (1991); SIAM J. Appl. Math. **51**, 1602 (1991); R. E. Meyer and M. C. Shen, SIAM J. Appl. Math. **52**, 730 (1992).
6. S. C. Creagh, J. Phys. A **27**, 4969 (1994).
7. M. Wilkinson, Physica **21**D, 341 (1986).
8. M. Wilkinson, J. Phys. A **20**, 635 (1987).
9. S. Takada and H. Nakamura J. Chem. Phys. **100** 98 (1994); *ibid.* **102** 3977 (1995).
10. S. Takada, P. N. Walker and M. Wilkinson, Phys. Rev. A **52**, 3546 (1995).
11. S. Takada, J. Chem Phys. **104**, 3742 (1996).
12. W. A. Lin and L. E. Ballentine Phys. Rev. Lett. **65**, 2927 (1990); Phys. Rev. A **45**, 3637 (1992); A. Peres, Phys. Rev. Lett. **67**, 158 (1991).
13. J. Plata and J. M. Gomez Llorente, J. Phys. A **25**, L303 (1992).
14. O. Bohigas, S. Tomsovic and D. Ullmo, Phys. Rep. **223** (2), 45 (1993).

15. O. Bohigas, D. Boosé, R. Egydio de Carvalho and V. Marlvulle, Nucl. Phys. A **560**, 197 (1993).
16. S. Tomsovic and D. Ullmo, Phys. Rev. E **50**, 145 (1994).
17. R. Utermann, T. Dittrich and P. Hänggi, Phys. Rev. E **49**, 273 (1994).
18. R. Roncaglia, L. Bonci, F. M. Israilev, B. J. West and P. Grigolini, Phys. Rev. Lett. **73**, 802 (1994).
19. M. Latka, P. Grigolini and B. J. West, Phys. Rev. E **50** 596 (1994); Phys. Rev. A **50**, 1071 (1994).
20. J. U. Nöckel, A. D. Stone and R. K. Chang, Opt. Lett. **19**, 1693 (1994); A. Mekis et al, Phys. Rev. Lett. **75**, 2682 (1995); J. U. Nöckel and A. D. Stone, Nature **385**, 45 (1997).
21. E. Doron and S. D. Frischat, Phys. Rev. Lett. **75**, 3661 (1995).
22. E. M. Zanardi, J. Gutiérrez and J. M. Gomez Llorente, Phys. Rev. E **52**, 4736 (1995); E. M. Zanardi and J. M. Gomez Llorente, preprint.
23. F. Leyvraz and D. Ullmo, J. Phys. A **29** 2529 (1996).
24. J. Ortigoso, Phys. Rev. A **54**, R2521 (1996).
25. H. J. Korsch, B. Mirbach and B. Schellhaaß, J. Phys. A **30** 1659 (1997).
26. S. D. Frischat, PhD Thesis, Heidelberg (1997).
27. S. D. Frischat and E. Doron, preprint chao-dyn/9707005
28. L. S. Schulman, *Techniques and applications of path integrals* (Wiley, New York, 1981).
29. D. W. McLaughlin, J. Math. Phys. **13**, 1099 (1972).
30. W. H. Miller and T. F. George, J. Chem. Phys. **56**, 5668, (1972); *ibid.* **57**, 2458, (1972).
31. T. Banks, C. M. Bender and Tai T. Wu, Phys. Rev. D **8**, 3346 (1973); T. Banks and C. M. Bender, Phys. Rev. D **8**, 3366 (1973).
32. R. Balian and C. Bloch, Ann. Phys. **85**, 514 (1974).
33. S. Coleman "The uses of Instantons," in *The Whys of Subnuclear Physics*, A. Zichichi ed. (Plenum, New York 1979); J. Zinn-Justin, *Quantum Field Theory and Critical Phenomena* (Oxford University Press, (1989).
34. W. H. Miller, J. Phys. Chem. **83**, 960 (1979).
35. S. Adachi, Ann. Phys. **195** 45 (1989).
36. J. M. Robbins, S. C. Creagh and R. G. Littlejohn, Phys. Rev. A **39**, 2838 (1989); *ibid.* **41**, 6052 (1990).
37. P. Lebœuf and A. Mouchet Phys. Rev. Lett. **73**, 1360 (1994).
38. A. Shudo and K. Ikeda, Prog. Theor. Phys. Supplement No. 116, 283 (1994); Phys. Rev. Lett. **74**, 682 (1995); preprint, to appear in Physica D.
39. A. Shudo and K. Ikeda, Phys. Rev. Lett. **76**, 4151 (1996).
40. S. C. Creagh and N. D. Whelan, Phys. Rev. Lett. **77**, 4975 (1996).

41. S. C. Creagh and N. D. Whelan, in preparation.
42. S. C. Creagh and N. D. Whelan, in preparation.
43. R. Balian, G. Parisi and A. Voros, Phys. Rev. Lett. **41**, 1141 (1978).
44. A. Voros, Ann. Inst. H. Poincaré A **39**, 211 (1983); Ann. Inst. Fourier **43**, 1509 (1993); J. Phys. A **27**, 4653 (1994); in *Quasiclassical Methods*, J. Rauch and B. Simon Eds., The IMA Volumes in Mathematics and its Applications **95**, 189 (1997).
45. J. Ecalle, *Les fonctions résurgentes*, Publ. Math. d'Orsay, Université Paris-Sud, 1981-05, 1981-06, 1985-05; *Cinq applications sur des fonctions résurgantes*, Publ. Math. d'Orsay, Université Paris-Sud, 84T 62.
46. M. V. Berry and C. J. Howls, Proc. R. Soc. A **430**, 653 (1990); *ibid.* **434**, 657 (1991).
47. E. Delabære, H. Dillinger and F. Pham, Ann. Inst. Fourier **43**, 163 (1993); J. Math. Phys. **38**, 6127 (1997); E. Delabaere and F. Pham Ann. Phys. **261**, 180 (1997).
48. V. P. Maslov and M. V. Fedoriuk, *Semiclassical Approximations in Quantum Mechanics* (Reidel, Dordrecht, 1981).
49. V. I. Arnold, *Mathematical Methods of Classical Mechanics* (Springer-Verlag, 1989).
50. J. J. Morehead, Phys. Rev. A **53**, 1285 (1996).
51. A. J. Lichtenberg and M. A. Lieberman, *Regular and Stochastic motion* (Springer-Verlag, 1983).
52. C. E. Porter ed. *Statistical Theories of Spectra: Fluctuations.*, (Academic Press, New York (1965); M. L. Mehta *Random Matrices and the Statistical Theory of Energy Levels.* (Academic Press, New York, (1967).
53. I. C. Percival, J. Phys. B **6**, L229 (1973).
54. CHAOS focus issue on periodic orbit theory, CHAOS **2** (1992), P. Cvitanović, I. Percival and A. Wirzba, eds.
55. A. Voros, Colloques Internationaux CNRS no. 237, 277 (1975); W. H. Miller, J. Chem. Phys. **63**, 996 (1975); P. J. Richens, J. Phys. A **15**, 2101 (1982); A. Voros, J. Phys. A, **21**, 685 (1988).
56. S. Tomsovic and E. J. Heller, Phys. Rev. E **47**, 282 (1993).
57. M. V. Berry and K. E. Mount, Rep. Prog. Phys. **35**, 315 (1972).
58. J. M. Greene and I. C. Percival, Physica **3**D, 530 (1981); I. C. Percival, Physica **6**D, 67 (1982).
59. A. J. Dragt and J. M. Finn, J. Math. Phys. **17**, 2215 (1976); R. Scharf, J. Phys. A **21**, 2007 (1988).
60. V. F. Latzutkin, I. G. Schachmannski and M. B. Tabanov, Physica **40**D, 235, 1989; P. Holmes, J. Marsden and J. Scheurle, Contemp. Math. **81**, 213, (1988); E. Fontich and C. Simó, Ergod. Theor. Dynam. Syst. **10**,

295 (1990); V. G. Gelfreich, V. F. Latzutkin and M. B. Tabanov, Chaos **1** 137, (1991). V. Hakim and K. Mallick, Nonlinearity **6**, 57 (1993); V. G. Gelfreich, V. F. Latzutkin and N. V. Svanidze, Physica **71**D, 82, (1994).

61. L. D. Landau and L. F. Lifshitz, *Quantum Mechanics* (Butterworth and Heinemann, Oxford, (1977).
62. J. Bardeen, Phys. Rev. Lett. **6**, 57 (1961)
63. C. Herring, Rev. Mod. Phys. **34**, 631 (1962).
64. R. G. Littlejohn, J. Math. Phys. **31**, 2953 (1990).
65. E. Bogomolny, Nonlinearity **5**, 805 (1992).
66. F. Grossmann, T. Dittrich, P. Jung and P. Haänggi, Phys. Rev. Lett. **67**, 516 (1991); Z. Phys. B **84**, 315 (1991); F. Grossmann, T. Dittrich and P. Haänggi, Physica B **175**, 293 (1991); T. Dittrich, F. Grossmann, P. Jung, B. Oelchlagel and P. Haänggi, Physica B **194**, 173 (1993).
67. G. Hackenbroich, E. Narimov and D. Stone, preprint.
68. E. Doron and U. Smilanski, Nonlinearity **5**, 1055 (1992).
69. J. P. Eckmann and C. A. Pillet, Comm. Math. Phys. **170**, 283 (1995).
70. W. H. Miller, Adv. Chem. Phys. **25**, 69 (1974).
71. T. Geisel, G. Radons and J. Rubner, Phys. Rev. Lett. **57**, 2883 (1986); G. Radons, T. Geisel and J. Rubner Adv. Chem. Phys. **73**, 891 (1988).
72. P. A. Boasman and J. P. Keating, Proc. Roy. Soc. Lond. A **449**, 629 (1995).
73. R. G. Littlejohn, J. Math. Phys. **31**, 2953 (1990).
74. R. C. Brown and R. E. Wyatt, Phys. Rev. Lett. **57**, 1 (1986).
75. J. M. Robbins, Phys. Rev. A **40**, 2128 (1989).
76. L. Leadbeater, F. W. Sheard and L. Eaves, Semicond. Sci. Technol. **6**, 1021 (1991); T. M. Fromhold et al, Phys. Rev. Lett. **72**, 2608 (1994); Phys. Rev. B **51**, 18029 (1995). P. B. Wilkinson et al, Nature **380**, 608 (1996).
77. T. S. Monteiro et al, Phys. Rev. E **53**, 3369 (1996); Phys. Rev. B **56**, 3913 (1997).
78. E. E. Narimov, A. D. Stone and G. S. Boebinger, cond-mat/9707073; E. E. Narimov and A. D. Stone, cond-mat/9705167; E. E. Narimov and A. D. Stone, cond-mat/9704083.
79. E. Bogomolny and D. Rouben, preprint.
80. A. Ozorio de Almeida, Nonlinearity **2**, 519 (1989).
81. E. Bogomolny and J. Keating, Nonlinearity **8**, 1115 (1995); *ibid.* **9**, 911 (1996); Phys. Rev. Lett. **77**, 1472 (1996).
82. M. Wilkinson and J. H. Hannay, Physica **27**D, 201 (1987).
83. S. C. Creagh, J. M. Robbins, and R. G. Littlejohn, Phys. Rev. **A** 42, 1907 (1990); J. M. Robbins, Nonlinearity **4**, 343 (1991).

84. R. A. Jalabert, A. D. Stone and Y. Alhassid, Phys. Rev. Lett. **68**, 3468 (1992); A. M. Chang et. al., Phys. Rev. Lett. **76**, 1695 (1996); J. A. Folk et. al., *ibid.* **76**, 1699 (1996).

QUANTUM ENVIRONMENTS: SPINS BATHS, OSCILLATOR BATHS, and applications to QUANTUM MAGNETISM

P.C.E. STAMP
Physics Department, and Canadian Institute for Advanced Research,
University of British Columbia, 6224 Agricultural Rd., Vancouver, B.C., Canada
V6T 1Z1
and
L.C.M.I./ Max Planck Institute, Ave. des Martyrs, Grenoble 38042, France
***
email: stamp@physics.ubc.ca

Abstract

The low-energy physics of systems coupled to their surroundings is understood by truncating to effective Hamiltonians; these tend to reduce to a few canonical forms, involving coupling to "baths' of oscillators or spins. The method for doing this is demonstrated using examples from magnetism, superconductivity, and measurement theory, as is the way one then solves for the low-energy dynamics. Finally, detailed application is given to the exciting recent Quantum Relaxation and tunneling work in nanomagnets.

1 Introduction

These lecture notes are taken from lectures given in April 1997 at the Institute for Nuclear Theory in Seattle (University of Washington). Their main point was to explain how condensed matter theorists deal with problems in which some degree/degrees of freedom of interest, are coupled to a background "quantum environment". In almost all cases the interesting degrees of freedom are showing quantum behaviour, and often they are mesocopic or even macroscopic.

The tactic adopted in the lectures was to explain the general framework within which the theory operates, and then to demonstrate its operation with the help of examples. These examples are taken from various fields (including superconductivity, the Kondo problem, and one-dimensional systems), but the emphasisis is on the very active new field of "Quantum Nanomagnetism". Great scientific interest has been generated (as well as some rather misleading articles in the more popular press) by recent experiments in crystals of magnetic macromolecules [1,2,3,4,5,6,7], which show resonant tunneling of their magnetisation. Other experiments which have also generated considerable press include coherence experiments on the ferritin biological macromolecule

[8,9], and tunneling experiments on domain walls in magnetic wires [10,11]. I thus spent some time explaining how the theoretical methods may be applied to these real systems. For another very important example of large-scale quantum behaviour, see the lectures of A.J. Leggett in this volume, who discusses macroscopic quantum phenomena in superconducting systems.

The notes are divided into 3 parts. The first part (Chapter 2) deals with the essential theoretical step in which one sets up a low-energy description of the system of interest. The result of this is an "effective Hamiltonian", which claims to accurately describe both the quantum degrees of freedom of interest, and the quantum environment (and their coupling), below a certain energy cut-off. We end up with 2 kinds of quantum environment. One is the oscillator bath model, which is very old - in its modern form it goes back to papers of Feynman et al. [12], at the beginning of the 1960's. Except in rather unusual cases, it well describes an environment of *delocalized* modes (eg. phonons, photons, conduction electrons, magnons, etc). It was introduced in this form into the discussion of tunneling by Caldeira and Leggett [13] in 1981. The second environment is the "spin bath" model, which has been developed recently by myself and Prokof'ev [14,15,16,17,18] (see also refs. [19,21,20,22]), stimulated particularly by problems in quantum nanomagnetism (although it can be applied outside this domain [21,22,23,24]; to illustrate this, brief space is given to superconducting SQUIDs and spin chains). It describes environmental modes like paramagnetic or nuclear spins, or other similar cases in which each environmental mode has only a few (often only two) levels of interest, and where they are very weakly coupled. The main point of this chapter is that readers may see clearly how the "truncation" to an effective Hamiltonian works, and to carry it out explicitly on some model examples in a tutorial way.

Once one has an effective Hamiltonian, the next task (chapter 3) is to solve for the dynamics of the degrees of freedom we are interested in, by "integrating out" (averaging over) the environmental modes. Doing this for an oscillator bath is standard, so I don't spend too much time on it, except to delineate the main ideas and results. More time is spent on the spin bath, where some new tricks are needed, and where the results are often radically different from the oscillator bath. Technical details are kept to a minimum, and ideas are emphasized. Again, it is then shown how these results apply to some real systems.

In both Chapters 2 and 3, I try to show how a very wide variety of physical problems tends to reduce to just a very few "canonical models" at low energies. Many readers will not be surprised by this- it is of course the idea underlying the whole philosophy of "universality classes" and the renormalisation group.

Finally in the 3rd part of these lectures (chapter 4) I turn to some real down-to-earth physics, and discuss recent advances in our understanding of quantum magnets, particularly at the nano/mesoscopic level, where all sorts of interesting tunneling phenomena occur. These include simple dissipative tunneling and quantum relaxation; one also gets resonant tunneling relaxation and there is evidence for quantum coherence at the mesoscopic level. Although not all experiments are properly understood at present, it is becoming very clear that much of the physics is controlled by the combined effect of nuclear spins and magnetic dipolar fields on the tunneling entities; moreover, it is *essential* to take proper account of the nuclear dynamics, which justifies *a posteriori* the interest in the spin bath model. There is a rich experimental literature here, on magnetic molecules and particles, and on magnetic wires, as well as related experiments on spin chains and on the "colossal resistance" Mn- perovskite materials.

From this summary the reader will see that I neglect discussion of other currently-used models of quantum environments such as the "Landau-Zener" model [19,25,26]. This is unfortunate, because it is much easier to connect this model with studies in quantum chaos, then the "bath" models I have described. My only excuses are (i) lack of space, and (ii) lack of experience - condensed matter physicists are typically forced to deal with real solids, liquids, or glasses, in which statistical averages over environmental modes are unavoidable.

2 EFFECTIVE HAMILTONIANS: SPIN BATHS AND OSCILLATOR BATHS

In this first lecture I discuss, from both the physical and mathematical points of view, a rather extraordinary simplifying feature of almost all physical problems, which is fundamental to our understanding of complex systems having many degrees of freedom. It is simply this - that provided we are only interested in a certain *energy scale* (or a limited range of energies), we may *throw away* most of the degrees of freedom - they do not play any direct role in the physics in the energy range of interest. Of course this can't be done willy-nilly; it must be done carefully, and those degrees of freedom which have been dumped will still play an *indirect* role in the physics.

Readers will of course be familiar with this idea in many forms - one's first encounter with it, as a student, is often in the form of the "coarse-graining" assumptions inherent in statistical mechanics or thermodynamics. In this lecture we begin by showing how one actually carries out the truncation of the Hamiltonian of a complex system down to a low-energy "effective Hamiltonian". This is done explicitly for one real physical example, in some detail, so that readers can see what is involved.

Another important idea, also rather extraordinary, is that if one carries out such truncations for many different physical systems, the same kinds of low-energy effective Hamiltonian recur again and again. These "universality classes" thus become very important, and the corresponding effective Hamiltonians become worthy of study in their own right. In this lecture we study 2 such Hamiltonians. One is the "oscillator bath" model, in which some simple central system is coupled to an environment of harmonic oscillators. The simplest such central system is a single 2-level system; the model is then the famous "spin-boson" model. We will also look at a model in which *two* central spin systems couple to the oscillator bath (the "PISCES" model, where "PISCES" is an acronym for "Pair of Interacting Spins Coupled to an Environmental Sea"); and I mention a number of other such "canonical" models.

The other model effective Hamiltonian we shall study is the "spin-bath" model, where now the environment itself is represented by a set of 2-level systems. The simplest model of this kind is the "central spin" model, where the central system is also represented by a 2-level system. In the same way as for the oscillator bath environments, one can also develop a number of canonical spin bath models. Typically the spin bath models arise for environments of *localised* modes, whereas the oscillator baths represent delocalised modes.

I will discuss various examples in this lecture, to illustrate how these simple archetypal models really do represent the low-energy physics of a huge variety of real systems. What I will *not* do here is discuss the dynamics of these models- this is saved for the next lecture (section 3).

2.1 Truncation to Low Energies

The idea of a low-energy "truncated" effective Hamiltonian (or Lagrangian) goes back a long way - in classical mechanics to hydrodynamics, and in quantum theory to the old spin Hamiltonians. In its modern form (partially inspired by Landau's treatment of turbulence) it is often discussed in the renormalization group (RNG) framework [27]. For the purposes of these lectures, we can formulate the problem as follows.

Typically one is presented with a reasonably accurately known "high-energy" or "bare" Hamiltonian (or Lagrangian) for a quantum system, valid below some "ultraviolet" upper energy cut-off energy E_c, and having the form

$$\tilde{H}_{\text{Bare}} = \tilde{H}_o(\tilde{P}, \tilde{Q}) + \tilde{H}_{\text{int}}(\tilde{P}, \tilde{Q}; \tilde{p}, \tilde{q}) + \tilde{H}_{\text{env}}(\tilde{p}, \tilde{q}) \quad (E < E_c) , \tag{1}$$

where $\tilde{Q}$ is an $\tilde{M}$-dimensional coordinate describing that part of the system we are interested in (with $\tilde{P}$ the corresponding conjugate momentum), and $(\tilde{p}, \tilde{q})$

are $\tilde{N}$-dimensional coordinates describing all other degrees of freedom which may couple to $(\tilde{P}, \tilde{Q})$. Conventionally one refers to $(\tilde{p}, \tilde{q})$ as environmental coordinates.

What is important about $\tilde{H}_{\text{Bare}}$ is that its form is known well (it can of course be regarded as a low-energy form of some other even higher-energy Hamiltonian, in a chain extending ultimately back to quarks and leptons). If, however, one is only interested in physics below a much lower energy scale Ω_o, then the question is - how can we find a new effective Hamiltonian, having the form

$$H_{\text{eff}} = H_o(P,Q) + H_{\text{int}}(P,Q;p,q) + H_{\text{env}}(p,q) \quad (E < \Omega_o) \,, \tag{2}$$

in the truncated Hilbert space of energies below Ω_o? In this H_{eff}, P and Q are generalised M-dimensional coordinates of interest, and p, q are N-dimensional environmental coordinates coupled to them. Since we have truncated the total Hilbert space, we have in general that $M < \tilde{M}$ and $N < \tilde{N}$. This truncation is illustrated in Fig. 1.

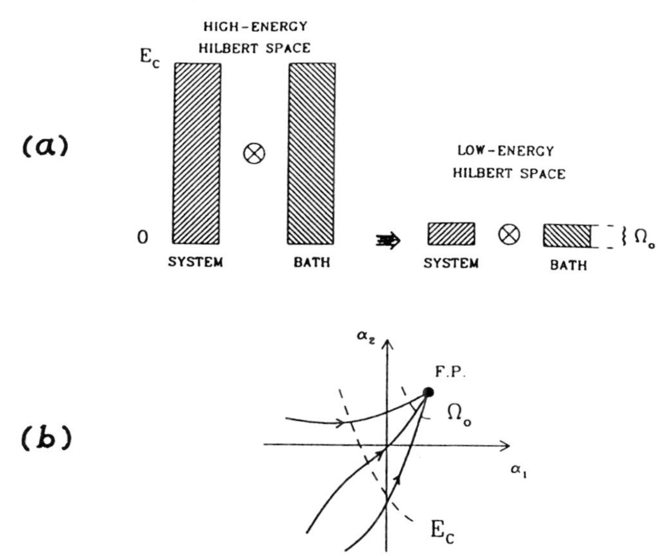

FIG. 1: The truncation Procedure (schematic). In (a) we see how the Hilbert space of the high- energy Hamiltonian (with UV cut-off E_c) is truncated down to low energies (UV cut-off Ω_o), both for the system and the environment. In (b) the corresponding flow in the coupling constant space of effective Hamiltonians (here shown for 2 couplings α_1 and α_2).

At first glance it is not at all obvious why anyone would want to make this truncation, because its inevitable effect is to generate various couplings between the low-energy modes which were not there before. However there are 2 very good reasons for truncating. First, in spite of the new couplings appearing at low energy, it almost always turns out that H_{eff} is easier to handle than $\tilde{H}_{\text{Bare}}$ (particularly in predicting the dynamics of the low-energy variables P, Q of interest), simply because there are fewer variables to deal with. Second, and much more fundamental, the truncation pushes the new H_{eff} towards some low-energy "fixed point" Hamiltonian (Fig. 1 again). Moreover, many different physical systems may "flow" to the same fixed point. This is particularly true for that part of H_{eff} which describes the environment, and this allows theorists to speak of "universality classes" of quantum environment, each describing many different physical systems. Each of these different physical systems will then flow towards the same fixed point, and they will all be described by the same form for $H_{\text{env}}(p,q)$, albeit with different values for the couplings. In fact the low-energy couplings parametrize the position of a given system, with a given UV cut-off, in the space of available couplings ("effective Hamiltonian" space). As one varies the UV cut-off, the couplings change and any given system moves in the coupling space; but all systems in a given universality class move towards one fixed point as the UV cutoff Ω_0 is reduced.

In this context one understands the various coupling terms in H_{eff}, as a parametrisation of the *path taken* by H_{eff}, as it approaches the fixed point.

From this point of view it is not so surprising that the description of quantum environments reduces to the discussion of just a few "universality classes". I emphasize this just because physicists are often quite surprised that one can discuss such a wide variety of, eg., tunneling problems with such a restricted class of models. Of course, in asserting that these fixed points exist, I have simply swapped one mystery for another - instead of scratching our heads over the way in which so many physical systems resemble each other in low energies, we are now left wondering *why* there must be a flow of H_{eff} towards one or other fixed point (ie., why fixed points?). This is often how physics proceeds, by substituting one conundrum for another (albeit a more precisely formulated one!).

Let us leave this question hanging for now, and simply note here the enormous *pragmatic* interest of this result. We shall see, in the 2 universality classes of interest, that we can parametrise very simply the form of the low-energy physics. In the case of oscillator bath models, this will be in terms of a function $J(\omega)$, where ω is a frequency - this is the famous Caldeira-Leggett spectral function [13], which convolves the coupling to the bath with the bath density of states. In the case of spin bath models, one has a density of states function

$W(\epsilon)$ for the spin bath (usually Gaussian), and parameters which describe both the coupling to the spin bath, and spin diffusion within the spin bath. At low temperatures and low applied fields, we will only care about the low-energy behaviour of ω or ϵ. This will allow us in some cases to make rather sweeping statements about whole classes of physical systems (as well as calculating their detailed dynamics, for which see the next lecture, in Ch. 3). To take an example, we will be able to make remarks about quantum measurements which apply to large classes of measuring devices and measured systems, and yet which are still *realistic* (in the sense that they apply to real physical systems, instead of just to some idle toy model).

Having said this, I emphasize that a lot of the physics is in the details (a fact often ignored in the quantum measurement literature!). Thus, rather than giving some (necessarily rather abstract) attempt at a general discussion of the derivation and validity of these 2 models, I will instead show how it is done for some particular real systems. After this, some at least of the general features will be evident (as well as the pitfalls in a too general approach!).

2.2 Spin Baths and Nanomagnets- a worked example

Let's begin our formal approach to the truncation procedure, by picking a particular example of a physical system which reduces to a spin bath model at low energies. We shall look at a "nanomagnet", by which we mean a magnetically-ordered particle or molecule which is sufficiently small that all the electronic spins causing the magnetism are aligned along one axis (note that a *macroscopic* magnet does *not* satisfy this criterion - it is often full of domains, with spins pointing in various directions; and spin waves or other spin fluctuations can also cause the spin axis to wander around a macroscopic sample). The size of such monodomain particles does not often exceed 2000Å; yet a particle this large can contain as many as 10^8 electronic spins! Because all microscopic spins are aligned, the total system then behaves like a "giant spin", denoted by $\vec{S}$.

As we shall see, the most important environment to which $\vec{S}$ is coupled is the set of *nuclear* spins inside it (and often to nuclear and paramagnetic impurity spins outside it as well). This is how the spin bath arises. Our aim in this section is to truncate this physical system to low energies. Recall that the reason for going through this is so you can see how it works on a realistic example. You don't have to care about nanomagnets to follow the exercise - the point is to understand the truncation procedure, and see how to apply it elsewhere.

2.2(a) HIGH ENERGY HAMILTONIAN

Suppose we start by looking for a high-energy effective Hamiltonian, analogous to (2.1), but for a magnetic system which is coupled to a set of weakly-coupled spin-1/2 spins (ie., a "spin bath", instead of an "oscillator bath"). If the magnetic degrees of freedom are described by some continuous variable Q, and the oscillator coordinates $(\tilde{p}, \tilde{q})$ are replaced by a set of N spin-1/2 variables $\{\hat{\vec{\sigma}}_k\}$, i.e., two-level systems, we get a high-energy Hamiltonian of general form

$$H = H_o(P, Q) + H_{\text{int}}(P, Q; \{\hat{\vec{\sigma}}\}) + H_{\text{env}}(\{\hat{\vec{\sigma}}\}) \; ; \tag{3}$$

$$H_{\text{int}}(P, Q; \{\hat{\vec{\sigma}}\}) = \sum_{k=1}^{N} \left[F_{\|}(P, Q)\hat{\sigma}_k^z + \left(F_{\perp}(P, Q)\hat{\sigma}_k^- + h.c.\right) \right] ; \tag{4}$$

$$H_{\text{env}}(\{\hat{\vec{\sigma}}\}) = \sum_{k=1}^{N} \vec{h}_k . \hat{\vec{\sigma}}_k + \frac{1}{2} \sum_{k=1}^{N} \sum_{k'=1}^{N} V_{kk'}^{\alpha\beta} \hat{\sigma}_k^\alpha \hat{\sigma}_{k'}^\beta \; , \tag{5}$$

for energy scales $E < E_c$. In this Hamiltonian, we have a "central system", moving in a space described by the continuous coordinate Q, which couples simultaneously to the environmental spin variables $\{\hat{\vec{\sigma}}_k\}$. An example of such a Hamiltonian, which has been worked out in detail recently, is that for a magnetic soliton, like a domain wall, at position Q, coupled by hyperfine interactions to a set of N spin-1/2 nuclear spins [28]; this example will be discussed in section 2.3(b). In the case of nuclear spins $V_{kk'}^{\alpha\beta}$ describes the extremely weak internuclear dipolar coupling; typically $|V_{kk'}^{\alpha\beta}| \leq 10^{-7}\; K$; and $\vec{h}_k$ is any external field that might influence these nuclei.

However in the case of nanomagnetic grains or magnetic macromolecules things simplify a great deal, because the continuous coordinate Q can be replaced by a "giant spin" vector moving in the compact space of the surface of a sphere; we get a high-energy Hamiltonian of form

$$H(\vec{S}; \{\hat{\vec{\sigma}}\}) = H_o(\vec{S}) + \frac{1}{S} \sum_{k=1}^{N} \omega_k \vec{S} \cdot \hat{\vec{\sigma}}_k + H_{\text{env}}(\{\hat{\vec{\sigma}}\}) \; ; \tag{6}$$

where $H_o(\vec{S})$ is a "giant spin" Hamiltonian, describing a quantum rotator with spin quantum number $S = |\vec{S}| \gg 1$, and $H_{\text{env}}(\{\hat{\vec{\sigma}}\})$ is the same as in (5).

The assumption here is that we have a "monodomain" magnetic particle or macromolecule, in which the very strong exchange interactions lock the microscopic electronic spins $\vec{s}_j$ into either a giant ferromagnetic (FM) moment, with $\vec{S} = \sum_j \vec{s}_j$ (summed over ionic spin sites) or a giant antiferromagnetic

(AFM) Néel vector, with $\vec{N} \equiv \vec{S} = \sum_j (-1)^j \vec{s}_j$ (for a simple staggered Néel order). With this assumption, the usual hyperfine coupling to the nuclear spins $\vec{I}_k$ can be written

$$H_{Hyp} = \sum_{k=1}^{N} \omega_k \underline{s}_k \cdot \underline{I}_k \rightarrow \frac{1}{S} \sum_{k=1}^{N} \omega_k \underline{S} \cdot \underline{I}_k \tag{7}$$

(or a more complicated tensor interaction if we wish). In these lectures I will assume that $|\vec{I}_k| = I = \frac{1}{2}$, and write $\vec{I}_k \rightarrow \sigma_k$, i.e. the nuclear spins will be described by spin-$\frac{1}{2}$ Pauli matrices. In many cases even if $I \neq \frac{1}{2}$, the low-energy nuclear spin dynamics is well described by a 2-level system; the effect of the other levels appears in the field $\vec{h}_k$ in (5).

The "giant spin" Hamiltonian $H_o(\vec{S})$ itself can either be written as a continuous function of the vector $\vec{S}$ (assuming that S is constant), or in terms of the usual spin operators for $\vec{S}$, for which we have the general form

$$H_o(\underline{S}) = \sum_{l=1}^{S} \frac{{}^{\parallel}K_l}{S^{l-1}} \hat{S}_z^l + \frac{1}{2} \sum_{r=1}^{S} \frac{{}^{\perp}K_r}{S^{r-1}} (\hat{S}_+^r + \hat{S}_-^r) \tag{8}$$

This form of the Hamiltonian separates the longitudinal part ${}^{\parallel}\mathcal{H}(\hat{S}_z)$ (which conserves S_z) from the transverse part ${}^{\perp}\mathcal{H}(\hat{S}_+, \hat{S}_-)$. In the symmetric case (all l even) a pioneering semi-classical study of (1.8) was carried out by van Hemmen et al. [29,30]. A very simple example of the giant spin Hamiltonian in the symmetric case is the biaxial form:

$$H_o(S) = \frac{1}{S} \left[{}^{\parallel}K_2 \hat{S}_z^2 + {}^{\perp}K_2 \hat{S}_y^2 \right] \tag{9}$$

in which tunneling between the classical minima (at $S_z = \pm S$) is accomplished by the symmetry-breaking transverse ${}^{\perp}K_2$ term, if ${}^{\parallel}K_2$ is negative.

I refer here to the giant spin Hamiltonian for a nanomagnet as a "high-energy" Hamiltonian, but of course it is obvious that Hamiltonians like (8) or (9) are themselves the result of the truncation of an even higher energy (more "microscopic") electronic spin Hamiltonian like, eg.,

$$H_o(\{\underline{s}_j\}) = \frac{1}{2} \sum_{<ij>} J_{ij}^{\alpha\beta} \hat{s}_i^\alpha \hat{s}_j^\beta + \frac{1}{2} \sum_j K_j^{\alpha\beta} \hat{s}_j^\alpha \hat{s}_j^\beta \tag{10}$$

If $|s_j| = s$ (so there are $L = S/s$ electronic spins), this Hamiltonian acts in a huge Hilbert space of dimension $(2s+1)^L$.

I would like to strongly emphasize here a practical point of some importance (which will be familiar to anyone working in mesoscopic or nanoscopic physics). Even if we knew all the couplings $J_{ij}^{\alpha\beta}$, $K_j^{\alpha\beta}$ (which is hardly likely given the internal complexity of most nanomagnets), truncation of $H_o(\underset{\sim}{S})$ at low energies is *practically impossible* if $L > 5 - 10$, even with supercomputers. Instead one attempts to *measure* the parameters in (8) or (9), thereafter treating them as "fundamental" (similar situations are encountered in most problems involving strong interactions, ranging from QCD to Fermi liquid theory). Moreover, even this is not easy - experiments such as ESR (Electron Spin Resonance) can only parametrize terms with small ℓ or r in (8). Unfortunately the tunneling matrix element Δ_S, between levels$| \; S_z > = \pm \mid S >$ of the longitudinal part of (8) is just as much influenced by higher transverse couplings as by lower ones (a point to be discussed in more detail in the last section). Thus in any practical situation, we will probably *not* know all of the important terms in our giant spin Hamiltonian.

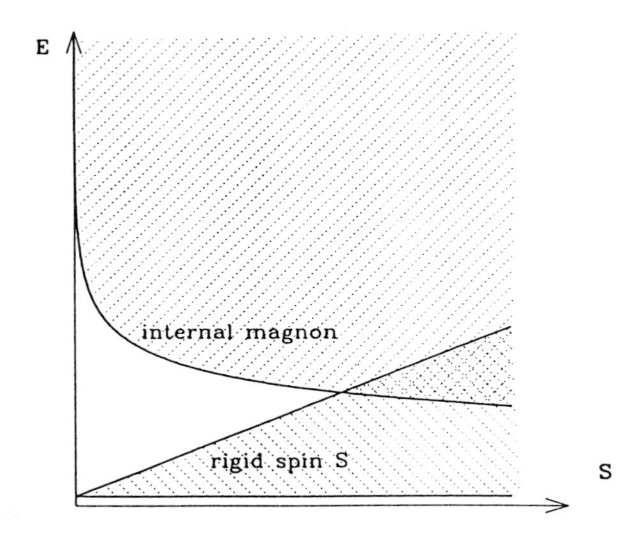

FIG. 2: The excited states of a nanomagnet (schematic): The $2S+1$ states of the giant spin appear at low energies. All states that break the constraint $\sum_j \vec{s}_j = \vec{S}$, with $|\vec{S}| = s$, are referred to as internal magnon states. As $S \to 0$ the energy at the bottom of the magnon band tends to the interspin exchange J, but it falls as S increases, because the magnons can spread out. When it reaches the top of the giant spin manifold, the two manifolds mix strongly in the overlap region shown.

In any case it is clear that $H_o(\vec{S})$ is only strictly meaningful if the higher states in $H_o(\{\vec{S}_j\})$ are not excited; these "internal magnon" states (of which there are a huge number) break the constraint $\sum_j \vec{s}_j = S$ (or $\sum_j (-1)^j \vec{s}_j = \vec{N}$). What this implies is shown schematically in Fig. 2; as the nanomagnet gets bigger, internal magnon states (where internal spin flips overcome the powerful J_{ij} exchange) can lower their energy by spreading over the whole sample. Once these start impinging on the $(2S+1)$ - dimensional giant spin manifold, we may be in trouble. Note that the giant spin manifold is spread over $\sim SK$ in energy (where K is the dominant anisotropy term in $H_o(\vec{S})$); values of K are typically in the range 0.1-10 K.

2.2(b) TRUNCATION to LOW ENERGIES

Suppose however that one is interested in energies or temperatures where only the 2 lowest levels of $H_o(\vec{S})$ are involved. This arises in particular when one is interested in quantum phenomena, since the dynamics of the system go over into a temperature-independent quantum regime once the higher levels can no longer be thermally activated. We are then at liberty, in the absence of the nuclear spins, to truncate $H_o(\vec{S})$ to a simple 2-level Hamiltonian $H_o(\vec{\tau})$, which is typically written as

$$H_o(\vec{\tau}) \; = \; \Delta_o \hat{\tau}_x \; + \; \xi \hat{\tau}_x \tag{11}$$

where $\vec{\tau}$ is a Pauli spin. It is very important, in the context of macroscopic quantum phenomena, that $H_o(\vec{\tau})$ is meaningful below a UV cut-off energy $\Omega_o \sim K$ *which is independent of* S (at least until the monodomain assumption fails, and we return to (10)). This means that we enter the quantum regime at temperatures which are often as high as several degrees Kelvin, even for very large spins.

Our essential task now becomes clear - we need to continue our truncation of (6) down to low T. For the simple giant spin Hamiltonian (no nuclear spins), the passage from (8) to (11) has been discussed by many authors for $S \gg 1$. Korenblit and Shender [31] used perturbation expansions (valid if ${}^{\perp}\mathcal{H}/\,{}^{\parallel}\mathcal{H} \ll 1$); van Hemmen and Sutő [29] used WKB methods (valid for arbitrary ${}^{\perp}\mathcal{H}/\,{}^{\parallel}\mathcal{H}$, provided S is large enough), and Enz and Schilling introduced instanton methods [32] (in many cases, formally equivalent to the WKB method, but not so general - one requires a semiclassical path or paths). Quite a lot can (and has been) said about this kind of truncation, but the problem is now solved, and you can read about it in the literature. What interests us here is the fact that such calculations are basically *irrelevant* to the real world, because of the strong coupling to the nuclear spins. This is a much more complicated problem, involving many more degrees of freedom- yet such problems are quite generic in this game, since there is always going to be some environment which relaxes the system.

In what follows, I will describe how the truncation can be done using instanton methods. This is certainly not the only way one could imagine it being done, and in fact it would be very interesting to see it done otherwise. Since this course is pedagogical, I will not try to to the reduction for a general Hamiltonian; instead it is done for the simple biaxial Hamiltonian given above.

(i) Free Giant Spins: To warm up, lets briefly recap how the instanton analysis works for the simple *isolated* biaxial Hamiltonian (1.9) (cf. ref [32,33,34,35]). We start by choosing a basis in the truncated (2-level) space such that the eigenstates of $\hat{\tau}_z$ correspond to the 2 semiclassical minimum states of $H_o(\vec{S})$, defined by coherent state vectors $|\vec{n}_1\rangle$ and $|\vec{n}_2\rangle$, such that $\vec{n}_1|\vec{S}|\vec{n}_1\rangle = S\vec{n}_1$ and $\vec{n}_2|\vec{S}|\vec{n}_2\rangle = S\vec{n}_2$; the eigenstates of $H^o_{eff}(\vec{\tau})$ are then linear combinations of $|\vec{n}_1\rangle$ and $|\vec{n}_2\rangle$, which we can determine once we have found the four matrix elements $\langle\vec{n}_\alpha|H^o_{eff}|\vec{n}_\beta\rangle$ with $\alpha,\beta = 1,2$.

Formally one can do this as follows, for the free spin. Consider the path integral expression for the transition amplitude $\Gamma^o_{\alpha\beta}(t)$, during the time t; this is given by :

$$\Gamma^o_{\alpha\beta}(t) = \langle\vec{n}_\alpha|e^{-iH_o(\vec{S})t}|\vec{n}_\beta\rangle = \int_{\vec{n}(\tau=0)=\vec{n}_\beta}^{\vec{n}(\tau=t)=\vec{n}_\alpha} \mathcal{D}\vec{n}(\tau)\exp\left\{-\int_0^t d\tau\mathcal{L}_o(\tau)\right\}\,, \tag{12}$$

where the free spin Euclidean Lagrangian is

$$\mathcal{L}_o = -iS\dot{\theta}\varphi\sin\theta + H_o(\vec{n})\,. \tag{13}$$

Here θ and φ are the usual polar and azimuthal angles for the unit vector $\vec{n}(\tau)$.

Now in the semiclassical approximation there are two fundamental time scales in the paths $\vec{n}(\tau)$ in (12); these are Ω_o^{-1}, the time required for the instanton traversal to be made between states, and Δ_o^{-1}, the typical time elapsing between instantons. By definition, an effective Hamiltonian is supposed to reproduce the slow dynamics of the system in the truncated Hilbert space of the two lowest levels, i.e., for long time scales an evolution operator is approximated as

$$\Gamma^o_{\alpha\beta}(t) \approx \left(e^{-iH_{eff}t}\right)_{\alpha\beta}\,. \tag{14}$$

Since Δ_o is exponentially smaller than Ω_o, and the nondiagonal elements are $\sim\Delta_o$, we can write

$$\Gamma^o_{\alpha\beta}(t) \quad = \quad \langle\vec{n}_\alpha|e^{-iH_{eff}t}|\vec{n}_\beta\rangle$$

$$\approx \quad \delta_{\alpha\beta} - it\langle\vec{n}_\alpha|H^o_{eff}|\vec{n}_\beta\rangle \; ; \qquad (\Omega_o^{-1} \ll t \ll \Delta_o^{-1}) \; ; \tag{15}$$

Then we immediately find the matrix elements of $H^o_{eff}(\vec{\tau})$ for $\alpha \neq \beta$ as

$$\left(H^o_{eff}(\vec{\tau})\right)_{\alpha\beta} = \frac{i}{t}\Gamma^o_{\alpha\beta}(t) \; ; \qquad (\Omega_o^{-1} \ll t \ll \Delta_o^{-1}) \; . \tag{16}$$

As a concrete example, consider the easy-axis/easy-plane Hamiltonian (9), where

$$H_o(\vec{n}) = SK_{\|}\left[\sin^2\theta + \frac{K_\perp}{K_\|}\sin^2\theta\sin^2\varphi\right] , \tag{17}$$

The two lowest states are $\vec{n}_1 = \hat{z}$ and $\vec{n}_2 = -\hat{z}$; henceforth we write these states as $|\Uparrow\rangle$ and $|\Downarrow\rangle$. In the usual case where $K_\perp/K_\| \gg 1$ (so that the tunneling amplitude is appreciable) one has only small oscillations of φ about the semiclassical trajectories $\varphi = 0$ or π, and by eliminating φ one has

$$\mathcal{L}_o(\theta) = \frac{S}{4K_\perp}\dot{\theta}^2 + SK_\|\sin^2\theta \; , \tag{18}$$

giving a classical equation of motion $\dot{\theta} = \Omega_o \sin\theta$, and instanton solution [36,37], going from $|\Uparrow\rangle$ to $|\Downarrow\rangle$

$$\sin\theta(\tau) = 1/\cosh(\Omega_o\tau) \tag{19}$$

(with the instanton centered at $\tau = 0$); in this system the "bounce" or "small oscillation" frequency is

$$\Omega_o = 2(K_\|K_\perp)^{1/2} \; . \tag{20}$$

The frequency Ω_o then sets the ultraviolet cut-off for the Hilbert space of $H^o_{eff}(\vec{\tau})$, and one finds, by substituting the semiclassical solution into (12) and evaluating the determinant over the quadratic fluctuations around the semiclassical solution [38] (the zero mode contribution gives a factor it, in the usual way), that from (16) we get [29,32,33,34,36,37,39]:

$$\hat{H}^o_{eff}(\vec{\tau}) = \frac{\Delta_o(S)}{2}\hat{\tau}_x \; , \tag{21}$$

$$\Delta_o(S) = -\sum_{\eta=\pm}\sqrt{\frac{2}{\pi}ReA_o^{(\eta)}}\,\Omega_o\exp\{-A_o^{(\eta)}\} \equiv 2\Delta_o\cos\pi S \; , \tag{22}$$

$$A_o^{(\eta)} = 2S(K_\|/K_\perp)^{1/2} + i\eta\pi S \; , \tag{23}$$

where the action $A_o^{(\eta)}$ is that for the transition between the two semiclassical minima, either clockwise ($\eta = +$) or anticlockwise ($\eta = -$); the phase $\eta\pi S$

is the Kramers/Haldane phase, coming from the linear in time derivatives kinetic term in (13). Without this phase, we would simply have a splitting $|\Delta_o| = \sqrt{2ReA_o/\pi}\Omega_o \exp\{-A_o\}$ with $A_o = 2S(K_{\|}/K_{\perp})^{1/2}$.

For this symmetric problem (where the 2 semiclassical states $|\Uparrow\rangle$ and $|\Downarrow\rangle$ are degenerate), the *diagonal* matrix elements in $\hat{H}^o_{eff}$ are zero (in fact if we computed them directly, again by evaluating the derminant, we would find a value Ω_o for each- but this is just the energy of the 2 states measured from the bottom of each well, and it makes sense to redefine the energy zero to make these diagonal elements zero). If instead of considering the symmetric problem, I had also added an external bias field, causing an energy difference ξ in the absence of tunneling, then the diagonal elements would simply be $\pm\xi/2$, and we would end up with an effective Hamiltonian

$$H^o_{eff} = \frac{\Delta_0}{2}\hat{\tau}_x + \frac{\xi}{2}\hat{\tau}_z \tag{24}$$

with eigenvalues

$$E_\pm = \pm|E| = \pm\frac{1}{2}\sqrt{\Delta^2 + \epsilon^2} \tag{25}$$

and eigenfunctions

$$\psi_\pm = A_\pm \left[(E_\pm + \epsilon)|\Uparrow\rangle - \Delta_0|\Downarrow\rangle \right], \tag{26}$$

$$A_\pm = \left[\frac{1}{(E_\pm + \epsilon)^2 + (\Delta_0)^2} \right]^{1/2} \tag{27}$$

Suppose this 2-level system starts off at $t = 0$ in the state $|\Uparrow\rangle$. Then after a time t, the system is described by a density matrix in the same basis, given by

$$\rho(t) = \begin{pmatrix} 1 - \frac{\Delta_0^2}{E^2}\sin^2 Et & -i\frac{\Delta_0}{E}\sin Et \\ i\frac{\Delta_0}{E}\sin Et & \frac{\Delta_0^2}{E^2}\sin^2 Et \end{pmatrix} \tag{28}$$

In the equivalent 2-well problem, the initial "wave-packet" $|\Uparrow\rangle$ partially oscillates between the 2 wells- the diagonal elements give the occupation probability of the wells, and the off-diagonal elements describe oscillatory quantum interference between them, which is suppressed by the bias; when $\xi \gg \Delta_0$, the system stays in one well, in the absence of any coupling to the environment, even if this means it is in an excited (high-energy) state. To remove this "blocking" of transitions to the lower energy state (and to give irreversible *relaxation*, as opposed to just oscillations), we need some kind of *dynamic environment*.

Notice in passing the limitations of this instanton derivation - it would not work in the absence of a few well-defined semiclassical paths. Thus if we replaced the transverse term ${}^{\perp}K_2S_y^2$ by the rotationally invariant ${}^{\perp}K_2(S_x^2 + S_y^2)/2$, there would be no favoured semiclassical path, and we would be forced back onto the WKB or perturbative analyses. However when we can use the instanton formalism, it is easily adaptable to include the spin bath as well; we now turn to this task.

(ii) Including the Spin Bath: I will not give all the details here (for which see Tupitsyn et al. [35]), just the main ideas. First, the basic physics. Before we couple the giant spin to the spin bath, the spin bath spectrum, containing 2^N lines, is almost completely degenerate - only the tiny internuclear coupling splits these lines. However the hyperfine coupling is very large (it ranges from 1.4 mK, in Ni, to nearly 0.5 K, in Ho, *per nuclear spin*, and it drastically alters the environmental spectrum. The nuclear levels, for a given giant spin state, now find themselves spread in a Gaussian multiplet of half-width $E_o \sim \omega_o N^{\frac{1}{2}}$ in energy (where ω_o is a typical hyperfine coupling) around the giant spin state; E_o is many orders of magnitude greater than the width $T_2^{-1} \sim 10^4 - 10^5$ Hz of the nuclear multiplet before coupling. In this sense the nuclear spin bath degrees of freedom are slaved to the giant spin. Notice also that if the giant spin was formerly able to tunnel (because of near resonance between states $\mid S_z >$ and $\mid -S_z >$), this is unlikely now, because an extra *internal* bias field $\epsilon = \tau_z \Sigma_k \omega_k \sigma_k^z$ acts on $\underline{\tau}$ (indeed on $\vec{S}$ itself), and typically $\epsilon \gg \Delta_o$, pushing the giant spin way off resonance. Thus the hyperfine coupling to even a single nuclear spin drastically alters the giant spin dynamics as well. before doing anything formal, lets see qualitatively what must happen.

(i) In the absence of any nuclear dynamics this "degeneracy blocking" would only allow a tiny fraction of giant spins to make any tunneling transitions at all [14,16,17,40] (Fig. 3). We will see in Chapter 4 that in the low temperature limit, it has been recently discovered that such giant spins can have very fast dynamics, which is (to me at any rate!) a very convincing demonstration of the role that the nuclear spin *dynamics* must play in the relaxation (since at these temperatures this is the only environment left with any dynamics at all!). This justifies *a posteriori* the theoretical effort that has been put into understanding these effects!

(ii) A second effect of the spin bath is "topological decoherence" [14,15,16]. Each transition of the giant spin causes a time-dependent perturbation (via H_{hyp}) on the nuclear bath, which can cause transitions in the bath - moreover, this perturbation causes a *phase change* in the nuclear bath state. Since this phase change varies from one transition of $\vec{S}$ to the next, the net effect is to *randomize* the phase of $\vec{S}$, ie., to cause phase decoherence. From the measurement point of view the nuclear spins are measuring time-varying fields due to $\vec{S}$

(ie., inhomogeneous in time rather than in space, as in the usual Stern-Gerlach experiment); this is a kind of "reverse Stern-Gerlach measuring apparatus" [22].

(iii) A final effect of the nuclear spins on the tunneling is that of "orthogonality blocking". Suppose that, semiclassically speaking, all nuclear spins are parallel/antiparallel to the net field acting on them (due to $\vec{S}$ and/or an external field), *before* $\vec{S}$ tunnels. What happens to them afterwards? The answer is that if they are still parallel/antiparallel to the new field, nothing at all. But in general the new field is not parallel/antiparallel to the old one, and so the bath spins must now precess (or quantum mechanically, make transitions) in the new field. The "mismatch" between old and new nuclear eigenstates causes a severe suppression of the tunneling of $\vec{S}$ (as well as decohering it).

Well, this all sounds very nice; but how do we deal formally with this physics? We wish to truncate to a low energy Hamiltonian $H_{eff}(\vec{\tau}, \{\vec{\sigma}_k\})$, valid for energies $\ll \Omega_o$ (cf. Fig. 1), where $\Omega_o \sim K$. First, let's cheat and look at the final answer! One finds for the general problem of a giant spin coupled to a spin bath that [14,16,35]

$$\begin{aligned} H_{\text{eff}} &= \left\{ 2\Delta_o \hat{\tau}_- \cos\left[\pi S - i \sum_k (\alpha_k \vec{n} \cdot \hat{\vec{\sigma}}_k + \beta_o \vec{n} \cdot \vec{H}_o)\right] + H.c. \right\} \\ &+ \frac{\hat{\tau}_z}{2} \sum_{k=1}^{N} \omega_k^{\parallel} \vec{l}_k \cdot \hat{\vec{\sigma}}_k + \frac{1}{2} \sum_{k=1}^{N} \omega_k^{\perp} \vec{m}_k \cdot \hat{\vec{\sigma}}_k + \sum_{k=1}^{N} \sum_{k'=1}^{N} V_{kk'}^{\alpha\beta} \hat{\sigma}_k^{\alpha} \hat{\sigma}_{k'}^{\beta} \quad . \end{aligned} \tag{29}$$

I have also added an external field $\vec{H}_o$. This is the *general* form; we shall see below that for the biaxial nanospin, with the simple contact hyperfine interaction in (7), it simplifies somewhat. This effective Hamiltonian looks forbidding (although we shall see it is not so difficult to use); let us now look at each term in turn.

The first thing to notice is the separation into a diagonal term (in $\hat{\tau}_z$) and a non-diagonal one (in $\hat{\tau}_+, \hat{\tau}_-$). The non-diagonal term operates during transitions of $\vec{S}$, and it also causes transitions in the nuclear bath, because of the time-dependent field $\omega_k \vec{S}/S$ acting on each $\hat{\vec{\sigma}}_k$ during a transition. The diagonal term operates when $\vec{S}$ is in one of its two quiescent states $\vec{S}_1$ and $\vec{S}_2$. Defining as before the basis states $\mid \vec{n}_1 > \equiv \mid \Uparrow >$ and $\mid \vec{n}_2 > \equiv \mid \Downarrow >$, we have corresponding fields $\vec{\gamma}_k^{(1)}$ and $\vec{\gamma}_k^{(2)}$ acting on $\vec{\sigma}_k$. We define the *sum* and the *difference* terms as

$$\begin{aligned} \omega_k^{\parallel} \vec{l}_k &= \vec{\gamma}_k^{(1)} - \vec{\gamma}_k^{(2)} \\ \omega_k^{\perp} \vec{m}_k &= \vec{\gamma}_k^{(1)} + \vec{\gamma}_k^{(2)} \; . \end{aligned} \tag{30}$$

where the $\vec{l}_k$ and $\vec{m}_k$ are unit vectors. Then the truncated diagonal interaction takes the form (we project on states $|\Uparrow\rangle$ and $|\Downarrow\rangle$ using standard $(1+\hat{\tau}_z)/2$ and $(1-\hat{\tau}_z)/2$ operators)

$$H^D_{eff} = \sum_{k=1}^{N} \left\{ \vec{\gamma}_k^{(1)} \frac{1+\hat{\tau}_z}{2} + \vec{\gamma}_k^{(2)} \frac{1-\hat{\tau}_z}{2} \right\} \cdot \hat{\vec{\sigma}}_k \equiv \frac{1}{2} \left\{ \hat{\tau}_z \sum_{k=1}^{N} \omega_k^{\parallel} \vec{l}_k \cdot \hat{\vec{\sigma}}_k + \sum_{k=1}^{N} \omega_k^{\perp} \vec{m}_k \cdot \hat{\vec{\sigma}}_k \right\} , \tag{31}$$

i.e., one term which changes when $\vec{S}_1 \to \vec{S}_2$, and one which does not. Usually $\omega_k^{\parallel} \gg \omega_k^{\perp}$, and $\omega_k^{\parallel} \sim \omega_o$. Thus we get the diagonal term in (29). For the biaxial system with contact hyperfine interaction, the diagonal term is trivial; we get

$$\omega_k^{\parallel} = \omega_k \ ; \quad \omega_k^{\perp} = 0 \ ; \quad (biaxial\ system) \tag{32}$$

Notice that it is this diagonal term which is responsible for the degeneracy blocking just mentioned- if we look at the *combined* density of states for the giant spin plus the nuclear levels, we see that there will be a huge multiplet of 2^N levels around each giant spin level. We will see later (section 3.4(a)) under what circumstances this takes the Gaussian form mentioned above.

Turning now to the non-diagonal term, we notice that unless $\omega_k \ll \Omega_o$, the hyperfine coupling itself can mediate interactions between the 2 lowest giant spin levels, and the higher levels. In practise $\omega_k/\Omega_o \ll 1$ almost always and we have a small parameter. Then instead of the bare transition matrix $\Gamma^o_{\alpha\beta}$ in (14), we calculate

$$\Gamma_{\alpha\beta}(\{\sigma_k^{(\alpha)}, \sigma_k^{(\beta)}\}; t) =$$

$$\prod_{k=1}^{N} \int_{\vec{\sigma}_k^{(\alpha)}}^{\vec{\sigma}_k^{(\beta)}} \mathcal{D}\vec{\sigma}_k(\tau) \int_{\vec{n}_\alpha}^{\vec{n}_\beta} \mathcal{D}\vec{n}(\tau) exp \left\{ - \int d\tau \left[\mathcal{L}_o(\tau) + \sum_{k=1}^{N} \mathcal{L}_k^o(\tau) + \delta\mathcal{L}_\sigma(\tau) \right] \right\} \tag{33}$$

where $\mathcal{L}_k^o(\tau) = -(i/2)\dot{\Theta}_k \varphi_k \sin\Theta_k$ is the nuclear spin Lagrangian, and $\delta\mathcal{L}_\sigma(\tau) = \Sigma_{k=1} \vec{\gamma}_k(\tau) \cdot \vec{\sigma}_k(\tau)$, with $\vec{\gamma}_k(\tau) = \omega_k \vec{S}(\tau)/2S$ the time-dependent field, from $\vec{S}$, acting on $\vec{\sigma}_k$. During the transition, $\vec{S}(\tau)$ varies over a timescale Ω_o^{-1}, and so we get a time-dependent perturbation $\sim \omega_k/\Omega_o$ (ie., a sudden perturbation).

To calculate $\Gamma_{\alpha\beta}$ one first finds the new instanton trajectory by minimizing $\mathcal{L}_o(\tau) + \delta\mathcal{L}(\tau)$, and then calculates the transition amplitude from $|\vec{n}_\alpha>$ to $|\vec{n}_\beta>$; for details see Tupitsyn et al. [35]. For the simple biaxial Hamiltonians (1.9), one gets for the transition element $\Gamma_{\Uparrow\Downarrow}(t)$ (the "non-diagonal" amplitude) after an "intermediate" time t, (ie., for $\Omega_o^{-1} \ll t \ll \Delta_o^{-1}$):

$$\hat{\Gamma}_{\Downarrow\Uparrow}(t) = it \sum_{\eta=\pm} \sqrt{\frac{2}{\pi} ReA_o \Omega_o} \exp \left\{ -A_o^{(\eta)} - \eta \sum_k (\alpha_k \vec{n} \cdot \hat{\vec{\sigma}}_k + \beta_o \vec{n}_o \cdot \vec{H}_o) \right\} \tag{34}$$

where the vectors $\alpha_k \vec{n}$ and $\beta_o \vec{n}_o$ have to be determined in terms of the parameters in the original high-energy Hamiltonian, in the course of minimizing the action. For the case of the biaxial giant spin Hamiltonian with contact hyperfine interactions, one easily gets

$$\alpha_k \vec{n} = \frac{\pi \omega_k}{2\Omega_o} (\hat{\vec{x}},\ i\sqrt{K_{\|}/K_{\perp}}\ \hat{\vec{y}})\ ; \tag{35}$$

$$\beta_o \vec{n}_o = \frac{\pi \gamma_e S}{\Omega_o} (\hat{\vec{x}},\ i\sqrt{K_{\|}/K_{\perp}}\ \hat{\vec{y}})\ , \tag{36}$$

From this result the non-diagonal term in (29) follows immediately. The dimensionless parameter α_k is particularly interesting, since it parameterizes the inelastic effect, on the nuclear spin $\underline{\sigma}_k$, of transitions made by $\vec{S}$ (and it leads to topological decoherence [14]). Note that it is *complex*, and that for $\omega_k/\Omega_o \ll 1$, we have $|\alpha_k| \sim \pi\omega_k/2\Omega_o$. Since the *probability* that $\vec{\sigma}_k$ will flip is $\frac{1}{2}|\alpha_k|^2$, this result is what we would expect (it follows from time-dependent perturbation theory in the sudden approximation). The only subtlety is that we cannot apply perturbation theory directly, since (at least in the instanton formalism), the tunneling takes place in *imaginary* time τ (not real time t); this is why the more detailed treatment of Tupitsyn et al. [35] is necessary. Roughly speaking, the real part of α adds an extra phase to the existing giant spin Berry phase (leading to topological decoherence), and the imaginary part is a renormalisation of the tunneling action caused by the nuclear spins.

This completes the discussion of how this particular spin bath effective Hamiltonian is derived - I have given details so you can see some of what is involved. In Chapter 3 we will see how to extract the dynamics of the giant spin (and we will then see the quantitative realisation of the concepts of topological decoherence, degeneracy blocking, and orthogonality blocking, discussed above). But first we look at a few other examples of systems coupled to a spin bath.

2.3 Some Other Systems Coupled To Spin Baths

The previous section attempted to explain the details of a particular truncation procedure in some pedagogical detail. However there are other important physical examples of systems coupled to spin baths, which can be profitably studied if you wish to understand the truncation procedure better. I shall briefly recount how two such systems work, and then mention some other examples which can be found in the literature. The first example involves a Josephson superconducting ring, coupled to nuclear and paramagnetic spins; and the second involves a magnetic soliton (a domain wall) coupled to nuclear spins.

2.3(a) MQC in SUPERCONDUCTORS, and the SPIN BATH

Let us start with a system which, in the *absence* of the spin bath, is thought to truncate at low energies to a 2-level system, just like the giant spin system. I am thinking here of a superconducting "ring" with a single Josephson "weak link"; a typical geometry is shown in Fig.3.

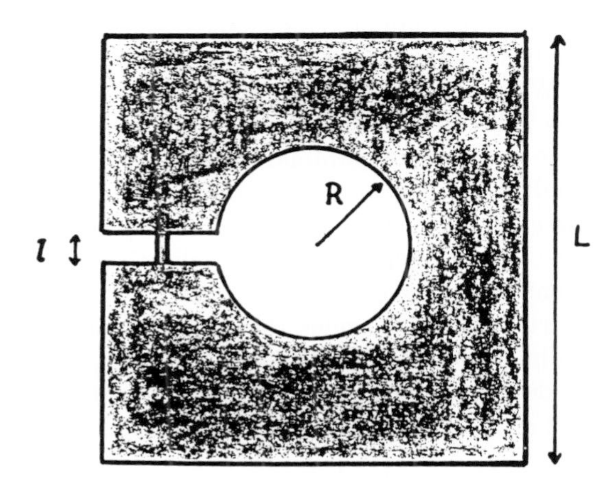

FIG. 3: The geometry chosen for the analysis of the Josephson ring. Topologically, the system is a torus, with the weak link allowing the passage of a fluxon between the inside and the outside of the ring, if the current through the weak link exceeds I_c. The geometry is discussed in the text.

Following the work of a number of authors, it is understood that the flux Φ passing through the ring moves in an adiabatic ("high-energy") potential

$V(\Phi)$ of form

$$V(\Phi) = \frac{1}{2L}(\Phi - \Phi_e)^2 - 2\pi E_J \cos(2\pi\Phi/\Phi_o) \tag{37}$$

where Φ_o is the flux quantum $h/2e$, L is the ring inductance, Φ_e is the external imposed flux, Φ the total flux (both of these through the ring) and $E_J = I_c\Phi_o/2\pi$ is the weak link "Josephson coupling energy", with I_c the critical current through the ring. This "RF SQUID", and the possibility of macroscopic tunneling of Φ through the Josephson cosine potential, has been discussed in great detail in the literature[13,41] (and see also Leggett's chapter in this volume). There is also a kinetic term $T = \frac{1}{2}C\dot{\Phi}^2$ in the system Hamiltonian; and as a result at low energies it finds itself near the bottom of a potential well, able both to oscillate (small oscillation frequency $\Omega_o \sim 2\pi(E_j/\pi C)^{1/2}/\Phi_o$) or to tunnel to the nearest potential well. If these 2 wells are the 2 lowest in energy (with others at energies higher by Ω_o or more) then when $kT < \Omega_o/2\pi$, we can again model the system by a 2-level system.

There is also a resistive coupling to normal electrons and Bogoliubov quasiparticles, which can be understood in terms of a coupling to a bath of oscillators (see next section). What we are interested in here is - what will be the coupling of the flux Φ to any spins in the ring? These may be either nuclear (of which there is a very large number) or paramagnetic spins (coming from magnetic impurities). We shall see in this and the next chapter that this question is of fundamental importance for the search for "Macroscopic Quantum Coherence" in RF SQUID rings. In what follows I give a qualitative discussion only, based on a detailed calculation by myself and Prokof'ev. [18]

Suppose to start with we consider the geometry shown in Fig. 3. We assume a cube of dimension 1x1x1 cm ($L = 1\ cm$) of type-I superconducting material, with London penetration depth $\lambda_L = 5 \times 10^{-6}\ cm$, and a hole in the center of radius $R = 0.2\ cm$, and surface area $S = \pi R^2 = 0.1\ cm^2$. The magnetic field inside the hole corresponding to a half-flux quantum is

$$B_o = \frac{\pi\hbar c}{e\pi R^2} = 2 \times 10^{-6}\ G\ . \tag{38}$$

There is a slit in the cube, bridged by the wire-shaped junction, of length $l = 10^{-4}\ cm$ and diameter $d = 2 \times 10^{-5}\ cm$. The current density in the junction is enhanced over that elsewhere on the system surface by a factor $L/d = 5 \times 10^4$, and correspondingly the magnetic field in the junction is as high as $B_{jun} \sim B_o L/d = 10^{-1}\ G$.

There are both nuclear spins and paramagnetic impurities in the spin bath. Consider first the nuclear spins; assuming all nuclei have spins, we find that in the bulk of the ring, within a penetration depth of the surface, there is a number

$$N_{ring} = (L \times 2\pi R \times \lambda_L) \times 10^{23} \approx 5 \times 10^{17} \ , \tag{39}$$

of nuclear spins coupling to the ring current; and in the junction itself, a number

$$N_{jun} = (l \times \pi d \times \lambda_L) \times 10^{23} \approx 3 \times 10^{9} \ . \tag{40}$$

coupling to the junction current. In the "high-temperature" limit the average nuclear polarisation is $M = \sqrt{N}$, and so the typical coupling between the junction current and the nuclear spins is $\Gamma_{ring} \sim \omega_{ring} M = 2 \times 10^{-13} K \times 7 \times 10^8 \approx 10^{-4} K$, where $\omega_{ring} = \mu_n B_o$, and μ_n is the nuclear Bohr magneton. For the junction nuclei one has a coupling $\Gamma_{jun} \sim \omega_{jun} \sqrt{N_{jun}} \approx 5 \times 10^{-4} K$, where $\omega_{jun} = \mu_n B_{jun}$, which is actually similar, because there is a linear increase in the magnetic field (since $B_{jun} \sim 1/d$), but a quadratic suppression in the number of spins (ie., $N_{jun} \sim ld$). Notice the magnetic coupling is $\ll kT$, justifying the high-T assumption. Notice also we have taken no account of a possible coupling to substrate spins (the ring is in superfluid He-4!), which would have a much larger coupling again to the current. We have also assumed perfect screening from external fields.

This coupling gives a longitudinal bias energy $\xi \sim 10^{-4} - 10^{-3} K$, acting on the tunneling flux coordinate Φ, which is rather bigger than any presently realistic numbers for the SQUID tunneling matrix element Δ_o; we are again in a situation of strong degeneracy blocking. However, it also turns out that we have *very* strong orthogonality blocking! This is because each nuclear spin feels the dipolar fields from the other nuclear spins; in general there will be some component perpendicular to the field from the SQUID, of strength $\sim 1G$, and associated energy $\omega_k^{\perp} \sim V_{NN} \sim 10^{-7} K$, where V_{NN} is the nearest neighbour internuclear dipolar coupling energy. Physically, when the SQUID flips, the field on each nuclear spin hardly changes its direction, being dominated by the more slowly varying (but much stronger) nuclear dipolar field.

We may summarize this analysis of the nuclear spin effects on the flux dynamics in the form of the effective Hamiltonian

$$H_{\text{eff}} = \Delta_o \hat{\tau}_x + \frac{1}{2} \sum_{k=1}^{N} (\omega_k^{\perp} \hat{\sigma}_k^z + \hat{\tau}_z \omega_k^{\parallel} \hat{\sigma}_k^z) \tag{41}$$

where $\omega_k^{\parallel} = \omega_{ring}$ or ω_{jun} (with values $\omega_{ring} \sim 2 \times 10^{-13} K$ and $\omega_{jun} \sim 10^{-8} K$ respectively), whilst $\omega_k^{\perp} \sim 10^{-7} K$ as above. Thus $\omega_k^{\parallel}/\omega_k^{\perp} \ll 1$, which is the opposite limit considered to that for the giant spin! Notice further that both these couplings are much less than Δ_o (ξ is bigger than Δ_o only because there are so many nuclei involved). Thus the nuclear bath is no longer slaved to the central system at all! In the next chapter we will show that this allows us to map this problem to that of an oscillator bath, coupled to Φ.

Consider now the effect of paramagnetic impurities, with concentration n_{pm}, a single-spin coupling $\omega_{pm} \sim 10^3 \times \omega_{ring}$ to the current, and hence a total coupling energy $\Gamma_{pm} \sim 10^6 n_{pm} \Gamma_{ring} \sim n_{pm} \times 100\ K$. This is obviously bigger still, unless the superconductor is very pure indeed! The dynamics of these will come either from the Kondo effect (in the superconductor), or from flip-flop processes between these impurities. The former go at a rate $\sim T_K$, the Kondo temperature (which varies over many orders of magnitude depending on the impurity; the latter would go at a rate $(T_2^{-1})_{pm} \sim 10^9 n_{pm}$ Hz, except that in pure samples these flip-flop processes will themselves be blocked by the local dipolar coupling between the impurity and nearby nuclear spins (of strength $\sim 10^{-4} K$); thus this will happen once $n_{pm} \ll 10^{-3}$.

Just as for the Giant spins, to properly understand the flux dynamics we must then consider the nuclear bath dynamics, which can in principle relieve this blocking. We will deal with this in the next chapter.

2.3(b) MAGNETIC DOMAIN WALL TUNNELING

Let us now move to a quite different example, in which at low energies the central system does not behave at all like a 2-level system, but instead like a particle moving continuously in a one-dimensional potential. Of course we expect that many physical systems will behave like this at low energies, and many of them will couple to spin environments - but the following example is the only one which has so far been worked out in detail[28,42].

The reason that a magnetic domain wall moves like a 1-dimensional particle in many realistic situations is that it is a membrane-like soliton whose "flexural" oscillations are hindered by the strong magnetic dipolar field, which tries to keep the wall flat. As an example we take the famous "Bloch wall", which has a Hamiltonian

$$\begin{aligned} \mathcal{H} &= \int d\mathbf{r}[J(\nabla \mathbf{m})^2 - K_{\|} m_z^2 + K_{\perp} m_x^2] \\ &= \int d\mathbf{r}[J((\nabla\theta)^2 + \sin^2\theta(\nabla\phi)^2) - K_{\|}\cos^2\theta + K_{\perp}\cos^2\phi\sin^2\theta] \end{aligned} \quad (42)$$

representing a ferromagnet with easy axis along the z axis and the $z-y$ plane being easy. In 3-dimensions, the units of J are J/m and the units of the anisotropy constants are in J/m^3.

The domain wall corresponding to this Hamiltonian is perpendicular to the x axis, with the magnetisation rotating in the $z-y$ plane. We refer to the wall by its center, located at a position Q along the x axis The new frame of reference is thus $(x_1, x_2, x_3) = (z, y, x)$. This is represented in Fig.4 below.

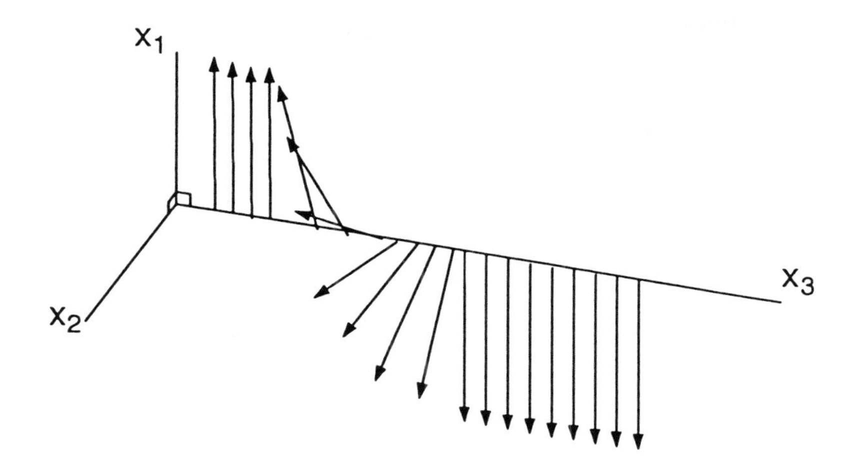

FIG. 4: A standard Bloch wall in a uniaxial ferromagnet, showing the direction of the electronic spins as one passes along a path perpendicular to the wall plane. The chirality is defined by the sense in which the spins turn, and the topological charge by the difference in magnetisation along the easy axis, between the 2 sides of the wall.

The components of the magnetisation are given by:

$$\begin{aligned}
\hat{m}_1^B &= C \tanh\left(\frac{x_3 - Q(t)}{\lambda_B}\right) \\
\hat{m}_2^B &= \chi \left(1 - \frac{\dot{Q}^2(t)}{8c_0^2}\right) \operatorname{sech}\left(\frac{x_3 - Q(t)}{\lambda_B}\right) \\
\hat{m}_3^B &= C \frac{\dot{Q}(t)}{2c_0} \operatorname{sech}\left(\frac{x_3 - Q(t)}{\lambda_B}\right)
\end{aligned} \tag{43}$$

$C = \pm 1$ is the "topological charge" of the wall and $\chi = \pm 1$ is the "chirality". The topological charge corresponds to the direction along which the wall moves under the application of an external magnetic field in a direction parallel to the easy axis, while the chirality refers to the sense of the rotation of the magnetisation inside the wall. A static Bloch wall only rotates in the easy plane.

However, as soon as it moves it creates a demagnetising field which causes the spins to precess and the appearance of a component of the magnetisation out of the plane, directly proportional to the wall velocity. The precession of the spins also causes the appearance of an inertial term, the Döring mass, given by

$$M_w = \frac{S_w M_0^2}{\gamma_g^2 (JK_{\parallel})^{1/2}} \left[\frac{1}{(1 + K_{\perp}/K_{\parallel})^{1/2} - 1} \right]^2 \tag{44}$$

where S_w is the surface area of the wall.

Now from this phenomenological picture of the wall, one can proceed to a treatment which eliminates the details of the wall profile altogether, and simply describes everything in terms of the coordinate Q of the wall centre (assuming a wall held flat by the dipolar or "demagnetisation fields- in realistic situations wall curvature is small and has little effect on what follows). In this case the wall will have a kinetic energy (parametrised by the Döring mass), and there will also be potential terms. There are 2 obvious sources for these:

(i) If defects are present in the sample, it will be energetically favourable for the magnetisation to rotate at this site since there is no associated energy cost. We will assume that the radius R_d corresponding to the defect volume is much smaller than λ_B, the domain wall width. The wall thus becomes pinned by a potential of the form [42]

$$V(Q) = V_0 \,\mathrm{sech}^2(Q/\lambda_B) \tag{45}$$

with V_0 proportional to the volume of the defect. We further assume that there is a very small concentration of defects, so that there is only 1 important pinning center for the wall. This would correspond to an ideal experimental situation.

(ii) The application of an external magnetic field $\mathbf{H}_e$ in the direction of the easy axis couples to the magnetisation to give a potential term linear in Q.

We can then write a "bare" Hamiltonian (ie., neglecting the environment) for the wall [42]

$$H_w = \frac{1}{2} M_w \dot{Q}^2 - V(Q) - 2 S_w \mu_B M_0 H_e Q \tag{46}$$

where we put the topological charge $C = 1$ for brevity.

What now of the environment? In the literature you will find extensive discussion of the effect of magnons [37,42] (ie., spin waves),electrons [43], and phonons [28] on the wall dynamics. However these are all oscillator baths. What we are interested in here is any spin bath effects. It turns out here that, just as for the nanomagnets, these are far more important than any oscillator bath effects. To see why this is we recall again the strength of hyperfine interactions to nuclear spins. Nuclear hyperfine effects vary enormously between magnetic systems. The weakest is in Ni, where only 1% of the nuclei have spins, and the hyperfine coupling is only $\omega_0 = 28.35\,\mathrm{MHz}$ ($\sim 1.4\,\mathrm{mK}$). On the other hand, in the case of rare earths, ω_0 varies from 1 to 10 GHz (0.05 to 0.5 K). In this latter case the hyperfine coupling energy to a *single* nucleus may be comparable to the other energy scales in the problem !

To derive an effective interaction Hamiltonian, we consider our system of ferromagnetically ordered spins to be coupled locally to N nuclear spins $\mathbf{I}_k$ at positions $\mathbf{r}_k$ ($k = 1, 2, 3, \ldots N$), which for a set of dilute nuclear spins (where only one isotope has a nuclear spin) will be random. The total Hamiltonian for the coupled system is then

$$H = H_m + \sum_{k=1}^{N} \omega_k \mathbf{s}_k \cdot \mathbf{I}_k + \frac{1}{2} \sum_k \sum_{k'} V_{kk'}^{\alpha\beta} I_k^\alpha I_{k'}^\beta \tag{47}$$

where H_m is the electronic Hamiltonian for the magnetisation (Eq. (42)), written in terms of the electronic spins $\mathbf{s}_k$ at the sites where there happen to be nuclear spins, and ω_k is the hyperfine coupling at $\mathbf{r}_k$; $V_{kk'}^{\alpha\beta}$ is the internuclear dipolar interaction,, with strength $|V_{kk'}^{\alpha\beta}| \sim 1-100\,\mathrm{kHz}$ ($0.05-0.5\,\mu\mathrm{K}$). In terms of the continuum magnetisation $\mathbf{M}(\mathbf{r})$, we have

$$\begin{aligned} H &= H_m + \sum_{k=1}^{N} \omega_k \int \frac{d^3r}{\gamma_g} \delta(\mathbf{r} - \mathbf{r}_k)[M_z(\mathbf{r}) I_k^z + (M_x(\mathbf{r}) I_k^x + M_y(\mathbf{r}) I_k^y)] \\ &+ \frac{1}{2} \sum_k \sum_{k'} V_{kk'}^{\alpha\beta} I_k^\alpha I_{k'}^\beta \end{aligned} \tag{48}$$

I would like to draw your attention to the similarity between this equation and that quoted at the very beginning of section 2.2. We have in essence reduced our problem to the coupling of a single 1-dimensional coordinate to a bath of spins. I will not go into any further detail here as to how one can rewrite this Hamiltonian in terms of "degeneracy blocking" diagonal terms, plus non-diagonal terms operating when the domain wall moves (for which see the original paper[28]). Suffice it to say that the longitudinal term gives a total "hyperfine potential field" acting on the wall, of diagonal form

$$U(Q) = \frac{\omega_0 M_0}{\gamma_g} \sum_{k=1}^{N} \int d^3r \delta(\mathbf{r} - \mathbf{r}_k) \left(1 - C \tanh\left(\frac{x_3 - Q}{\lambda_B}\right)\right) I_k^w \tag{49}$$

where the nuclear spins now have their axis of quantisation defined according to the *local* magnetisation orientation; the axis w is along $\vec{M}(r)$, and so the component I_k^w is diagonal. This potential has the spatial form of a *random walk* (Fig.5); however it also fluctuates in time (because of nuclear spin diffusion). There are of course non-diagonal terms as well, for which see Dubé and Stamp [28].

As discussed in detail in the original paper, the effect of this potential on the wall dynamics can be very large indeed, particularly for rare earth magnets.

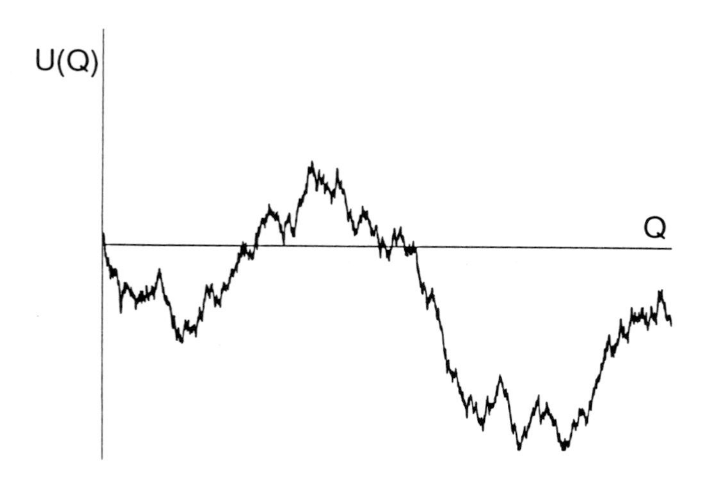

FIG. 5: Typical form of the "random walk" potential $U(Q)$ acting on a magnetic domain wall, due to the hyperfine coupling to disordered nuclear spins. We assume that the temperature is higher than the hyperfine coupling to the nuclei. In reality this potential also fluctuates in time, because of nuclear spin diffusion.

This concludes our detailed discussion of the spin bath models, and where they come from. I should emphasize that there are other "canonical" spin bath models which can be studied (and which for the most part have not been, at least not at the time of writing). Two obvious ones are

(i) A model in which a simple oscillator is coupled to a spin bath. There are many obvious systems for which such a model applies- two of them are immediately apparent, for we can simply put either the SQUID flux or the domain wall mentioned above, into a harmonic potential. This would describe the SQUID trapped in one well, near the bottom, or a domain wall trapped deep in a pinning potential.

(ii) The "Landau-Zener plus spin bath model", in which a 2-level system, coupled to a spin bath, is also *driven* by some external (c- number) field. This problem is of considerable interest for experiments on quantum nanomagnets (where one applies an AC field to the magnets, and looks at the absorption), and it is also of theoretical interest, as an example of a dissipative landau-Zener model. The simplest such model involves the single passage of the 2-level system through resonance.

The reader can certainly think of more such examples (particularly after referring forward to the discussion of canonical oscillator bath models, which have been much more thoroughly studied- see section 2.4(c)).

2.4 Coupling to The Oscillator Bath

Let us now turn to that class of effective Hamiltonians H_{eff} which can be written in terms of an environment of oscillators (the "oscillator bath" models). These models are not only very old (early examples include QED, as well as Tomonaga's model of 1-dimensional electrons, and early theories of both spin waves and the electron gas [44]), but also of great generality - they apply to almost any problem in which the central system is coupled to a set of *delocalized modes* (whether these be bosonic or fermionic, by the way). The lengthy history of this model means that I will refrain from giving a detailed treatment of truncation to its various low-energy forms - instead, I refer the reader to relevant original references or reviews. The reader is urged to consult these - the truncation is not without interesting subtleties, particularly where non-linear interactions to the environmental modes are involved.

Many physicists (and even more chemists) are suspicious of the claimed generality for the oscillator bath model for quantum environments, despite the justifications of Feynman and Vernon [12], and Caldeira and Leggett [13]. The essential claim is that at low energies, our general effective Hamiltonian (2) can very often be written as

$$\begin{aligned} H_{\rm eff}^{\rm osc} &= H_o(P,Q) + \sum_{k=1}^{N}\Big[F_k(P,Q)x_k + G_k(P,Q)p_k\Big] \\ &+ \frac{1}{2}\sum_{k=1}^{N}\left(\frac{p_k^2}{m_k} + m_k\omega_k^2 x_k^2\right) , \end{aligned} \tag{50}$$

where P, Q are the momentum and coordinate of interest, and the $\{x_k\}$ are harmonic oscillators with frequency $\omega_k < \Omega_o$, the UV cut-off. The crucial point is that the couplings F_k and G_k are weak ($\sim N^{-\frac{1}{2}}$), and this justifies one in stopping at linear coupling. In most cases the oscillators represent delocalized modes of the environment, and the $N^{-\frac{1}{2}}$ factor then just comes from normalisation of their wavefunctions.

Now in fact physicists should *not* be surprised at the generality of (1.31), since it embodies precisely the same assumptions as conventional response function theory [45] - that the effect of the central coordinates P, Q, on each environmental mode, is weak ($\sim N^{-\frac{1}{2}}$) and may be treated in 2nd order perturbation theory. Then, by an appropriate diagonalisation to normal modes [46,13], one may always write (1.31). From this point of view it is clear that this should work for, say electrons - although electrons are fermions, the low-energy modes of the Fermi liquid are incoherent particle-hole pairs, plus collective modes [45]. It also works nicely for superconductors [13,41,47] (although in this case one really needs 2 oscillator baths [47]), and in fact oscillator bath models have been applied to a vast array of electronic excitations, ranging from Luttinger liquid quasiparticles to magnons.

In what follows I will attempt to convince you that the "oscillator bath' models really are very generally applicable to environments of delocalised modes. The discussion will be fairly informal- readers wanting rigour should go to the literature.

2.4(a) SPIN-BOSON MODELS

Consider the case where $H_o(P, Q)$ describes a system moving in a 2-well potential, at which point, as with the giant spin, one truncates down to the 2 lowest levels therein, to get the celebrated "spin-boson" model [48]:

$$\begin{aligned} H_{\mathrm{SB}} &= \Delta_o \hat{\tau}_x + \xi_H \hat{\tau}_z + \sum_{k=1}^{N} \left[(c_k^{\parallel} \hat{\tau}_z) x_k + (c_k^{\perp} \hat{\tau}_- q_k + H.c.) \right] \\ &+ \frac{1}{2} \sum_{k=1}^{N} \left(\frac{p_k^2}{m_k} + m_k \omega_k^2 x_k^2 \right) , \end{aligned} \qquad (51)$$

with couplings $c_k^{\parallel}, c_k^{\perp} \sim O(N^{-\frac{1}{2}})$, and $\vec{\tau}$ again describing the 2-level central tunneling system. Often the term $c_k^{\perp}$ is dropped, since formally $c_k^{\perp}/c_k^{\parallel} \sim \Delta_o/\Omega_o$; however sometimes one has $c_k^{\parallel} = 0$ (eg. for symmetry reasons), and $c_k^{\perp}$ survives and is important [49].

Now at this point we could go through the same kind of exercise as before, and derive a spin-boson Hamiltonian at low energy for some model system. This has in fact been done by many authors for systems like SQUID's, or the

Kondo problem, etc. However, to give some continuity to the discussion, let's stick to the giant spin model for the moment, since we have already seen how it truncates to a 2-level system in isolation. To see how it also couples to oscillators, we will consider a giant spin in an environment of phonons and electrons. Readers interested in SQUID's should go to Leggett's lectures; and there is of course a huge literature on the spin-boson model.

(i) Giant Spin coupled to Phonons: We thus go back to our giant spin model (eg. (8)), and ask how one should now incorporate the coupling of phonons and electrons to $\vec{S}$. Once we have done this, we can also ask how one reduces to a form like (51), once we truncate down to the lowest 2 levels of $\vec{S}$.

Let us start with phonons - this case is almost trivial, since the phonon representation is already an oscillator one. The coupling between $\vec{S}$ and the phonon variables b_q, b_q^+ is just the standard magnetoacoustic one, which can either be understood from macroscopic considerations[50,51], or at a microscopic level[52]. In either case, one has couplings like

$$\mathcal{H}_2^{\phi} \sim \frac{\Omega_o}{S} \hat{S}_x \hat{S}_z \left(\frac{m_e}{M_a}\right)^{1/4} \sum_{\vec{q}} \left(\frac{\omega_q}{\Theta_D}\right)^{1/2} [b_{\vec{q}} + b_{\vec{q}}^{\dagger}] \,, \tag{52}$$

where m_e is the electron mass, M_a the mass of the molecule, and $\Theta_D \sim c_s a^{-1}$ is the Debye temperature (with a the lattice spacing in a molecular crystal of giant spins, and c_s the sound velocity). This interaction describes a non-diagonal process in which a phonon is emitted or absorbed with a concomitant change of ± 1 in S_z (since $S_x = \frac{1}{2}(S_+ + S_-)$). One also has diagonal terms of similar form, in which $S_z S_x$ is replaced by, eg. S_z^2; and there are also higher couplings to, eg, pairs of phonons. [50]. The bare phonon Hamiltonian is just the usual form

$$H_{\phi} = \sum_q \omega_q (b_q^+ b_q + \frac{1}{2}) \tag{53}$$

and can be written in the usual oscillator form with the transformation $x_q = (2m_q\omega_q)^{-1/2}[b_q^+ + b_{-q}]$. Armed with (1.33) and (1.34), one can discuss the dynamics of a tunneling giant spin coupled to phonons, as has been done recently by the Grenoble-Firenze group [53,54]. At low T ($kT \ll \Omega_o$) one can truncate the giant spin to a 2-level system as before. The general technique for doing this is straightforward[17,49].

(ii) The "Giant Kondo" Problem: Consider now a much more interesting example of a giant spin coupled to electrons. Many situations can be envisaged, depending on whether the nanomagnet, or the matrix/substrate in which it is embedded, or both, are conducting; and a great deal of physics revolves around how the electrons move across the boundary between the nanomagnet and the background matrix. Some of the this is discussed in ref. [17] (but a lot more is not!).

Here we take a simplified example which has the virtue of bringing out the essential physics. We assume that an electronic fluid freely permeates both the conducting nanomagnet (of volume V_o), and the background matrix; and we assume that the individual electronic spins couple to the mobile electrons via a Kondo exchange, so that at the giant spin level we have

$$H = H_o(\vec{S}) + \sum_{\vec{k},\sigma} \epsilon_k c^+_{\vec{k}\sigma} c_{\vec{k}\sigma} + \frac{1}{2} \sum_{i \epsilon V_o} J_i \hat{\vec{s}}_i . \hat{\vec{\sigma}}^{\alpha\beta} \sum_{\vec{k}\vec{q}} e^{i\vec{q}.\vec{r}_j} c^+_{\vec{k}+\vec{q},\alpha} c_{\vec{k}\beta} \quad (54)$$

where $H_o(\vec{S})$ is the usual Giant spin Hamiltonian, whereas $\vec{s}_i$ is an individual electronic spin at site i, and position $\vec{r}_i$ (cf. eq. (10)); $c^+_{k\sigma}$ creates a conduction electron in momentum state $| \vec{k} >$, with spin projection $\sigma = \pm 1$ along $\vec{z}$; and $\vec{\sigma}^{\alpha\beta}$ is a Pauli matrix. I have written this interaction in terms of the individual electronic spins in order to show its microscopic origin as an exchange interaction; but of course the locking together of all the $\vec{s}_i$ into a giant $\vec{S}$ means that we can immediately rewrite the interaction as

$$H^{GK}_{int} = \frac{1}{2} \bar{J} \underline{\hat{S}} . \underline{\hat{\sigma}}^{\alpha\beta} \sum_{\vec{k}\vec{q}} F_q \, c^+_{\vec{k}+\vec{q}_i\alpha} \, c_{\vec{k}_i\beta} \quad (55)$$

where $\bar{J}$ is the mean value of the J_i, and $F_q = \int (d^3r/V_o)\rho(\vec{r})e^{i\vec{q}.\vec{r}}$ is a "form factor" which integrates the number density $\rho(\vec{r})$ of electronic spins over the volume V_o of the nanomagnet.

This form of the coupling is very famous in condensed matter physics - it is a "Kondo coupling", but instead of being between a single spin-1/2 and conduction electrons, it couples a spin $\vec{S}$ (with $2S+1$ levels) to the electrons. It is an unusual form of what has come to be known as the "multi-channel Kondo problem". In the energy regime over which the giant spin description is valid, one may use the "Giant Kondo" interaction in (55). At low T, a truncation to 2 levels produces the simple "spin-boson" Hamiltonian, with only the diagonal coupling $\hat{\tau}_z c_k x_k$, and with a Caldeira-Leggett spectral function

$$J(\omega) = \pi \alpha_K \omega \quad (56)$$

$$\alpha_\kappa \sim 2g^2 S^2 \int_{V_o} \frac{d^3r d^3r'}{V_o^2} \left(\frac{\sin k_F \mid \underline{r} - \underline{r\prime} \mid}{k_F \mid r - r\prime \mid} \right)^2 \sim g^2 S^{4/3} \quad (57)$$

where k_F is the Fermi momentum, $g = \bar{J}N(0)$ is the dimensionless Kondo coupling, and $N(0)$ the Fermi surface density of states; typically $g \sim 0.1$, so that α_κ is not necessarily small. Consider, eg., a very small FM particle made from Fe, embedded in a conducting film or similar substrate/background matrix. Suppose it is only 15 Angstroms across; such a particle could still have a spin $S \sim 300$. If the conductor were, eg., Cu, then with a $g \sim 0.1$, we would have $\alpha_\kappa \sim 20$. As we will see in Chapter 3, the dynamics of this system would then be not only overdamped but *frozen*, in the quantum regime.

The reason for this remarkably simple effective Hamiltonian is that the number of microscopic electronic spins $\vec{s}_i$ in the nanomagnet $\sim O(S)$, whereas the number of interaction channels (essentially angular momentum channels for electrons scattering off $\vec{S}$) is $\sim S^{2/3}$ (proportional to the cross-sectional area of the nanomagnet), which means that the effective coupling *per microscopic electronic spin* is weak ($\sim S^{-1/3}$). A renormalization group analysis of the problem substantiates this result [17].

We see that, as previously advertised, a coupling to fermionic electrons has now been reduced to an effective coupling to bosonic oscillators. Why this happens can already be guessed from the bilinear form of the fermionic terms in (55). To an incoherent particle-hole superposition $\sum_{k,\sigma} c^+_{k+q,\sigma} c_{k\sigma}$ we associate a bosonic operator b^+_q. If we treat the coupling perturbatively (and we have just seen we can) then it is obvious that using the same transformation between b^+_q, b_q, and x_q as we need for the phonons, we can go immediately to the oscillator bath representation. This perturbative argument is of course just a special case of the general arguments given by Feynman and Vernon [12] and Caldeira & Leggett [13]. In the case of a stronger coupling (as in the single spin Kondo problem) the argument can actually be made in a modified form [48] (and likewise in the context of the quantum diffusion of defects in solids [55]).

From these examples it will perhaps be clear that a very large number of physical systems can be mapped, at low T, to a spin-boson model. Essentially all one requires is that the central system of interest may be truncated to a 2-level system, and that it be coupled to a set of extended modes. We also need to have some good reason for ignoring any couplings to a spin bath (as we shall discuss below, it is not so easy to ignore environmental spins, so that such reasons may be hard to find!). Thus, in the sense described at the beginning of this section, the spin-boson Hamiltonian constitutes an important example of a "universality class" of low-energy Hamiltonians.

2.4(b) TWO SPINS- The "PISCES" MODEL

I think it is worthwhile dwelling a bit on another such universality class, which also involves an oscillator bath environment. This one involves not just one but <u>two</u> spins, each coupled to an oscillator bath. The effective Hamiltonian is then in general of the form

$$H = H(\boldsymbol{\tau}_1, \boldsymbol{\tau}_2) + H_{osc}(\{\mathbf{x_k}\}) + H_{int}(\boldsymbol{\tau}_1, \boldsymbol{\tau}_2; \{\mathbf{x_k}\}) \tag{58}$$

where $\boldsymbol{\tau}_1$ and $\boldsymbol{\tau}_2$ are 2 Pauli spins and the $\mathbf{x_k}$, with $\mathbf{k} = 1, 2, ...N$, are the oscillator coordinates. This model has been studied in great detail by M. Dubé and myself[57]; it is interesting in a wide variety of physical situations, particularly at the microscopic level (examples include, eg., coupled Anderson or Kondo spins, coupled tunneling defects, coupled chromophores in biological molecules, 2-level atoms coupled through the EM field, or nucleons coupled via the meson field). Perhaps even more interesting are the applications at the macroscopic level, where $\vec{\tau}_1$ and $\vec{\tau}_2$ represent the 2 lowest levels of a pair of mesoscopic or macroscopic quantum systems. Some concrete examples include 2 coupled SQUID's, or 2 coupled nanomagnets[58]. I should add that the initial development of the model was really directed towards the understanding of *quantum measurements.* In this case one of the spins represents a measured system, and the other the low-energy subspace of the measuring apparatus degrees of freedom - we shall see in a minute how this can happen.

We shall be interested here in situations where $\vec{\tau}_1$ and $\vec{\tau}_2$ represent mesoscopic and/or macroscopic systems. In this case, our low-energy Hamiltonian will be as above, with the restricted form

$$H_0 = -\frac{1}{2}(\Delta_1 \hat{\tau}_1^x + \Delta_2 \hat{\tau}_2^x) + \frac{1}{2} K_{zz} \hat{\tau}_1^z \hat{\tau}_2^z \tag{59}$$

$$H_{osc} = \frac{1}{2} \sum_{\mathbf{k}=1}^{N} m_{\mathbf{k}} (\dot{\mathbf{x}}_{\mathbf{k}}^2 + \omega_{\mathbf{k}}^2 \mathbf{x_k}^2) \tag{60}$$

$$H_{int} = \frac{1}{2} \sum_{\mathbf{k}=1}^{N} (c_{\mathbf{k}}^{(1)} e^{i\mathbf{k}\cdot\mathbf{R}_1} \hat{\tau}_1^z + c_{\mathbf{k}}^{(2)} e^{i\mathbf{k}\cdot\mathbf{R}_2} \hat{\tau}_2^z) \mathbf{x_k} \tag{61}$$

where the N oscillator bath modes $\{\mathbf{x_k}\}$ are assumed for simplicity (but without real loss of generality) to be momentum eigenstates, and $\vec{R}_1$ and $\vec{R}_2$ are the "positions" of the two spins. The two spins are coupled through a *direct* interaction term K_{zz} having ferro- or antiferromagnetic form, depending on whether K_{zz} is negative or positive. The direct interaction itself comes from integrating out very high energy modes (with frequencies even greater than the UV cut-off Ω_0 implicit in (61); for frequencies much lower than Ω_0 it can be considered to be static. In the absence of the coupling to the environmental sea, the spins have their levels split by "tunneling" matrix elements Δ_1 and Δ_2. We work in the basis in which $|\uparrow\rangle$ and $|\downarrow\rangle$, eigenstates of $\hat{\tau}^z$, are degenerate until split by the off-diagonal tunneling.

We refer to Hamiltonians like this as "PISCES" Hamiltonians, where PISCES is an abbreviation for "Pair of Interacting Spins Coupled to an Environmental Sea". H_{PISCES} as written above is clearly not the most general model of this kind. A much wider range of direct couplings is possible; instead of $K_{zz}\hat{\tau}_1^z\hat{\tau}_2^z$ we could use

$$H_{int}^{dir} = \frac{1}{2}\sum_{\mu\nu} K_{\mu\nu}\hat{\tau}_1^{\mu}\hat{\tau}_2^{\nu} \tag{62}$$

We could also use more complicated indirect couplings (ie., couplings to the bath) like $\frac{1}{2}\hat{\tau}_\alpha^\mu \sum_{\mathbf{k}} c_{\mathbf{k}\mu}^{(\alpha)} e^{i\mathbf{k}\cdot\mathbf{R}_\alpha}\mathbf{x}_{\mathbf{k}}$, with $\mu = x, y, z$ and $\alpha = 1, 2$. In the present discussion we will stick to the diagonal coupling in (61) and keep only the direct coupling in $\hat{\tau}_1^z\hat{\tau}_2^z$. Our reasoning is as follows. Just as for the single spin-boson problem, we expect diagonal couplings to dominate the low-energy dynamics of the combined system, since the spins spend almost all their time in a diagonal state (only a fraction Δ/Ω_0 of their time is in a non-diagonal state, when the relevant system is tunneling under the barrier). In certain cases the diagonal couplings can be zero (usually for symmetry reasons), and then one must include non-diagonal couplings like $\frac{1}{2}\hat{\tau}_\alpha^\perp \sum_{\mathbf{k}}(c_{\mathbf{k}\perp}^\alpha e^{i\mathbf{k}\cdot\mathbf{R}_\alpha}\mathbf{x}_{\mathbf{k}} + H.c.)$.

Our reason for having a direct coupling in (62) of *longitudinal* form is then connected to our choice of diagonal couplings between the bath and the systems. In fact, one may quite generally observe that in a field-theoretical context, any direct longitudinal interaction $K_{zz}\hat{\tau}_1^z\hat{\tau}_2^z$ can be viewed as the result of a high-energy coupling of $\hat{\tau}_1^z$ and $\hat{\tau}_2^z$ to the high-frequency or "fast" modes of some dynamic field; "fast" in this context simply means much faster than the low-energy scales of interest, so that the interaction may be treated as quasi-instantaneous. In our case, integrating out these fast modes (those with a frequency greater than Ω_0) then produces the longitudinal static interaction in H_0. In the next section we see how at the frequencies of interest for the application of H_{PISCES}, a further longitudinal interaction appears, mediated by those bath modes with frequencies ω such that $\Omega_0 > \omega \gg \Delta_1, \Delta_2$.

One further term we have omitted from (61) is an applied bias acting on one or both of the spins, of the form $H_{ext} = \xi_j\hat{\tau}_j^z$, for example. We will analyse its effects in a moment; its origin is usually fairly obvious.

The crucial new ingredient, of course, in H_{PISCES} is that we are now interested in the *correlations* between the behaviour of 2 systems (microscopic or macroscopic) whilst they are both influenced by the environment; even if

K_{zz} is zero, there will be correlations transmitted via the bath, of a dynamic and partly dissipative nature. In the next section we will see how such physics comes out of the solution for the dynamics of H_{PISCES}.

In this section we shall stick firmly to the question of the *physical origin* of our models. It is then useful to mention 2 concrete examples of H_{PISCES}, where $\vec{\tau}_1$ and $\vec{\tau}_2$ represent macroscopic systems. The first is a simple generalization of the "Giant Kondo" model just discussed above - we imagine 2 nanomagnetic particles in a conducting background. The second involves the much more general "measurement problem", in which one spin represents the low-energy Hilbert space of some quantum system, and the other the low-energy Hilbert space of a measuring apparatus; both are coupled to a quantum environment of oscillators.

(i) Two Interacting Giant Kondo Spins: We imagine starting from a high-energy Hamiltonian analogous to that for the single Giant Kondo system, now with two nanomagnets coupled to the same conducting "substrate", ie., of form

$$H \;=\; H_o(\vec{S}_1, \vec{S}_2) \;+\; H_{bath} \;+\; H_{int} \tag{63}$$

where $H = (H_o(\vec{S}_1) + H_o(\vec{S}_2) + H_{dip}(\vec{S}_1, \vec{S}_2)$, describing 2 separate giant spins, coupled by a high-energy dipolar interaction; where $H_{bath} = \sum_{\vec{k},\sigma} \epsilon_k c^+_{k,\sigma} c_{k,\sigma}$ as for the giant Kondo model; and finally

$$H_{int} \;=\; \frac{1}{2} \sum_{\vec{k},\vec{q}} \hat{\vec{\sigma}}^{\alpha\beta} \cdot \left(\bar{J}_1 \vec{S}_1 F^{(1)}_{\vec{q}} \;+\; \bar{J}_2 \vec{S}_2 F^{(2)}_{\vec{q}} \right) \, c^+_{k+q,\alpha} \, c_{k\beta} \tag{64}$$

is the obvious generalization of the single giant Kondo coupling to the electron bath.

The messy details of the truncation of this Hamiltonian to the PISCES form are given in the original paper. The main point I wish to emphasize here is the way in which the truncation generates an *indirect* interaction between the 2 nanomagnets, of form $\epsilon(R)\hat{\tau}^{(1)}_z \hat{\tau}^{(2)}_z$, where R is the distance between the centres of the nanomagnets, and $\epsilon(R)$ is an indirect interaction which is calculated to be

$$\epsilon(R) \;\sim\; 2g_1 g_2 S_1 S_2 \int \frac{d^3r}{V_1} \int \frac{d^3r'}{V_2} \left(\frac{sink_F \mid \underline{r} - \underline{r} \mid}{k_F \mid r - r\prime \mid} \right)^2 \tag{65}$$

which, not surprisingly, has the usual RKKY form; for 2 identical nanomagnets one finds [58]

$$\epsilon(R) \;\sim\; 144\pi^3 \varepsilon_F g^2 S^2 \; \frac{cos(2k_F R)}{(2k_F R)^2} \left[\frac{cos(2k_F R_o) \;-\; sin(2k_F R_o)}{(2k_F R_o)^2} \right] \tag{66}$$

where ε_F is the Fermi energy of the substrate (ie., the bandwidth, in this simple calculation).

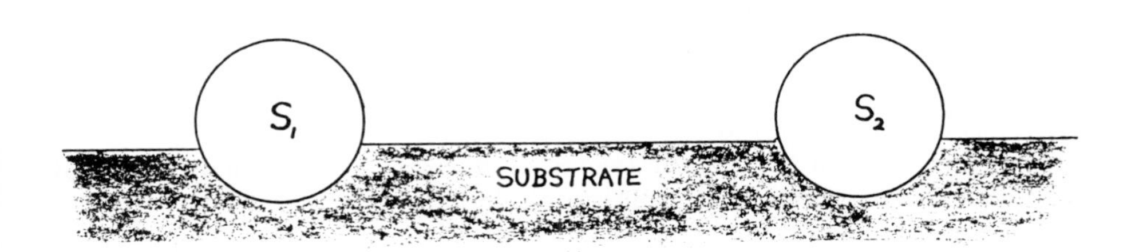

FIG. 6: How the problem of coupled Kondo impurities might arise in practise. We imagine 2 conducting nanomagnets (perhaps part of an array) embedded in a conducting or even semiconducting substrate. Electrons can pass across the boundary between the nanomagnets and the substrate. The nanomagnets are a distance R apart, and have radius R_o.

As remarked above, the generation of such a bath-mediated coupling is quite typical. In this case the relative strengths of the dipolar and RKKY couplings depends very much on the distance between the particles, as well as the electron density- we will discuss numbers in more detail in Chapter 3.

Another example of the PISCES model is provided by a pair of SQUID rings - in this case we generate an interaction (non-Ohmic) between the rings once we integrate out high-frequency EM field modes. This is of course nothing but a fancy way of deriving the usual low-frequency inductive coupling between them.

(ii) Model for Quantum Measurements:

This brings us to a 2nd example, directly connected to the quantum theory of measurement. The standard set-up in a quantum measurement involves some "measuring coordinate" (necessarily macroscopic), and a quantum system which may be microscopic or very rarely macroscopic, which is being measured. Both are in general coupled to their surrounding environment. Now of course there are many different ways in which this set-up can be realised, but the essential features are already revealed in a model in which the quantum "system" of interest exists in one of 2 interesting quantum states, and the apparatus does the same. In this case the apparatus works as a measuring device if (and only if) its final state is correlated with the initial state of the system.

Quite how this works has been discussed *qualitatively* for a few models, which tend to fall in one or another of 2 classes. In the first class, the apparatus coordinate of interest can exist in one or other of 2 stable (or metastable) states, which are "classically distinguishable". The best-known example of this is the Stern-Gerlach experiment (where atomic beams are widely separated according to their microscopic spin state). Similar examples are provided by SQUID magnetometers, and one may imagine a large number of "Yes/No" experiments, in which the measuring system ends up with the relevant collective coordinate (the "measuring coordinate") in one of the 2 quite different states. By "quite different", one usually means that (a) the overlap between the 2 quantum states is very small, and (b) that the classical counterparts of the quantum states are sufficiently different that one is justified in calling them "macroscopically distinguishable". Thus, in the case of Stern-Gerlach experiment, the beams are sufficiently widely separated that they no longer overlap, *and* they can be effectively separated by, eg., a microscope when they hit a screen or counter. In the case of a SQUID magnetometer, the 2 relevant states differ by a single flux quantum through the SQUID ring - the overlap between the states is via tunneling, and hence exponentially small, and the currents flowing in the 2 states can differ by values $\sim$ mA.

A second class of models is typified by the Geiger counter. In this case we again have a "Yes/No" measuring system, designed to discriminate between 2 states of some central system under investigation. In this case the central system is typically subatomic, involving, eg., nuclear decay which has/has not occured. However the measuring system clearly involves a measuring coordinate which is not microscopic - the Geiger counter is metastable, and designed so that a large potential change (kV) occurs between the electrodes if a microscopic (subnuclear) decay product interacts with atoms inside the counter. As has been repeatedly stressed in the literature, this example well illustrates the massive *irreversibility* inherent in many measuring systems. From this point of view any further coupling of the Geiger counter to another macroscopic system (eg., to Schrodinger's cat) seems almost superfluous as far as the measurement paradox is concerned - we are already at the classical level!

The important difference between these 2 sorts of measurement is that in the latter case there is a large energy difference between the initial and final states of the apparatus coordinate (the energy having been fed irreversibly into an "environment" of other coordinates).

However, given our experience of the truncation procedure, we can ask whether it is possible to strip these models down to a truncated form involving only 4 states (2 for the central system, 2 for the apparatus). All irreversible effects then arise from a coupling of these states to a bath - which if modelled by oscillators gives the PISCES model.

This is part of the rationale behind recent work by M. Dubé, E. Mueller, and myself[59]. One aim of this was to see how valid are some of the quantum measurement models, by showing how they may be derived from underlying "high energy" microscopic theory (and finding corrections to them).

Before describing the work, let us first see what sort of "measurement model" we are aiming at. One very commonly used model (known as the "Bell/Coleman-Hepp" model[60]) takes the form

$$H_o^{(A)}(\vec{\tau}) \; + \; H_o^{(s)}(\vec{\sigma}) \; + \; \frac{1}{2}K \; \hat{\tau}_x(1 - \hat{\sigma}_z) \tag{67}$$

where $\vec{\tau}$ is a 2-state "apparatus", and $\vec{\sigma}$ a 2-state "system". The coupling is such that if $\vec{\sigma}$ is initially $\mid \uparrow>$, then there is no effect on $\vec{\tau}$, whereas if $\vec{\sigma}$ is initially $\mid \downarrow>$, then a transition occurs in the apparatus (with no change in the system). For this scheme to function properly, one requires the initial state of $\vec{\tau}$ to be $\mid \Uparrow>$ or $\mid \Downarrow>$. Typically K will depend on time (ie., we turn the coupling on and off), but this is not crucial for what follows. If we let $K = \pi$, then we have the classic "ideal measurement" scheme , ie.,

$$\begin{aligned} &\mid \Uparrow> \mid \uparrow> \longrightarrow \mid \Uparrow> \mid \uparrow> \\ &\mid \Uparrow> \mid \downarrow> \longrightarrow \mid \Downarrow> \mid \downarrow> \end{aligned} \tag{68}$$

ie., the final state of the apparatus is uniquely correlated with the initial state of the system.

Now in view of our earlier discussion of universality classes and fixed points for low-energy Hamiltonians, it is clear that such a scheme can be obtained starting with a large variety of initial high-energy Hamiltonians. It is rather illuminating to consider one such starting point- we begin from a Hamiltonian

$$\begin{aligned} H = \frac{1}{2}(M_A q_A^2 + M_s q_s^2) + V_A(q_A) + V_s(q_s) - \xi_A q_A \\ -K_o q_s q_A + \frac{1}{2}\sum_{k=1}^{N} m_k(\dot{x}_k^2 + \omega_k^2 x_k^2) + \sum_{k=1}^{N}(c_k^{(s)} q_s + c_k^{(A)} q_A) x_k \end{aligned} \tag{69}$$

It is useful to visualize the "2-dimensional" potential here, in the space of the coordinates q_s, q_A (referring to system and apparatus).

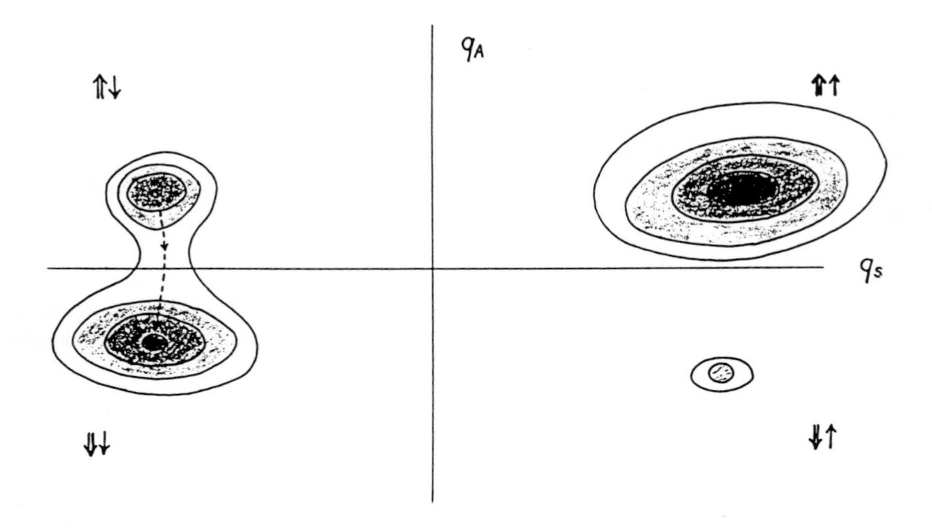

FIG. 7: The effective potential (shown as a contour map, with low energies shaded) for a system of 2 coupled two-well systems, with a mutual "ferromagnetic" coupling and a bias applied to the apparatus coordinate q_A (but not the system coordinate q_s). On truncation to low energies this produces the biased PISCES model of eqtn. (61).

We assume potentials $V_s(q_s)$ and $V_A(q_A)$ which describe symmetric 2-well systems, with minima at $q_s = \pm 1, q_A = \pm 1$; however we also assume that the action required for the tunneling between the 2 apparatus states is considerably smaller than for the tunnelling between the 2 system states. The coupling $-K_o q_A q_s$ is a "ferromagnetic" coupling, of quite generic form (produced by expanding the interaction in powers of q_A and q_s).

When we truncate this down to low energies, we get a PISCES Hamiltonian of form

$$H \sim \Delta_A \hat{\tau}_x + \Delta_s \hat{\sigma}_x - \xi_A \hat{\tau}_z - \mathcal{K} \tau_z \sigma_z + \frac{1}{2} \sum_{k=1}^{N} \left[m_k (\dot{x}_k^2 + \omega_k^2 x_k^2) + (c_k^{(A)} \hat{\tau}_z + c_k^{(s)} \hat{\sigma}_k^z) x_k \right] \tag{70}$$

where $\mathcal{K}$ is a renormalized (but still high-energy) coupling, which includes the effect of oscillators above a UV cut-off frequency Ω_o. We will make 2 assumptions here, viz.,

$$(i)\ \Delta_A \gg \Delta_s \tag{71}$$

$$(ii)\ \mathcal{K} > \xi_A \gg \Delta_n, kT \tag{72}$$

That $\Delta_A \gg \Delta_s$ implies that the apparatus reacts quickly to any change in the system state. The purpose of ξ_A is to hold the apparatus in state $|\Uparrow>$ when there is no coupling; when the renormalized coupling $\mathcal{K}$ acts, it must overcome this bias. We also want the resulting net bias between initial and final states of the apparatus to be much greater than either Δ_A or kT, so that there is no chance the apparatus, once it has made a transition, can return back to its former state.

Without doing any calculations, we can immediately see from the Figure that this system will behave like an ideal measuring device. The combined system-apparatus starts off either in $|\Uparrow\uparrow>$ (and stays there) or in $|\Uparrow\downarrow>$ (in which case it can tunnel inelastically to $|\Downarrow\downarrow>$; provided we wait long enough, it will always do this, no matter how weak is the coupling of $\vec{\tau}$ to the bath). Typically the apparatus-bath coupling is strong, so that the apparatus will relax quickly. Notice, however, that the relaxation rate must depend also on the size of the transition matrix element Δ_A, and the strong coupling to the bath will renormalize Δ_A down to a considerably smaller value Δ_A^*. Thus we must take care that $\Delta_A^* \gg \Delta_s^*$, after both have renormalised (and the weaker coupling of the system to the bath means that Δ_s will decrease much less than Δ_A).

In the next Chapter we will discuss the *dynamics* of models like these more thoroughly. Notice already, however, one obvious remark that can be made about these models. This is that any interesting dynamics possessed by the *system* before the coupling to the apparatus is switched on, is *frozen* by this coupling. Thus, if the system is originally tunneling between $|\uparrow>$ and $|\downarrow>$ at a frequency Δ_s, the large coupling $\mathcal{K}$ blocks this completely. In this sense the measurement drastically interferes with the system dynamics. We shall see in the next section that this is a quite general feature of the PISCES model, and most likely of measurements in general. What is more, a good part of $\mathcal{K}$ comes from the oscillator bath itself (indeed K_o already arises from truncation of even higher energy modes), and the *dissipative* effect of the remaining oscillators (having energy $< \Omega_o$) also strongly influences the dynamics of the system once $\mathcal{K}$ is switched on, in a way which is *quite different from that produced by the oscillators when* $\mathcal{K} = 0$. This latter feature, discussed in detail in Dubé and Stamp[57] in the context of the PISCES model, is quite new as far as we know. We thus see that models of this kind, which include both apparatus *and* system (as well as the environment), have interesting light to shed on measurement theory. The PISCES model appears to be the first attempt to discuss all 3 partners in the measurement operation (system, apparatus, environment) on an equal and fully quantum-mechanical level.

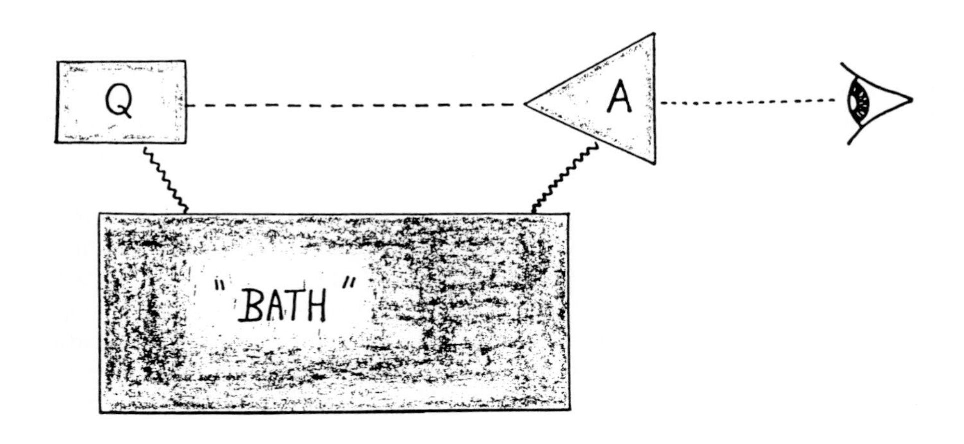

FIG. 8: The relationship between the measuring apparatus, the measured coordinate(s) Q of interest, and the environment (ie., all other relevant degrees of freedom), involved in a typical measurement. The disciplines of mesoscopic and nanoscopic physics often involve similar situations, in which all variables must be treated quantum-mechanically.

It is the experience of this author that many physicists are somewhat averse to the discussion of abstract quantum measurement problems in the context of down-to-earth questions of solid-state physics. This is unfortunate, since even if one is uninterested in the foundations of quantum mechanics, there is no question that one learns a great deal about these down-to-earth problems by simply placing them in the more general context that the measurement problem requires. This is becoming really very obvious in the field of nanophysics, where both theorists and experimentalists must deal every day with the emergence of classical properties from quantum ones, and the relationship between the quantum and classical domains. In fact with a bit of hindsight one can see the steady evolution of condensed matter theory towards an ever-closer examination of the interaction between observer (or probe), and a quantum system of interest, in the presence of dissipation and/or decoherence (cf Fig.8). In modern mesoscopic or nanoscopic physics there is often not a huge difference in the size of the probe and the quantum system of interest. Condensed matter physics has come along way since the early days of statistical physics, when only the *equilibrium* behaviour of a *macroscopic* system was of interest!

2.4(c) CANONICAL OSCILLATOR BATH MODELS

Just as for the spin bath models, we have seen that it is possible to reduce the description of a large number of systems, at low energies, to that of a few effective oscillator bath models.

In this subsection I just briefly note some of the other "canonical" oscillator bath models that have also been studied. In contrast with the more recent spin bath models, many of the following are rather old, in some cases with a distinguished history - I will refer to reviews for historical details and references.

(i) Harmonic Oscillator coupled to Oscillators : A very thorough review of this model was given a few years ago by Grabert et al [61]; we are interested in a low-energy Hamiltonian of the standard form in (50), where now

$$H_o(P,Q) = \frac{1}{2}[P^2/2M_o + M_o\Omega_o^2 Q^2] \tag{73}$$

and where in most cases one simply uses the bilinear coupling $H_{int} = \sum_k c_k x_k Q$. In this case the Hamiltonian is obviously exactly diagonalisable, and so the dynamics of the central oscillator is exactly solvable.

There are many obvious physical examples; some were already discussed by Feynman and Vernon, since many measuring systems are based on resonant absorption by one or more oscillators. An example which has stimulated a lot of recent work is that of a gravity wave detector - this is usually a very heavy bar of sphere, made from a conducting alley, in which Ω_o represents the principal oscillator (phases) mode. The coupling to higher phonons, or to electron-hole pairs, is deliberately made very weak. The purpose of these detectors is to absorb a single quantum of energy in this mode, from a gravitational wave. An enormous literature exists on this kind of application; see in particular the books by Braginsky and collaborators [62], which are particularly interesting for the generality of their coverage. It has indeed been remarked on many occasions that one can understand much of Quantum Mechanics with reference to 2 simple models (the 2-level system and the harmonic oscillator); in these lectures I am simply coupling these to an environment!

One can obviously also consider systems of coupled oscillators, in analogy with the PISCES model.

(ii) Tunneling/Nucleating System, coupled to Oscillators: Another canonical model which has been studied in hundreds if not thousands of papers is that of a particle tunneling from a potential well into an open domain - usually the problem is designed so that this domain is semi-infinite, and the relevant potential is 1-dimensional. This restriction is actually rather weak, since many multi-dimensional tunneling or nucleation problems can be reduced to 1-dimensional one, as already noted in our discussion of domain wall tunneling in the spin bath. In the context of tunneling of systems coupled to oscillators, this question has been discussed in great detail by A. Schmid[63].

The relevant Hamiltonian is exactly that in (2.1), with V(Q) representing the tunneling potential - this is the famous "Caldeira-Leggett" model[13]. So many discussions have been given of how models like this can be derived, that I do not propose adding to them here.

(iii) Free Particle coupled to Oscillators. If we let $V(Q) = 0$ in (2.1), we get a model which is of somewhat academic interest but which can be solved for a variety of couplings - it is particularly interesting for the study it allows of different initial conditions imposed on the environmental state, and an initial correlations between bath and particle[61].

This concludes our survey of the various ways in which one may truncate high-energy Hamiltonians down to one or other of the universality classes mentioned in the introduction. We now proceed to a survey of the *dynamics* of these models.

3 Tunneling Dynamics: Resonance, Relaxation, Decoherence

3.1 Generalities

There are a number of features common to any problem in which some central quantum system tries to tunnel (or otherwise show coherent dynamics), whilst coupled to an environment. Let us first note 2 important mathematical features. One is the division of coupling terms in H_{eff} into "diagonal terms" (operating when the system is *not* tunneling), and "non-diagonal terms" (active *during* the tunneling). The difference is very clear when the central system is a 2-level one, and easily shown diagramatically, for either spin or oscillator baths. A second feature appears when one comes to average over the bath variables, to produce a reduced density matrix for the central system. Mathematically this averaging supposes some distribution of probability over the possible bath states, which is assumed *invariant in time* - often it is simply a thermal average, assuming a thermal equilibrium distribution for the bath states. For the time invariance to be physically reasonable, it is necessary that the energy transferred from the central system, while it is relaxing, be rapidly distributed over the bath states, via some kind of mixing. This assumption is usually reasonable for the oscillator bath (although one has to be careful sometimes about non-linear effects). For the spin bath, often containing a finite number N of spins (no thermodynamic limit), the assumption can be wrong, and then one has to think more carefully (see later).

All these problems also have obvious physical features in common. The environmental wave-function "entagles" with that of the central system, thereby destroying coherence in the latter; and if we start the central system in an excited state, it decays or relaxes incoherently by exciting bath modes. In a large system this may occur via tunneling into a quasi-continuum of states ("Macroscopic Quantum Tunneling"), but at low energies it is often more common to have tunneling between discrete levels on either side of a barrier. In this case, if the bias $\xi = (\varepsilon_L - \varepsilon_R)$ between these 2 levels is substantially larger than the tunneling matrix Δ, resonance is lost and tunneling is suppressed; the overlap integral between $|L>$ and $|R>$ is now $\sim \Delta/(\Delta^2 + \xi^2)^{\frac{1}{2}}$, and so the new effective "tunneling amplitude" is $\Delta^2/(\Delta^2 + \xi^2)^{\frac{1}{2}}$ (compare the off-diagonal matrix elements in the density matrix of eqtn. (28), for the simple 2-level system). However at this point the bath can come to the rescue, by absorbing the energy difference ξ, and thereby mediate inelastic (and thus incoherent) tunneling. Other related processes of interest are quantum diffusion [55] (in which the diffusion rate of a system in the quantum regime *increases* as one lowers the temperature), and quantum nucleation (which is essentially barrier tunneling).

Quite generally we may characterise the dynamics of the central system via the *reduced* density matrix; for the general Hamiltonian (1.2) this is

$$\rho(Q,Q';t) \;=\; Tr_{\{q_k\}}\left[\rho_{tot}(Q,Q';\{q_k\},\{q_k'\};t)\right] \tag{74}$$

where the trace is over the environmental variables (usually with a thermal weighting). Our problem is to evaluate the *propagator* K_{12} for ρ, defined by

$$\rho(Q,Q';t) \;=\; \int dQ_2 \int dQ_2' K(1,2)\rho(Q_2,Q_2';t_2) \tag{75}$$

In this equation, K(1,2) $= K(Q_1,Q_1';Q_2,Q_2';\,t_1,\,t_2)$, and is the 2-particle Green function describing the propagation of the central system, first from Q_1 (at time t_1) out of Q_2 (at time t_2); and then back from Q_2' (at time t_2) to Q_1' (at time t_1). K(1,2) simply describes the "flow" of the reduced density matrix in (Q,Q') space (Fig.9). "Decoherence" in the "Q-space representation" then corresponds to the suppression of "off-diagonal" parts of $\rho(Q, Q')$, for which Q and Q' differ significantly.

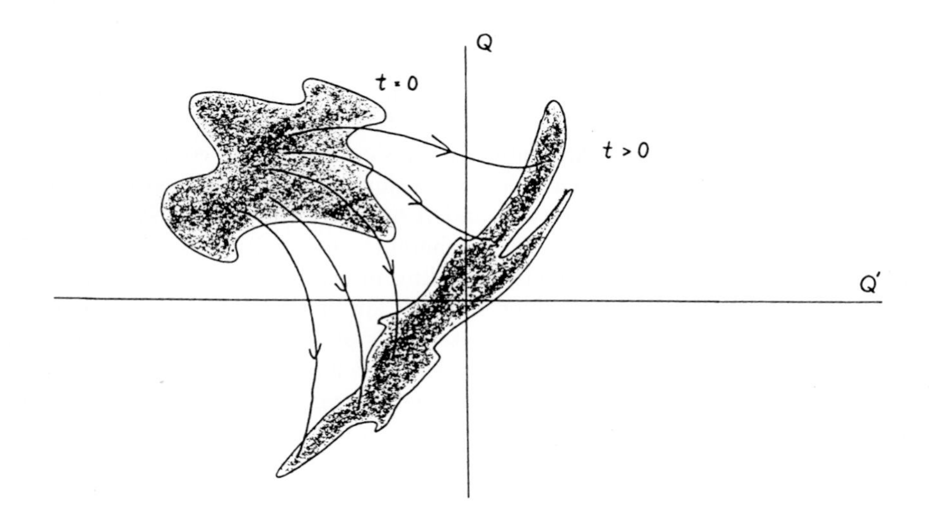

FIG. 9: The flow of the reduced density matrix in the space of generalised coordinate variables (Q, Q'). The propagator $K(1,2)$ then simply describes this flow. If there is strong decoherence in the coordinate variables, the suppression of off-diagonal interference then tends to compress the flow along the streamlines, as in the figure.

All of these features are true of tunneling systems, regardless of the bath. However in what follows we shall see in detail how going between a spin and oscillator bath can change some very important details.

3.1(a) OSCILLATOR BATHS

As noted by Feynman and Vernon [12], the oscillator bath environment is easily integrated out because the oscillator actions are *quadratic*. Thus the central system density matrix has a propagator

$$K(1,2) = \int_{Q_1}^{Q_2} dQ \int_{Q_1'}^{Q_2'} dQ' \; e^{-i/\hbar(S_o[Q] \; - \; S_o[Q'])} \mathcal{F}(Q, Q') \tag{76}$$

Here $S_o[Q]$ is the free central system action, and $\mathcal{F}(Q, Q')$ is the famous "influence functional", defined in general by

$$\mathcal{F}(Q,Q') = \prod_k \langle \hat{U}_k(Q,t)\hat{U}_k^\dagger(Q',t)\rangle \ , \tag{77}$$

where $\hat{U}_k(Q,t)$ describes the evolution of the k-th environmental oscillator, given that central system follows the path $Q(t)$ on its "outward" voyage, and $Q'(t)$ on its "return" voyage. Now we can always write this as

$$\mathcal{F}(Q,Q') = e^{-i\Phi(Q,Q')} = e^{-i\sum_{k=1}^{N}\phi_k(Q,Q')} \ , \tag{78}$$

where $\Phi(Q,Q')$ is a *complex* phase, containing both real (reactive) contributions, and imaginary damping contributions. Thus $\mathcal{F}(Q,Q')$ acts as a weighting function, which weights different possible paths $(Q(t),Q'(t'))$ differently from what would happen if there was no environment (thus, if the oscillators couple directly to the coordinate via a coupling $F_k(Q)x_k$, then $\mathcal{F}_k(Q,Q')$ will typically tend to suppress "off-diagonal" paths in which Q and Q' differ strongly).

The crucial result of weak coupling to each oscillator is that an expression for each $\mathcal{F}_k(Q,Q')$ can be written down (and thence for $\mathcal{F}(Q,Q') = \prod_k \mathcal{F}_k(Q,Q')$), in which $\phi_k(Q,Q')$ is expanded to 2nd order in the coupling only [12,13]. Since the couplings are $O(N^{-1/2})$, summing over the N oscillators then gives an expression for $\Phi(Q,Q')$ which is independent of N and exact in the thermodynamic limit.

The second-order (in $F_k(Q)$) graphs contributing to $\Phi(Q,Q')$ are shown in Fig.10(a) below; the exponentiation of these gives graphs like those in Fig. 10(b). The bosonic propagator for each oscillator, in equilibrium, is just

$$g_k(t) = \frac{1}{2m_k\omega_k}\left[e^{i\omega_k t} + 2\frac{\cos\omega_k t}{\exp\{\beta\hbar\omega_k\}-1}\right] \qquad (\beta = 1/k_BT) \ . \tag{79}$$

Summing over all bosonic modes allows us to subsume all environmental effects into a spectral function [12,13], whose form I quote here in the case where the coupling $F_k(Q)$ is linear in Q, ie., $F_k(Q) = c_kQ$ (so that the system-oscillator coupling is bilinear):

$$J(\omega) \ = \ \frac{\pi}{2}\sum_{\kappa=1}^{N}\frac{|\ c_\kappa\ |^2}{M_\kappa\omega_\kappa}\,\delta(\omega \ - \ \omega_\kappa) \tag{80}$$

which of course just has a "Fermi golden rule" form, characteristic of response functions. As emphasized by Leggett, at low ω this function will often be dominated by an "Ohmic" term $J(\omega) \sim \eta(\omega)$. This is partly because of course a linear form dominates over higher powers at low ω. Moreover, in the classical dissipative dynamics it gives Ohmic friction. In this context it is important that even if one finds, in some calculation, a non-Ohmic form for $J(\omega)$, it is almost certain that if one pursues the calculation to higher orders, an Ohmic coupling will be found. However this coupling will usually be *temperature-dependent*, ie., $J(\omega) \rightarrow J(\omega, T)$. For a more detailed discussion of this point, see, eg., Dubé and Stamp [28], and refs. therein.

The function $J(\omega)$ is usually called the "Caldeira-Leggett spectral function".

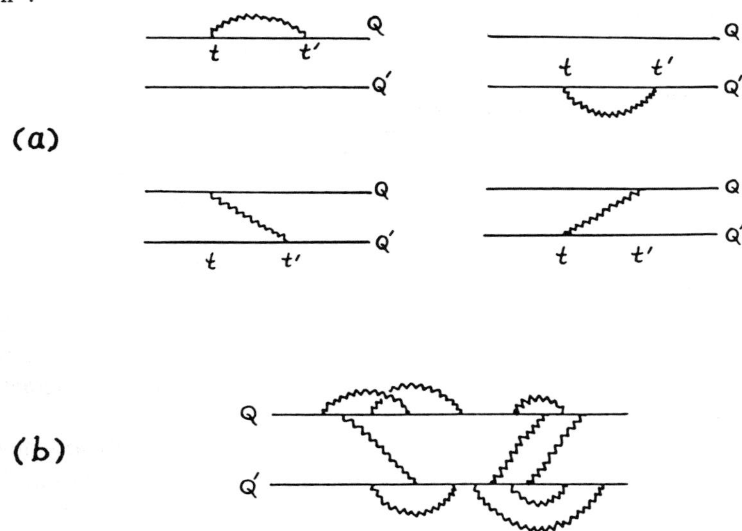

FIG. 10: Graphical representation of the influence functional phase $\Phi(Q, Q')$. In (a) I show the 4 contributions to 2nd order in the coupling $F_k(Q)$; the wavy line is the propagator $g_k(t - t')$ for the k-th bath oscillator. Exponentiation gives graphs to all orders; a typical example is shown in (b).

3.2 Dynamics of the Spin-Boson Problem

As our first example of the dynamics of a system coupled to an oscillator bath, we go to the famous spin-boson problem. There are several detailed reviews of the spin-boson dynamics [48,41,64] (see also Leggett's notes in this volume). Here I summarize results relevant to the present topic.

Although one can imagine a whole variety of forms for the spectral function $J(\omega)$ in this problem, in many systems it is the Ohmic coupling which dominates at low energies, and I will only talk about this case here. Readers interested in the finer details can go to the reviews, particularly if they are interested in non-Ohmic spin-boson systems. For the Ohmic spin-boson problem one writes $J(\omega) = \pi\alpha\omega$, and enormous effort has been devoted to understanding the phase diagram and dynamics of the central spin, in contexts ranging from particle theory, the Kondo problem, defect tunneling and quantum diffusion, flux tunneling in SQUID's, and 1-dimensional fermion problems. From these studies we recall several results. First, the well known phase diagram for Ohmic coupling[48] (Fig.11).

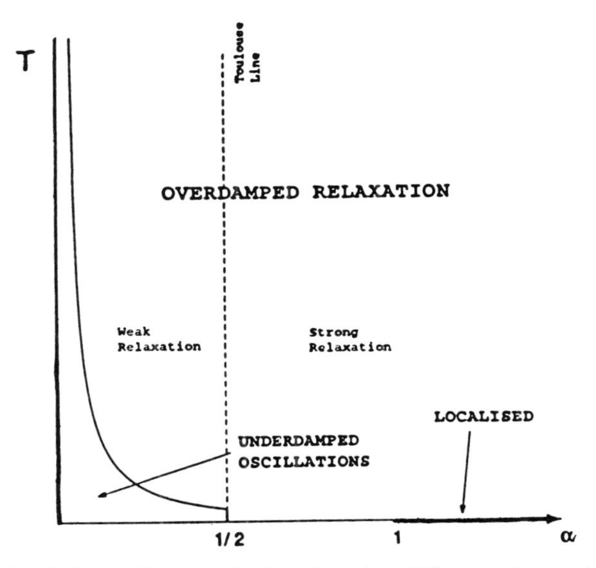

FIG. 11: The "phase diagram" showing the different dynamic behaviours possible for an *unbiased* 2-level system coupled to an Ohmic bath, as a function of the dimensionless coupling α to the bath, and the temperature T. At $T = 0$ and $\alpha > 1$, the bath oscillators "quantum localise" the system- it freezes. For $\alpha \gg 1$, the system is quasi-localised for temperatures up to a crossover temperature $T_c \sim \Omega_o/2\pi$. Elsewhere the system shows overdamped relaxation (exactly exponential along the Toulouse line $\alpha = 1/2$), except for low T and low α.

The most important feature here is how delicate quantum coherence turns out to be, as a function of the dimensionless parameters α and T/Δ_R, where

$\Delta_R(\alpha)$ is a renormalized tunneling amplitude. This is of course because (i) the oscillators suppress Δ_o to Δ_R, via a standard Franck-Condon effect, and (ii) the coupling c_κ to $\hat{\tau}_z$ allows the oscillators to sense or "measure" the state of $\hat{\tau}$, and thereby suppress coherent oscillations.

A second important result, which concerns many oscillator bath models (not just the spin-boson model), is the relationship between the classical and quantum behaviour of the system coupled to the oscillator bath. It is of course trivially obvious that if the spectral function $J(\omega; T)$ contains all information about the system/bath coupling, then both the classical frictional dynamics and the low-T tunneling dynamics are determined by the same $J(\omega, T)$. In this sense knowledge of the classical motion may help us in predicting the quantum dynamics (albeit at a different T). However in those (not uncommon) cases where $J(\omega) = \eta\omega$, with η independent of T, Caldeira and Leggett showed that knowledge of η obtained from the classical friction entirely determined the quantum dynamics.

One should, nevertheless, refrain from pushing this result too far. In particular, it is a serious mistake to imagine that one can understand the *decoherence* in the quantum dynamics, solely from a knowledge of the classical dynamics. A related mistake is the idea that there is some well-defined "decoherence time", referring to exponential relaxation of off-diagonal matrix elements - as a rule this is incorrect, even in the oscillator bath models.

To better appreciate the above results, a brief word on the mathematical techniques is in order. To solve for the dynamics of the spin-boson model, Leggett et al. [48] employed an approximation dubbed the "NIBA" ("Non-Interacting Blip Approximation"), in order to evaluate the influence functional. This approximation simply assumes that the system density matrix (with oscillators already averaged over) spends most of its time in a *diagonal* state (as opposed to an off-diagonal one). It is useful to understand this approximation in 3 different ways.

(i) diagramatically, the system is in a diagonal state (dubbed a "sojourn" by Leggett et al. [48]), when both paths contributing to the density matrix evolution happen to be in the same state. An off-diagonal (or "blip") state, where the 2 paths at a given time are in opposite states, is then considered to be rare.

(ii) mathematically, the suppression of the off-diagonal elements arises because the propagator $g_k(t)$, for the oscillators, is complex, and the imaginary part damps out off-diagonal blip states. Notice that by "diagonal" we mean diagonal in $\hat{\tau}_z$; and the environment couples to $\hat{\tau}_z$.

Formally we can express this point in terms of the influence functional as follows. For an arbitrary path of the system, the complex phase $\Phi(Q, Q')$ in

(78) can be written

$$\Phi(Q,Q') = \int_0^t d\tau \int_0^\tau ds (i\Sigma(\tau - s)\xi(\tau)\chi(s) - \Gamma(\tau - s)\xi(\tau)\xi(s)) \tag{81}$$

where the double path is parametrised by sum and difference variables $\xi(s) = Q(s) - Q'(s)$ and $\chi(s) = Q(s) + Q'(s)$ (again we assume a coupling of the oscillators to Q).

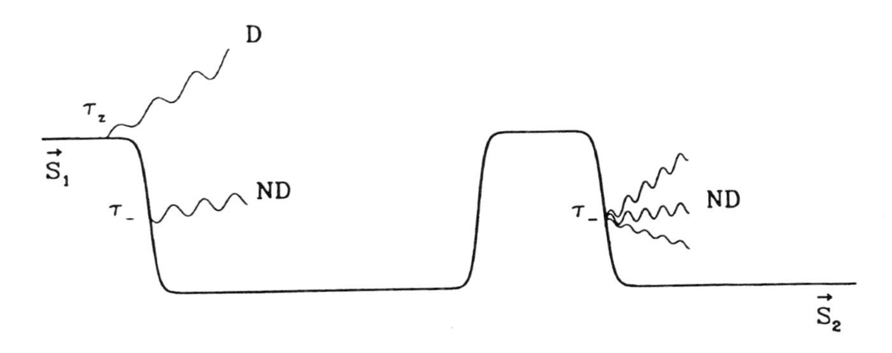

FIG. 12: A typical path $Q(t)$ for a 2-state system coupled to an environment. The 2 levels result from truncating a higher-energy Hamiltonian containing many levels (here a giant spin $\vec{S}$). The fast high-energy fluctuations are invisible, and only the rapid (on a timescale Ω^{-1}) transitions between the 2 classically stable states $\vec{S}_1$ and $\vec{S}_2$ are seen. The system interacts diagonally (D) or non-diagonally (ND) with the environment; in the latter case a non-linear coupling to an environmental triplet is shown (as in the coupling to the spin bath).

Now for the specific case of the spin-boson system system, the paths are very simple; they are nothing but sums over outgoing or return paths of the form

$$Q_{(n)}(s) = 1 - \sum_{i=1}^{2n} \left[sgn(s - t_{2i-1}) + sgn(t_{2i} - s) \right] , \tag{82}$$

where $sgn(x)$ is the sign-function, n is the number of transitions of the central system, occuring at times $t_1, t_2, \ldots, t_{2n}$ (and here for definiteness we assume trajectories starting and ending in the same state). Notice that the paths have sudden jumps because once we truncate our Hamiltonian to energies $\ll \Omega_o$, we have no way of resolving processes occurring over the time scale Ω_o^{-1} of the instanton jump- to all intents and purposes they are instantaneous! I have shown in Fig.12 a typical such path (including the couplings to the environmental modes, which in a path integral formalism take place at particular times).

Notice that now the off-diagonal "blip" states are simply the ones having finite $\xi(s)$ (and $\chi(s) = 0$). How and when are these suppressed by the oscillator bath? Well, we can see this by looking at the forms of the reactive and dissipative contributions to the phase $\Phi(Q, Q')$, which can easily be calculated in terms of the spectral function $J(\omega)$ for the bilinear coupling model (which just integrates over the complex propagator $g_k(t)$):

$$\Sigma(\tau - s) = \int_0^\infty d\omega\, J(\omega)\, \sin\omega(\tau - s) \tag{83}$$

$$\Gamma(\tau - s) = \int_0^\infty d\omega\, J(\omega)\, \cos\omega(\tau - s) \coth(\omega/2T) \tag{84}$$

From this we see that whether blips are suppressed relative to sojourns depends very much on the form of $J(\omega)$ (for a detailed discussion see Leggett et al. [48]), but if they are, it is because of the imaginary part Γ of the complex phase.

(iii) Finally and most physically, we can see that the diagonal state of the density matrix is favoured because the environment is in effect continually "measuring" $\hat{\tau}_z$ (the coupling $\sum_k c_k x_k \hat{\tau}_z$ distinguishes between $|\uparrow>$ and $|\downarrow>$); this suppresses quantum interference between the two states (ie., off-diagonal elements). Notice that we can think of this as a "dynamic localisation" of the system in one or other state for long periods, and the opportunity for tunneling is reduced to periods when resonance between the 2 wells persists long enough for tunneling to occur. Note that this is not really the same as the "degeneracy blocking" we encountered before, which comes from a *static* bias field. Nevertheless it does bring us to perhaps the most important question about the spin-boson model, which is, what happens to the spin when there is an *external* bias field ξ, and a finite temperature? I will simply quote the (practically very useful) answer for you here. Suppose the system starts up in a state $|\uparrow\rangle$ at $t = 0$. Then at time t later the probability $P_{11}(t)$ that it will still be $|\uparrow\rangle$ is given by the simple *incoherent relaxation* expression $P_{11}(t) = e^{-t/\tau(T,\xi)}$, where for the most interesting case of Ohmic coupling, the rate τ^{-1} is given by (assuming energy scale less than Ω_o as usual):

$$\tau^{-1}(\xi,T) = \frac{\Delta^2}{2\Omega_0}\left[\frac{2\pi T}{\Omega_0}\right]^{2\alpha-1}\frac{\cosh(\xi/2T)}{\Gamma(2\alpha)}|\Gamma(\alpha+i\xi/2\pi T)|^2 \qquad (85)$$

where $\Gamma(x)$ is the Gamma function. This result is valid for all but the smallest biases, and even when the bias is $\leq \Delta$, it is valid throughout the incoherent relaxation region of the phase diagram in Fig. XXX. Thus even at very low bias we get relaxation at a rate

$$\tau^{-1}(T) = \frac{\Delta^2}{2\Omega_0}\frac{\Gamma^2(\alpha)}{\Gamma(2\alpha)}\left[\frac{2\pi T}{\Omega_0}\right]^{2\alpha-1} + O(\xi/T)^2 \qquad (\xi/T \ll 1) \qquad (86)$$

whereas in the opposite limit where we apply a strong bias $\xi \gg T$ to the system, we get incoherent fluctuations between the 2 states:

$$\tau^{-1}(\xi) = \frac{\Delta^2}{\Omega_0}\frac{1}{\Gamma(2\alpha)}\left[\frac{\xi}{\Omega_0}\right]^{2\alpha-1} \qquad (87)$$

These results have been applied to a large variety of problems in physics, particularly those involving conduction electron baths. Their application to SQUID dynamics has been particularly emphasized by Leggett [41,65].

Application to the Giant Kondo model: The application to the problem of a nanomagnet coupled to a conducting bath illustrates some of the dramatic features of the above results. Consider the example given previously of an Fe particle with $S = 300$ and dimensionless coupling $\alpha_\kappa \sim 20$ to the electronic bath. Then at zero bias, the incoherent fluctuation rate goes as the 40-th power of temperature, whereas at low T, the incoherent relaxation rate goes as the 40-th power of bias! Consider now an array of such particles, which we imagine to be functioning as a computer data storage bank. If we are at low T, and small bias, the giant spins are "quantum localised"; they only relax over eons of time (the crucial point is to stay below the crossover temperature $T_c = \Omega_o/2\pi$; if $T = T_c/2$, the relaxation time is already 10^6 years!). Thus even though these particles are no larger than small molecules, they can store information for astronomical time periods below T_c (and recall that T_c can easily be $1\ K$ if we have a reasonably strong anisotropy!). On the other hand suppose we keep T well below T_c, and raise the bias. Then over the time-window that might be of interest to data storage (ie., between $nsec$ and years), we will see a very sudden "switching-on" of the dynamics of the giant spins, in the small region of bias around $\xi \sim \Omega_o/2\pi$.

This result is obviously of practical importance [17])! In this context an experimental test is highly desirable. Note that one cannot of course extend these results above $T, \xi > T_c$, since our truncated Hamiltonian does not work there (in fact when $T > T_c$, we go into the thermally activated regime). Note also that we have neglected the nuclear spin bath; for the combined effects of the 2 baths, see Prokof'ev and Stamp [17].

3.3 Other Bosonic Models

3.3 (a) The PISCES MODEL

This is the simplest non-trivial generalization of the spin-boson model. As we saw in section 2.4, what is interesting here is the communication between the spins via the bath, and the way in which each affects the dynamics (including the coherence properties) of the other.

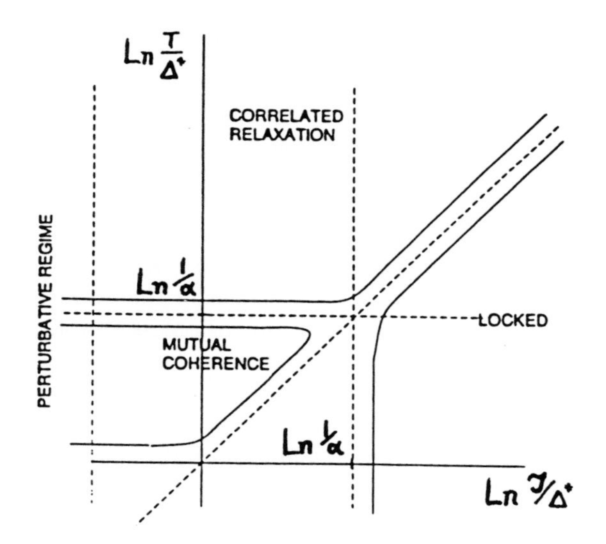

FIG. 13: The phase diagram for the symmetric PISCES model (both spins identical), for an Ohmic bath, as a function of the dimensionless parameters $ln(T/\Delta^*)$ and $ln(\mathcal{J}/\Delta^*$, where $\mathcal{J}$ is the full renormalised coupling between them. We assume the dimensionless coupling α of each to the bath is $\ll 1$, so the mutual coherence phase appears.

To characterize the dynamics of the model one calculates the 2-spin reduced density matrix $\rho(\tau_1\tau_2;\tau_1'\tau_2';t)$, which is a 4x4 time- dependent density matrix. Instead of discussing how these calculations were done (for which see the long paper of Stamp and Dubé [57]), I think it will be more useful here to give an intuitive feeling for the results, and some of their consequences for coupled mesoscopic or macroscopic systems.

The most interesting examples of the PISCES model have Ohmic coupling of each spin to the oscillator bath (cf. section 2.3). In this case one can delineate the behaviour of each spin by a "phase diagram" in the 4-dimensional space of couplings $\alpha_1, \alpha_2, \mathcal{J}(\mathcal{R})$, and T, where α_1, α_2 are the Ohmic coupling of each spin to the bath, and $\mathcal{J}$ the total static coupling between the spins, including the bath-mediated coupling $\epsilon(R)$ (section 2.4). The other 2 energy scales in the problem are the Δ_β^*, where $\beta = 1, 2$ labels the 2 spins, and the star superscripts indicate the *renormalised* tunneling matrix elements (after coupling to the bath). Then there are 4 dimensionless parameters in the problem, viz., $\Delta_\beta^*/\alpha_\beta, \mathcal{J}/T$, and T itself; this is our parameter space.

The simplest case to discuss is where the spins are identical (ie., $\Delta_1 = \Delta_2 = \Delta$, and $\alpha_1 = \alpha_2 = \alpha$). The behaviour can then be displayed in a simple way as a function of $\mathcal{J}(\vec{R})/\alpha$ and T (cf. Fig.13), and we can identify 4 different regions in this phase diagram, as follows:

(i) The "Locked Phase" ($\mathcal{J} \gg T, \Delta_\beta^*/\alpha_\beta$): In this regime, the effective coupling $\mathcal{J}$ is so strong that the 2 spins lock together, in either the states $\mid\uparrow\uparrow\rangle$ or $\mid\downarrow\uparrow\rangle$ depending on the sign of $\mathcal{J}$. One finds that the combined "locked spin" oscillates between $\mid\uparrow\uparrow\rangle$ and $\mid\downarrow\downarrow\rangle$ (for FM coupling), or between $\mid\uparrow\downarrow\rangle$ and $\mid\downarrow\uparrow\rangle$ (for AFM coupling), at a renormalised frequency $\Delta_c = \Delta_1\Delta_2/|\mathcal{J}|$. This result is not at all surprising- in fact one would get the same result just by coupling 2 tunneling spins in the complete absence of the environment! However the oscillations are now *damped* - in fact the locked spin now behaves like a single spin-boson system, with a new coupling $\alpha_c = \alpha_1 + \alpha_2 \pm 2\alpha_{12}$ to the oscillator bath, the + ($-$) corresponding to ferromagnetic (antiferromagnetic) coupling between the spins. For the dynamics of the locked spin one simply then refers back to the spin-boson results.

(ii) The "Mutual Coherence" phase ($\Delta_\beta/\alpha_\beta \gg T \gg \mathcal{J}$; $\mathcal{J} > \Delta_\beta$): Here, the thermal energy overcomes the mutual coupling; nevertheless if the dissipative couplings α_β's are sufficiently small ($\alpha_\beta \ll 1$), it is possible for the energy scale $\Delta_\beta^*/\alpha_\beta$ to dominate even if $\Delta_\beta < \mathcal{J}$. In this case, even though we are dealing with a strong coupling, and the bath dissipation is still important, some coherence in the motion of each spin is maintained - moreover, the small $\mathcal{J}$ causes "mutual coherence" between the two spins, ie., their damped oscillations are correlated to some extent. The analytic form of the correlation functions is extremely complex (see the original paper [57]), and moreover is not given simply in terms of the single spin-boson results; the mutual correlations are essential in this regime. However the physics is fairly easy to understand- we are seeing a kind of "beat" or "breather" oscillation between the 2 spins, which the damping has not quite succeeded in destroying. In addition, each spin is exhibiting its own weakly damped oscillations. The situation is reminiscent of 2 guitar strings, which couple through the air (and the guitar box), but where the air also causes weak damping.

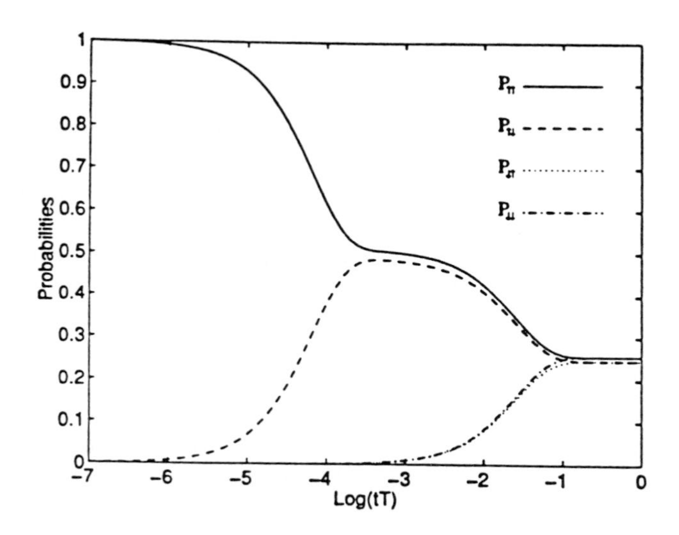

FIG. 14: The Probability $P_{\tau_1^z\tau_2^z}(t)$ for the 2 PISCES spins to be in state $|\tau_1^z\tau_2^z\rangle$ after a time t, if at $t=0$ they started in state $|\uparrow\uparrow\rangle$. The system is in the "Correlated relaxation" regime, with time shown logarithmically, in units of inverse temperature. We assume $\alpha_1 = 1.5$, $\alpha_2 = 2$, $T/\Delta_1 = 100$, $T/\Delta_2 = 300$, and a ferromagnetic coupling $\mathcal{J}/T = -0.04$.

(iii) The "Correlated Relaxation" or High-T phase ($T \gg \Delta_\beta/\alpha_\beta, \mathcal{J}$). In this regime, the bath causes each spin to relax incoherently; however, the relaxation of the two spins is still correlated (indeed each spin relaxes in the time-dependent bias generated by the other). This regime is, along with the locked regime, the most typical. The coupling between the 2 spins, via the bath oscillators, simply decoheres completely the spin dynamics of each, and they are both strongly damped. You can find the analytic expressions for their damping in the original paper ; what is perhaps remarkable is that the system exhibits 3 different damping rates, each of which is however a fairly simple algebraic function of the damping rates for 2 different single spins, each coupled to an oscillator bath and in a *static* bias field $\xi = \mathcal{J}$ (cf. equation (85)). In this sense the problem is not fundamentally different from the single spin-boson model. Here I just show a typical relaxation of the 4 diagonal elements of the 2-spin density matrix, which illustrates the vastly different relaxation timescales which can result during the relaxation of 2 initially parallel spins (Fig.14).

(iv) Finally, and much less interesting, the "perturbative regime" ($\mathcal{J} \ll \Delta_\beta^*$), in which the total coupling is so weak that the 2 spins relax almost independently; all correlations can be handled perturbatively, and only weakly affect the behaviour that one calculates from the standard spin-boson model.

The remarkable thing is that *analytic* results for the dynamics of the 2 spins can be found, although we shall not need these in this lightning survey. Readers familiar with some particular case of the PISCES model will doubtless recognise features of the phase diagram. Thus, eg., those who have looked at the problem of 2 coupled Kondo impurities will recognise the "locked regime" from the old studies of Wilkins et al [66]. The advantage here is that instead of just extracting a phase diagram (which is easily done using renormalisation group methods), we have been able, using instanton methods, to extract the dynamic properties as well.

Perhaps the most important thing to take away from these results is just how easily the quantum dynamics of each spin can be decohered or even blocked by the other. We should not be too surprised by this - recall how the environment causes a "dynamical localisation" of the spin in the single spin-boson model. However it is much more serious in the PISCES model, because to the weak random environmental bias we have added a much stronger dynamical bias coming from the other spin. From this point of view it is actually interesting to compare the PISCES dynamics with that of a much simpler "toy model", in which the environment is eliminated completely, and we just have 2 coupled spins (see Dubé and Stamp [57]).

Let us now fill out this picture by applying it to the 2 physical cases discussed in section 2.4.

(i) 2 coupled Giant Kondo spins: As mentioned in the previous subsections, the dynamics of a Giant Kondo spin is of potential technological importance. However it is also obvious that in a nanomagnetic array, interactions between the nanomagnets can alter the behaviour of a single nanomagnet. Since there is obviously a large number of different cases that can be discussed [17], I will simply give you a taste of the results. Consider again the example in which a Fe-based nanomagnet is embedded in a conducting film, whose conduction electron density can be varied. We again assume $S = 300$, and now also assume $\Omega_o = 1K, \Delta = 1mK$, and a typical nuclear bias energy $E_o \sim 6mK$ (roughly 2 per cent of the Fe nuclei have spins, and the hyperfine coupling for each is of order $2-3mK$). However we now assume a second such nanomagnet, at a distance R from the first. The effective bath-mediated interaction $\epsilon(R)$ between them depends strongly on R (as $1/R^3$), but even more strongly on the conduction electron density - it goes as $(N(0))^4$, where $N(0)$ is the Fermi surface density of states!

Let us suppose that $R = 300\AA$. Then the dipolar interaction between the spins is $V_{dip} \sim 10^{-5}K$. The conduction electron mediated interaction between the 2 nanomagnets in a typical metallic film, with $g \sim 0.1$ (so that $\alpha \sim 20$) is then $\epsilon(R) \sim 30mK$, ie., 3000 times larger then this (and 30 times larger than Δ). In this case it is clear that $\epsilon(R)$ has no real effect on the dynamics of the individual giant spins, which is completely controlled by the very large α. As we saw previously, the 2 spins relax independently at high T, but at low T, their dynamics are frozen into a quantum localised state. As discussed by Prokof'ev and myself[17], this localisation could be of great practical importance for future data storage at the molecular level. Notice however that if $R = 100\AA$, $\epsilon(R)$ climbs to $1K$, ie., $\epsilon(R) \sim \Omega_o$, and the interaction will completely change the dynamics of the 2 spins - apart from tending to unlock the spin dynamics of each, it will also take them out of the Hilbert space of the 2 lowest levels. The dynamics of an array of such spins would be very complicated, even at $T = 0$.

Now instead suppose that we drop the conduction electron density by a factor of ~ 10, so that $g \rightarrow 0.01$. As previously, this means that $\alpha \rightarrow 0.2$, thereby unlocking the giant spin dynamics at all temperatures. It also means that $\epsilon(R)$ is drastically reduced - it falls by a factor of 10^4 (so that for $R = 300\AA, \epsilon(R)$ is now $3 \times 10^{-6}K$), and is now smaller than V_{dip}! It is also much smaller than Δ, and in fact the dynamics of the spins, in zero applied field, will be controlled at low T by the random nuclear bias, since E_o is now the largest energy scale. The spins will again have essentially independent dynamics, when $R = 300\AA$; but now because we are in the perturbative regime.

Notice, however, that if instead $R = 100\AA$, then V_{dip} rises to $0.3mK$ (and $\epsilon(R)$ to $0.1mK$), ie., it is of similar size to Δ and E_o. Obviously by playing with α and R it is then possible to create a situation where mutually coherent oscillations can occur; otherwise, depending on the exact conditions, we will be either in the correlated relaxation or the locked phases. A nanomagnetic array in the regime might be expected to show very interesting quantum diffusive properties in the $T = 0$ limit.

Finally one can imagine further reducing the electron density and/or S, to produce a situation where $\alpha \ll 1$, but V_{dip} is of the same order as Δ (and both are larger than E_o). This is the "weak coupling" regime. In this case one's first naive guess might be that a nanomagnetic array in this regime should behave coherently, with an energy band of width $\sim \Delta$. It is clear that such an array is not beyond present technology to prepare. This would be a remarkable display of macroscopic coherence behaviour, if true- but have we left anything out of the model?

The answer is of course that we have. We have left out the nuclear dynamics, ie., the dynamics of the background spin bath. This is not so important when α is large, but if $\alpha \ll 1$, it is essential, and it will destroy the coherence of our giant spins, as we shall see in the next subsection (as well as in section 4).

(ii) <u>The Measurement Problem</u>: Let us quickly look at the analogous implications for our measurement model (which we have largely anticipated already, in introducing it above). Recall that the model discussed in section 2.3 for a quantum measurement involved an overdamped but rapidly responding apparatus (ie., Δ_A large, $\alpha_A > 1$) and an underdamped system with slow dynamics (Δ_s small, $\alpha_s \ll 1$). When a measurement is made, an interaction K(t) is switched on, which is sufficiently strong to allow the apparatus state to correlate with that of the system - however it is not strong enough to cause any transitions in the system state. What actually happens is that as soon as K(t) exceeds the small Δ_s, the system dynamics are frozen, and the apparatus state subsequently evolves to correlate with the frozen system state.

The quantitative calculations of the "Overdamped plus Underdamped" case [57] confirm this picture - I do not propose giving into the details here! There is obvious practical application of these results to the problem of the observation of macroscopic quantum coherent behaviour (on the part of, eg., a SQUID system, or a nanomagnet) - the most obvious way of determining the dynamics of such systems is by coupling in another SQUID magnetometer, or a MFM ("Magnetic Force Microscope") to the tunneling system. It is clear that the combined system/apparatus/ environment problem can be described by a PISCES model (provided we can ignore nuclear dynamics - see below)

(iii) <u>Miscellaneous Models</u>: To complete this discussion of results for the dynamics of various commercial oscillator bath models, a brief word on the dynamics of the models noted in section 2.4(c).

The simplest such model is of a "central" harmonic oscillator coupled to a bath of oscillators. If we assume the bilinear coupling $H_{int} = \sum_k c_k x_k Q$ between the two, the problem is exactly solvable; the details were discussed at great length by Grabert et al. [61]. Just as in the standard problem of a classical dissipative oscillator, the results are most easily discussed in the Laplace transform domain. Thus suppose we have a central oscillator of mass M_o and frequency Ω_o. The classical equation of motion then involves a dissipation coefficient $\eta(\omega)$, in general frequency dependent; if $\eta(\omega) \rightarrow \eta$, we have Ohmic dissipation. In terms of the Caldeira and Leggett spectral function $J(\omega)$, the Laplace-transformed function $\eta(s)$ is

$$\mu(s) = \int_0^\infty \frac{d\omega}{\pi} \frac{J(\omega)}{\omega} \frac{2s}{s^2 + \omega^2} \tag{88}$$

Then the retarded response function $\chi(t)$ of the central oscillator is most conveniently written as $\chi(t) = C(t)\Theta(t)$, where

$$\begin{aligned} C(s) &= \int_0^\infty dt C(t) e^{-st} \\ &= \frac{1}{M_o(\Omega_o^2 + s^2) + s\eta(s)} \end{aligned} \tag{89}$$

a result which demonstrates very nicely the correspondence, strongly emphasized by Caldeira and Leggett [13], and mentioned in section 3.2 above, between the classical and quantum behaviour.

In the case of the tunneling problem, there is a huge literature on the dynamics. This work is so extensive that I will simply refer you to the literature here - it is not central to the present discussion, and moreover is discussed in the chapter by Leggett in this book. The original long paper of Caldeira and Leggett [13] is a very good starting point. Hanggi et al. [67] discuss the connection with reaction rate theory, and a very fine set of reviews of the whole field appear in the book edited by Kagan and Leggett [56]. Finally, a good introduction to the theoretical techniques is given by Weiss [64].

For work on the dynamics of a free particle, coupled to an oscillator bath, see the review of Grabert et al. [61] again (as well as the paper of Hakim and Ambegaokar [68]).

So much for the dynamics of systems coupled to oscillator baths; now we turn to the very different effect of the spin bath.

3.4 Central Spin Model

In our discussion of oscillator bath models I occasionally warned that the results would be seriously compromised by any spin bath that coupled to the central system. Here we shall see how this happens, and review results for the dynamics of a central system coupled to a spin bath. It is crucial here to recall the essential difference between the oscillator and spin bath models- whereas the oscillators are only very weakly affected by the central system (so that their spectra and dynamics are hardly altered), the spins in the spin bath have their spectra and dynamics changed completely.

The discussion of the present section is based on papers by myself and Prokof'ev [14,16,17,18] (see also ref. [69]), for which see the gory details - here I attempt an intuitive perspective on these, and briefly note their application to nanomagnets (discussed in detail in Chapter 4), and SQUIDs.

3.4(a) THREE LIMITING CASES:

Let us first recall the low-energy effective Hamiltonian we derived for the central spin model, whose derivation was described in section 2.2. The Hamiltonian, in the absence of an external field, was

$$\begin{aligned} H_{\text{eff}} &= \left\{ 2\Delta_o \hat{\tau}_- \cos\left[\pi S - i\sum_k (\alpha_k \vec{n}\cdot\hat{\vec{\sigma}}_k + i\beta_o \vec{n}\cdot\vec{H}_o)\right] + H.c.\right\} \\ &+ \frac{\hat{\tau}_z}{2}\sum_{k=1}^{N}\omega_k^{\parallel}\,\vec{l}_k\cdot\hat{\vec{\sigma}}_k + \frac{1}{2}\sum_{k=1}^{N}\omega_k^{\perp}\,\vec{m}_k\cdot\hat{\vec{\sigma}}_k + \sum_{k=1}^{N}\sum_{k'=1}^{N} V_{kk'}^{\alpha\beta}\hat{\sigma}_k^{\alpha}\hat{\sigma}_{k'}^{\beta} \quad . \quad (90) \end{aligned}$$

where $\vec{\tau}$ represents the central spin, and the $\{\sigma_k\}$ the spin bath variables; the notation is that of section 2.2.

What we wish to do here is calculate the dynamics of $\vec{\tau}$ after integrating out the bath variables - our problem is similar to that addressed in section 3.2, only the bath has changed. I shall describe the physics in a tutorial manner, by first introducing you to 3 limiting cases, and then showing how the complete solution is an amalgam of these cases.

(i) Topological Decoherence Limit: This limiting case removes all of the static coupling between the central spin and the spin bath, and also removes the intrinsic spin bath dynamics, by supressing $V_{kk'}^{\alpha\beta}$. The resulting Hamiltonian is

$$H_{\text{eff}}^{top} = 2\Delta_o \hat{\tau}_x \cos\left[\pi S - i\sum_{k=1}^{N}\alpha_k \vec{n}_k\cdot\hat{\vec{\sigma}}_k\right] , \tag{91}$$

The essential feature of this model is that all of the spin bath dynamics is driven by the central spin - moreover, from the form of H_{eff}, we see that to the Berry phase πS of the central spin is added a complex phase $-i\Sigma_k\alpha_k\vec{n}_k\cdot\vec{\sigma_k}$ coming from the bath. Thus the main effect of the bath transitions is to add a *random phase* to the action incurred during each central spin transition - in effect the topological phase in the system dynamics is randomised. Both the physical and mathematical aspects of this case have been discussed in some detail in the original references , and I will not repeat this here. The main ideas can be entirely understood in the special case where we choose α_k to be entirely *imaginary*; and I will simply now refer to this imaginary quantity as α_k. The solution can then be written formally as

$$P_{11}(t) = \frac{1}{2}\left\{1 + \langle\cos\left[4\Delta_o t\cos\left(\Phi + \sum_{k=1}^{N}\alpha_k\vec{n}_k\cdot\hat{\vec{\sigma}}_k\right)\right]\rangle\right\} , \tag{92}$$

where $< ... >$ is an average over the environmental states, and we have made the replacement $\pi S \rightarrow \Phi$, to take account of any high-frequency ($> \Omega_o$) renormalization of the central spin Haldane phase, caused by the nuclear bath. In the present model all bath states are degenerate, and this average is trivial - it can be written as a weighted integration over topological phase φ, as

$$\begin{aligned} P_{11}(t) &= \sum_{m=-\infty}^{\infty} F_{\lambda'}(m)\int_0^{2\pi}\frac{d\varphi}{2\pi}e^{i2m(\Phi-\varphi)}\left\{\frac{1}{2}+\frac{1}{2}\cos(2\Delta_o(\varphi)t)\right\} & (93)\\ &= \frac{1}{2}\left\{1 + \sum_{m=-\infty}^{\infty}(-1)^m F_{\lambda'}(m)e^{i2m\Phi}J_{2m}(4\Delta_o t)\right\} , & (94)\end{aligned}$$

where we define

$$\lambda = \frac{1}{2}\sum_k |\alpha_k|^2 ; \qquad F_\lambda(\nu) = e^{-4\lambda\nu^2} \tag{95}$$

As mentioned in section 2.2, λ is the mean number of bath spins flipped each time $\vec{S}$ flips. If $\lambda > 1$, but not too large; I will call this the intermediate coupling limit. Then

$$F_\lambda(\nu) = \delta_{\nu,0} + \textit{small corrections} \qquad (\textit{intermediate}) . \tag{96}$$

so that, very surprisingly, we get a *universal form* in the intermediate coupling regime for $P_{11}(t)$ (here $\eta(x)$ is the step function):

$$P_{11}(t) \longrightarrow \frac{1}{2}\left[1 + J_0(4\Delta_o t)\right] \equiv \int\frac{d\varphi}{2\pi}P_{11}^{(0)}(t, \Phi = \varphi) \tag{97}$$

(compare the angular average of the coherent series in (94)), From this we can also compute the *absorption* of the 2-level giant spin, as a function of frequency, by Fourier transforming; we get

$$\chi''(\omega) \longrightarrow \frac{2}{(16\Delta_o^2 - \omega^2)^{1/2}}\eta(4\Delta_o - \omega) \quad (\textit{intermediate}) . \tag{98}$$

We plot this universal form in Fig.15.

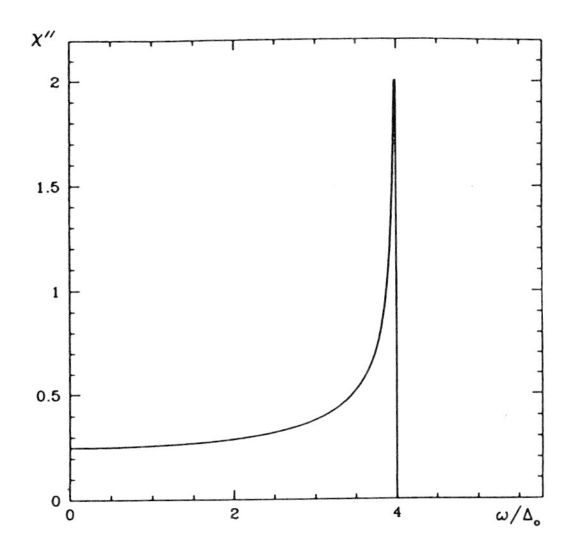

FIG. 15: The universal form for the power absorption $\chi''(\omega)$ in the case of pure topological decoherence.

The physics of this universal form is simply one of phase cancellation., As explained in refs.[14,15], this phase cancellation arises because successive flips of $\vec{S}$ cause, in general, a *different* topological phase to be accumulated by the spin environment, so that when we sum over successive instantons for $\vec{S}$, we get phase randomisation and hence loss of coherence. The only paths that can then contribute to coherent oscillations in $P_{11}(t)$ are those for which the number of clockwise and anticlockwise flips in the spin bath are *equal* - in this case the topological phase "eaten up" by the environmental spins is zero. By looking at expression (97) one sees another way of describing this- the universal behaviour comes from complete phase randomisation [15], so that all possible phases contribute equally to the answer! The final form shows decaying oscillations , with an envelope $\sim t^{-1/2}$ at long times. This decay can also be understood [16] by noting that the "zero phase" trajectories that contribute to P_{11} constitute a fraction $(2s)!/(2^s s!)^2 \sim s^{-1/2}$ of the total number of possible trajectories, where $s \sim \Delta_o t$. Because of this decay, the δ-function peak in the spectral function at $\omega = 4\Delta_o \cos\Phi$ is now transformed to the spectral function of a 1-dimensional tight-binding model.

One should also note that in the the *strong* coupling limit, when $\mid \alpha_k \mid \rightarrow$

$\pi/2$, so that each spin rotates adiabatically with $\vec{S}$, we simply have

$$P_{11}(t) = \frac{1}{2}[1 + \cos(4\Delta_o t \cos\tilde{\Phi})] \; , \tag{99}$$

where $\tilde{\Phi} = \pi S + N\pi/2$, i.e., the Haldane/Kramers phase is now $\tilde{\Phi}$, with the extra phase coming from the N bath spins which rotate rigidly with $\vec{S}$.

Recall that the above results are given for the special case where α_k is pure *imaginary*; for the general complex case see the original work.

(ii) Orthogonality Blocking Limit: Let us now consider the case where again all bath dynamics is suppressed ($V_{kk'}^{\alpha\beta} = 0$), but now we retain only the static diagonal part of H_{eff}, ie., we assume

$$H_{\text{eff}} = 2\Delta_\Phi \tau_x + \hat{\tau}_z \frac{\omega_o^{\|}}{2} \sum_{k=1}^{N} \vec{l}_k \cdot \hat{\vec{\sigma}}_k + \frac{\omega_o^{\perp}}{2} \sum_{k=1}^{N} \vec{m}_k \cdot \hat{\vec{\sigma}}_k \; . \tag{100}$$

$$\Delta_\Phi = \Delta_o \cos\Phi \; , \tag{101}$$

where $\omega_o^{\|} \gg \omega_o^{\perp}$; in (100) we use a "Kramers renormalised" matrix element Δ_Φ. The significance of the finite but small transverse term $\omega_o^{\perp}$, is that in any case where the initial and final fields $\vec{\gamma}_k$ acting on $\vec{\sigma}_k$ are not exactly parallel or antiparallel, the diagonal coupling must have a transverse part (cf. section 2.2(b)).

Now in this case no transitions occur in the spin bath when the central spin $\vec{S}$ flips. However the bath spins still play a crucial role, because the transverse "field" term $\omega_o^{\perp} \sum_k \hat{\sigma}_k^x$ changes the motion of $\vec{S}$. There is perhaps a temptation to treat this as nothing but the simple problem of the changed motion of $\vec{S}$ in this transverse field. However this is quite wrong - if we do this we forget that the transverse term is coming from a set of spin variables (the $\{\hat{\vec{\sigma}}_k\}$) to each of which is associated a *wave-function* $\mid \Phi_k >$. The crucial point here is that in the presence of the transverse coupling, there is a mismatch between the initial state $\mid \Phi_k^{in} >$ (before $\vec{S}$ flips) and the final state $\mid \Phi_k^{fin} >$ (after it flips); in general $|\langle\Phi_k^{in}|\Phi_k^{fin}\rangle| \; < 1$ if $\omega_o^{\perp} \neq 0$. When we average over all bath spins, this seriously alters the system dynamics. Since $\omega_o^{\perp} \ll \omega_o^{\|}$, we define an angle β_k, describing this mismatch between the initial and final states of $\vec{\sigma}_k$, according to

$$\cos 2\beta_k = -\hat{\vec{\gamma}}_k^{(1)} \cdot \hat{\vec{\gamma}}_k^{(2)} \; , \tag{102}$$

where $\hat{\vec{\gamma}}_k^{(1)}$ and $\hat{\vec{\gamma}}_k^{(2)}$ are unit vectors in the direction of the initial and final state fields acting on $\vec{\sigma}_k$ (and here we assume these are almost exactly antiparallel).

If we assume that $\beta_k \ll 1$, then the *total* mismatch, of the usual "Debye-Waller" or "Franck-Condon" form, is just $e^{-\kappa}$, where

$$\kappa = \frac{1}{2} \sum_k \beta_k^2 \, , \tag{103}$$

More generally, if the mismatch angles β_k are not so small, we just have

$$e^{-\kappa} = \prod_{k=1}^{N} \cos \beta_k \, . \tag{104}$$

Because of the similarity to the famous "orthogonality catastrophe" considered by Anderson (in which the electronic phase shift δ_k is analogous to the β_k considered here), we have called this limit that of "orthogonality blocking". However the physics here is rather different because the bath spins behave quite differently from electrons (which, we recall, map to an oscillator bath). As shown in the original work, the easiest way to understand and to solve for the dynamics of $\vec{S}$ is to notice that $\mid \Phi_k^{fin} >$ is produced from $\mid \Phi_k^{in} >$ by the unitary transformation:

$$\mid \vec{\sigma}_k^{fin} \rangle = \hat{U}_k \mid \vec{\sigma}_k^{in} \rangle = e^{-i\beta_k \hat{\sigma}_k^x} \mid \vec{\sigma}_k^{in} \rangle \, . \tag{105}$$

Thus, *mathematically*, the problem is identical to one in which there is an amplitude β_k for $\vec{\sigma}_k$ to flip each time $\vec{S}$ flips, and where typically $\beta_k \ll 1$. Physically, the mismatch between the 2 states arises because the fields $\vec{\gamma}_k^{in}$ and $\vec{\gamma}_k^{fin}$ acting on $\vec{\sigma}_k$ are neither parallel or exactly antiparallel. Thus if σ_k is initally aligned with $\vec{\gamma}_k^{in}$, it finds itself slightly misaligned with $\vec{\gamma}_k^{fin}$ after $\vec{S}$ flips (semiclassically, $\vec{\sigma}_k$ must start precessing in the field $\vec{\gamma}_k^{fin}$).

This problem is thus similar to that of topological decoherence (with β_k replacing α_k, and κ replacing λ), except that we now have the added complication that the spin bath states are not at all degenerate - in fact they are split by the large longitudinal coupling ω_o. Recall (end of section 2.2) that this spreads the nuclear bath states over a large energy range $E_o \sim \omega_o N^{\frac{1}{2}}$, so that in zero applied field, S cannot make any transitions at all unless the internal spin bath field $\epsilon = \omega_o \sum_k \sigma_k^z \tau_k^z$ bias energy is the same before and after $\vec{S}$ flips. Otherwise, since typically $\omega_o \gg \Delta$, the difference 2ϵ between the intial and final energies of $\vec{S}$ simply blocks all tunneling. This we immediately see that we require that either (i) the net polarisation $\tilde{M} = \sum_k \sigma_k^z$ of the spin bath is zero both before and after $\vec{S}$ flips (so $\epsilon = 0$); or (ii) $\tilde{M}$ changes from M to $-M$ when $\vec{S}$ flips - meaning that *at least* M spins flip (and $\epsilon = M\omega_o$ before and after $\vec{S}$ flips).

For details of the resulting calculations of $P_{\Uparrow\Uparrow}(t)$,and the resulting rather bizarre behaviour, I refer to the original papers [16,18].

(iii) Degeneracy Blocking Limit: Finally let us consider the case where again $V_{kk'}^{\alpha\beta} = 0$ (no spin bath dynamics), and both the non-diagonal terms *and* the transverse diagonal term $\omega_k^{\perp}$ are zero. Thus we have

$$H_{\text{eff}} = 2\Delta\tau_x + \frac{1}{2}\tau_z \sum_{k=1}^{N} \omega_k^{\parallel}\, \hat{\sigma}_k^z \; ; \tag{106}$$

with a spread of values of $\omega_k^{\parallel}$ of

$$\sqrt{\sum_k (\omega_k^{\parallel} - \omega_o)^2} \equiv N^{1/2}\delta\omega_o \, . \tag{107}$$

i.e., a distribution of width $\delta\omega_o$.

Clearly, this Hamiltonian is identical to the standard biased two-level system, with the bias energy ϵ depending on the particular environmental state; thus $\epsilon = \sum_{k=1}^{N} \omega_k^{\parallel}\sigma_k^z$. The introduction of this spread is to destroy the exact degeneracy between states in the same polarisation group. For coherence to take place, we require the initial and final states to be within roughly Δ of each other.

The crucial point is of course that $\omega_o^{\parallel} \gg \Delta$ in most cases, and in fact one often has $\delta\omega_o^{\parallel} > \Delta$ (thus, in a nanomagnet, $\delta\omega_o^{\parallel}$ comes not only from the variation of the local contact hyperfine couplings $\omega_j \vec{s}_j.\vec{I}_j$, but also from the large variety of weaker transfer hyperfine couplings between the $\vec{s}_j$ and nuclei on other non-magnetic sites). A suitable dimensionless parameter which describes the extent of the spread in the $\omega_k^{\parallel}$, around the "central" value ω_o, is $\mu = N^{\frac{1}{2}}\delta\omega_o/\omega_o$. In the limit where $\mu \to 0$, and then the 2^N nuclear levels are organised into a set of sharp lines, each one corresponding to a particular polarisation group- however this case is rather academic. More generally we define a *density of states* function $W(\epsilon)$ for the nuclear levels in the presence of the hyperfine coupling (Fig.16 below).

If $\mu < 1$ the lines still can be seen, with however the different polarization groups overlapping. If $\mu > 1$, however (and this is almost invariably the case in any real system), this structure disappears and we are left with a Gaussian form [14,16]:

$$W(\epsilon) = \left(\frac{2}{\pi E_o^2}\right)^{1/2} exp(-2\epsilon^2/E_o^2) \tag{108}$$

$$E_o = \omega_o^2 N \tag{109}$$

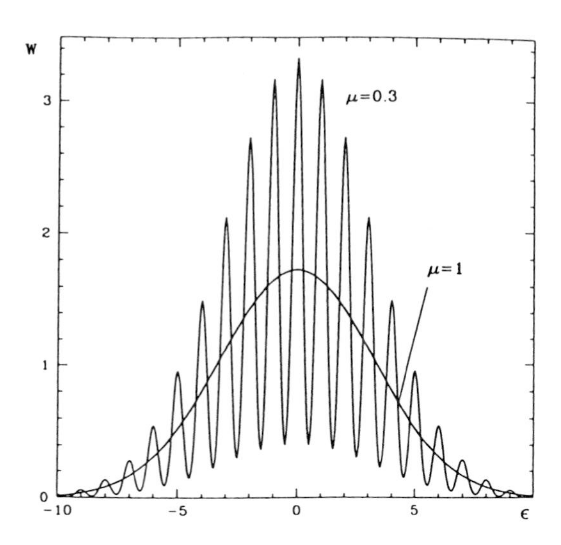

FIG. 16: The density of states function $W(\epsilon)$ for the nuclear multiplet for 2 different values of the parameter μ (see text). The case $\mu < 1$ is somewhat academic, and so $W(\epsilon)$ is Gaussian in practise.

This form was already mentioned in section 2.2. In nanomagnetic molecules μ is usually considerably greater than unity, even for small magnetic molecules, because of the large number of protons in the molecule and its ligands. In nanomagnetic grains the nuclei of O ions (as well as protons in hydrated systems) as well as other magnetic species, have the same effect.

The dynamics of an *ensemble* of nanomagnets, each described by the above Hamiltonian, but averaged over the nuclear distribution $W(\epsilon)$ in a thermal ensemble, is then quite trivial to obtain - it is just the *weighted ensemble average* over the correlation function of a simple 2-level system in this internal bias field ϵ. In the usual case where $kT \gg \Delta$, we then have

$$\begin{aligned} P_{11}(t) \quad &= \quad \int d\epsilon W(\epsilon) \frac{e^{-\beta\epsilon}}{Z(\beta)} \\ &\left[1 - \frac{2\Delta_\Phi^2}{\epsilon^2 + 4\Delta_\Phi^2}\left(1 - \cos(2t\sqrt{\epsilon^2 + 4\Delta_\Phi^2})\right)\right] \end{aligned} \tag{110}$$

$$\longrightarrow \quad 1 - 2A \sum_{k=0}^{\infty} J_{2k+1}(4\Delta_\Phi t) \; . \tag{111}$$

where the reduction factor A is just

$$A = 2\pi\Delta_\Phi / (\omega_o \sqrt{2\pi N}) \ll 1 \; . \tag{112}$$

and the spectral function is given by another universal form, viz:

$$\chi''(\omega) = A \frac{8\Delta_\Phi}{\omega\sqrt{\omega^2 - 16\Delta_o^2}} \eta(\omega - 4\Delta_\Phi) \; , \tag{113}$$

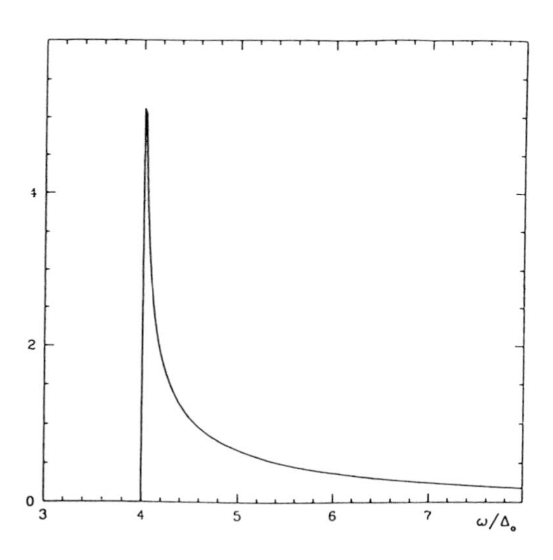

FIG. 17: The universal form for $\chi''(\omega)$ in the case of pure degeneracy blocking. Notice that absorption only occurs *above* the threshold defined by Δ_o.

The result here is very easy to understand. Only those nanomagnets near resonance (ie., with $\epsilon \sim \Delta$ or less) stand much of a chance of tunneling, since tunneling requires near degeneracy between intial and final states. Those systems which do not have this degeneracy are "degeneracy blocked"; and since $\Delta \ll E_o$, almost all of them are frozen. This result of course arises because the only dynamics in this limiting model comes from the tunneling term itself. As soon as we give the spin bath some dynamics (either by reintroducing the non-diagonal terms, or even more importantly, by making $V_{kk'}^{\alpha\beta}$ non-zero, so that the nuclear bath has independent dynamics), the blocking will be relieved, as we shall now see.

3.4(b) GENERAL SOLUTION The discussion of the 3 limiting cases given above allows us to give a simple intuitive picture of the general solution to the problem (readers wanting details, proofs, etc., should go to the original papers [17,18,69]).

(i) Independent Bath Spins: Suppose we start with the (often rather academic) case where the bath spins do not interact at all with each other (formally, this means $V_{kk'}^{\alpha\beta}$ is much less than all other energy scales, in particular Δ). The results are so pretty, and intuitively revealing, that I will quickly describe them to you.

Each of the 3 limiting cases above corresponds to one physical mechanism (already previewed back in our initial discusion of this model, in section 2.2(b)). Now the magic is that the general solution is simply obtained by combining them! Mathematically, each mechanism corresponds to a particular *statistical average*:

(a) A "topological phase average" given by
$$\sum_{\nu=-\infty}^{\infty} F_{\lambda'}(\nu) \int \frac{d\varphi}{2\pi} e^{i2\nu(\Phi-\varphi)} \; ; \tag{114}$$

(b) An "orthogonality average" given by
$$2\int_0^{\infty} dx x e^{-x^2} \; ; \tag{115}$$

(c) A "bias average"
$$\int d\epsilon W(\epsilon) \frac{e^{-\beta\epsilon}}{Z(\beta)} \; . \tag{116}$$

Apart from the orthogonality average (the derivation of which is too complicated to explain here), you have just seen these above. Now, consider a simple 2-level system in a bias ϵ, but with a *renormalised* tunneling splitting

$$\Delta_M(\varphi, x) = 2\tilde{\Delta}_o |\cos(\varphi) J_M(2x\sqrt{\gamma})| \; , \tag{117}$$

and associated energy splitting $E_M^2(\varphi,x) = \Delta_M^2(\varphi,x) + \epsilon^2$. This tunneling matrix element now depends explicitly on the topological phase, on the nuclear polarisation M, and on the orthogonality variable x. The parameter γ is just

$$\gamma = \begin{cases} \lambda & \text{if } \lambda \gg \kappa \text{ (topological decoherence regime)} \\ \kappa & \text{if } \kappa \gg \lambda \text{ (orthogonality blocking regime)} \end{cases} \; . \tag{118}$$

Recall now the very simple dynamics of an isolated biased 2-level system (eqtn. (28)); the probability that the system stays in state $|\uparrow\rangle$ after a time t is $P_{11}^{(0)}(t) = 1 - (\Delta/E)^2 sin^2(Et)$. Then, perhaps amazingly, the same function for the central system, after all averaging over the bath spins is carried out, is nothing but a generalisation of this; one gets

$$P_{11}(t;T) = 1 - \int d\epsilon W(\epsilon) \frac{e^{-\beta\epsilon}}{Z(\beta)} \sum_{M=-N}^{N} (1 - P_M(t, \epsilon - M\omega_o)) \; ; \qquad (119)$$

in which the probability relaxation function for a given polarisation group M is

$$\begin{aligned} P_M(t;\epsilon) \;\; &= \;\; 2\int_0^{\infty} dx x e^{-x^2} \sum_{\nu=-\infty}^{\infty} F_{\lambda'}(\nu) \\ & \int \frac{d\varphi}{2\pi} e^{i2\nu(\Phi-\varphi)} \left[1 - \frac{\Delta_M^2(\varphi, x)}{E_M^2(\varphi, x)} \sin^2(E_M(\varphi, x)t) \right] , \qquad (120) \end{aligned}$$

That is all! Perhaps even more amazing, these integrals and sums can be done *analytically* for almost all of the parameter regime (which, amusingly, is more than one can say for the spin-boson model, which just goes to show that a simple Hamiltonian does not always have simple dynamics!). The physical interpretation of this result is of course obvious, just by looking at our 3 limiting cases again; I will not go over it again. If you want to see analytic formulae and pictures, see Prokof'ev and Stamp [17,18,69]; you will find that in some cases they do not look anything like what you can get from a spin-boson model!

This of course raises the question- can this "central spin" model ever behave like a spin-boson model? The answer is yes. Lets first think about this physically- we wish the spins to somehow behave like a set of oscillators, *weakly coupled* to the central spin. If you recall the form of the Caldeira-Leggett spectral function $J(\omega)$, involving a coupling constant squared, divided by an energy denominator (cf. 2nd-order perturbation theory), then it is clear how this can happen, for in our case the 2 relevant parameters are $\omega_k^{\parallel}$ and $\omega_k^{\parallel}$, and each is capable of playing the role of the coupling c_k (with the other playing the role of the energy denominator). However *this only works if the coupling is small* (compared to Δ). An example is provided by the SQUID model we considered in section 2.3(a), summarized in the effective Hamiltonian (41); recall that there $\omega_k^{\parallel}/\omega_k^{\perp} \ll 1$, and both are very small. This problem (which is related to that studied by Caldeira et al. [20]), is easily solved by writing $P_{11}(t)$ as

$$P_{11}(t) = \sum_{nm}^{\infty} (i\Delta_o)^{2(n+m)} \int_0^t dt_1 .. \int_{t_{2n-1}}^t dt_{2n} \int_0^t dt'_1 .. \int_{t'_{2m-1}}^t dt'_{2m} \mathcal{F}(Q, Q')$$

$$\mathcal{F}(Q,Q') = \prod_k \langle \hat{U}_k(Q_{(n)},t)\hat{U}_k^\dagger(Q'_{(m)},t)\rangle \tag{121}$$

where as before $Q_{(n)}(t)$ refers to a path of the central system containing n instanton jumps, but now $\hat{U}_k(Q_{(n)},t)$ refers to the time evolution of the k-th environmental *spin* (not oscillator!), under the influence of the central spin (cf eqtn. (77)). In the present case this is

$$\hat{U}_k(Q_{(n)},t) = T_\tau \exp\left\{-i\int_0^t \frac{ds}{2}\left[\omega_k^\perp \hat{\sigma}_k^z + Q_{(n)}(s)\omega_k^\parallel \hat{\sigma}_k^x\right]\right\} , \tag{122}$$

In general this formula for the "influence functional" is intractable (hence our alternative approach above!). However in the present case, expanding to 2nd order in $\omega_k^\parallel$, one gets at $T \gg \omega_k^\perp$ that

$$\mathcal{F}(Q,Q') = \exp\left\{-\sum_k \frac{(\omega_k^\parallel)^2}{8}\left|\int_0^t ds e^{2i\omega_k^\perp s}\left[Q_{(n)}(s) - Q'_{(m)}(s)\right]\right|^2\right\} , \tag{123}$$

Now in general this is still hard to solve; but for the SQUID it is simple, because even though $\omega_k^\parallel$ is extremely small, there are so many nuclear spins that the parameter γ defined by $\gamma^2 = [(\sum_k \omega_k^\parallel)^2/2]^{1/2}$ is not (for our previous model parameters $\gamma \sim 5\times 10^{-4}K$, similar to the total longitudinal bias produced by the nuclei), and so we expect $\gamma \gg \omega_k^\perp$, Δ_o (unless Δ_o is very large indeed). In this case it is easy to see that $\mathcal{F}(Q,Q') \sim e^{-\gamma^2(t-t')^2}$, ie., the influence functional decays (and dephases everything) before anything else can move (which, incidentally, is why we can ignore the nuclear bath dynamics in studying this case). In path integral language, kink-antikink pairs are closely bound (justifying a "NIBA"); more prosaically, we get a fast "motional narrowing" relaxation, with $P_{11}(t) = 1/2(1-e^{-t/\tau_R})$, and with the relaxation time given just by the integral

$$\tau_R^{-1} = 2\Delta_o \int_0^\infty e^{-\Gamma^2(t_2-t_1)^2} \equiv \frac{\sqrt{\pi}\Delta_o^2}{\Gamma} . \tag{124}$$

This result does not give out much hope for the experimental search for MQC in SQUIDs; it indicates coherence will be destroyed by nuclear spins in a time scale $\ll \Delta_o^{-1}$.

(ii) Influence of Spin Bath Dynamics:

Suppose we now go back to the full spin bath problem. What will be the influence of the small dynamical term $V_{kk'}^{\alpha\beta}$ on the results. The answer is that it can be profound, because in a situation where the central system is more or less blocked from transitions by the strong degeneracy blocking field, the weak residual bath dynamics can "scan" the bias energy until the system finds a resonance. In general this can happen with the simultaneous flipping of some bath spins, so that the resonance condition must include the change in energy of the nuclear bath as well; ie., we will end up summing over polarisation groups M. Thus, we have an *external* bias ξ, an internal longitudinal bias $\epsilon(t) = \epsilon + \delta\epsilon(t)$ (where $\delta\epsilon(t)$ represents the time dependent fluctuations coming from the internal spin bath dynamics- or, at higher T, from external couplings which drive the spin bath, ie., T_1- processes in the case of a nuclear bath), and the condition for resonance is that $\xi + \epsilon + M\omega_o + \delta\epsilon(t) \sim 0$. Moreover, the system should stay in resonance long enough to give the central spin a chance to tunnel!

This point is of very great practical importance; in fact it is obviously going to be entirely responsible for the existence of any dynamics in the low-T limit (as was discussed in some detail by Prokof'ev and myself, a while ago, in the context of nanomagnets [17]), since every other relaxation mechanism is frozen! What is found is that for a *single isolated nanosystem* (ie., a nanomagnet), in the Quantum regime , the new correlation function should be given by the simple incoherent relaxation form

$$P_{11}(t) = \sum_M w(T,M) \int x dx e^{-x^2} \sum_{\nu=-\infty}^{\infty} \int \frac{\varphi}{2\pi} F_{\lambda'}(\nu) e^{i2\nu(\Phi-\varphi)} \left[1+e^{-t/\tau_M(x,\varphi)}\right] , \tag{125}$$

where the relaxation time $\tau_M(x,\varphi)$ for the M-th polarisation group is

$$\tau_M^{-1}(\xi,\varphi) = 2\Delta_M^2(x,\varphi)/\pi^{1/2}\xi_o \tag{126}$$

The derivation is simple but too long to be repeated here. The parameter $\xi_o \sim N^{1/2}\delta\omega_o$ is a measure of the total energy spread of each polarisation group in the density of states (recall Fig.16). In the absence of any such spread, we have $\xi_o \sim N^{1/2}T_2^{-1}$, where the residual couplings between the bath spins give the total bath spectrum a linewidth ξ_o; here $T_2^{-1} \sim V_{kk'}^{\alpha\beta}$. We use this notation because in the case of a nuclear spin bath, this linewidth is parametrised (and in principle defined experimentally) by the transverse nuclear spin diffusion time T_2. Typically $T_2^{-1} \sim 10 - 100kHz$ in this case, although it can vary a lot. Notice that these bath fluctuations are in the high-T limit for anyone except a nanoKelvin experimentalist- we cannot make them go away! This result is easiest to understand in the limit where both degeneracy blocking and topological decoherence are absent (no nuclei coflip with $\vec{S}$, and $H_o(\vec{S})$ has axial symmetry). Then the averages collapse, and we get a unique relaxation rate given by $\tau_o = 2\Delta_o^2/\pi^{1/2}\xi_o$. This is of course just the overlap between initial and final states in a typical bias ξ_o, but be careful- it is a *relaxation rate*, involving the *irreversible* passage of the sytem from one state to the other (with an energy absorption $\sim \xi_o$ by the spin bath). The difference with a simple overlap integral becomes clear if the calculation is redone in a finite bias ξ (see Prokof'ev and Stamp[17]); then we get instead that

$$\tau(\xi) = \tau_0^{-1} e^{-|\xi|/\xi_o} \tag{127}$$

and we see that the rate falls of *exponentially* with the bias, as we take the system away from any possibility of reaching the resonant tunneling window, during its peregrinations over an energy range ξ_o. This formula is essential to understanding the low-T dynamics of any system described by a central spin model.

This concludes the very long discussion of canonical models and their dynamics. Enough, then, of Olympian theory- let's now get down to a merely Homeric epic, with some real physics (including experiments)!

4 Quantum Nanomagnetism

Before plunging into a new field, it is certainly worth asking what fundamentally new ideas may come out of it. So why study quantum nanomagnets? Here are a number of reasons, each of which will appeal to different tastes:

(i) They constitute a (very rich) testing ground for theorists interested in tunneling in complex systems; one can attack on many different fronts (numerical, analytical, experimental), and there is an almost inexhaustible variety of real examples to work with. The spin Hamiltonians for magnetic nanomolecules lend themselves well to studies in quantum chaos.

(ii) Much more than superconductors, they allow detailed study of the crossover between quantum and classical properties of physical systems, as one changes the system size, temperature, and coupling constants. Understanding this crossover is not only important for the resolution of the notorious quantum measurement problem (and a better understanding of how decoherence works at this scale); it will also have enormous practical consequences in the next few decades, with the coming of quantum nanomagnetic devices. Unravelling the relevant physics promises to be one of the most exciting challenges in condensed matter physics.

(iii) In the field of magnetism itself, this crossover is of central interest; it has been highlighted by the discovery of high-temperature superconductors, heavy fermions, 1-dimensional magnets (spin chains), and a variety of "frustrated magnets", many of which show tendencies towards "spin liquid" behaviour (ie., quantum-disordered magnetic states), even at the macroscopic level. More recently there has been the discovery, (in Mn-perovskite layered magnets, having the same structure as the high-Tc La-perovskite superconductors), of "collosal magnetoresistance" phenomena; and the preparation of "spin ladder" materials, which are essentially sets of coupled spin chains. There is no question that the study of these quantum magnetic materials (and their remarkable superconducting properties when doped), is the central focus of most of the condensed matter research around the world at present. It is quite extraordinary that almost all of these materials (including the magnetic molecules to be discussed presently) are *Transition Metal Oxides*, at or near a metal-insulator transition, with magnetism controlled by superexchange, and great sensitivity to doping. Why all his should be, is one of the great mysteries in physics at the end of the 20th century.

(iv) Some magnetic systems offer the prospect of observing genuinely macroscopic quantum phenomena, in the sense discussed by Leggett [41]. The most likely possibility here is in the tunneling of domain walls (or possibly their quantum nucleation), which at least theoretically should occur on a large scale [42], particularly in systems isotopically purified of nuclear spins [28]. Some promising experiments in this direction have been done already [10,11].

From this short list we see that nanomagnets appeal to physicists (and even philosophers of science) interested in the mysteries of quantum mechanics and the measurement problem; and at the same time to industrialists and venture capitalists, investing in the technology of the next century. This curious mixture is not infrequent in subjects at the frontier of science.

In the following I will give an overview of 2 important themes in this relatively new field, aimed at the non-specialist. These show (sometimes rather dramatically) how the concepts discussed in Chapters 2 and 3 can be applied to real physical systems. Incidentally, although some of the experiments are understood theoretically, others are not, and there are some intriguing outstanding mysteries. I will start with the now quite extensive results (theoretical and experimental) on quantum relaxation in magnetic micromolecules and grains; I will argue that these are fairly well understood. Then we look at "macroscopic coherence" experiments in ferritin macromolecules, where the gap between theory and experiment is still controversial. I do not discuss other areas (eg., domain wall tunneling), simply because of a lack of space.

4.1 Resonant Tunneling Relaxation

Let us start with a set of experiments in which no coherence exists at all. Considerable press (some of it highly misleading) has been devoted recently to a series of experiments [1-7] which have shown, with steadily increasing clarity, the role of tunneling in the relaxation of magnetisation in an initially polarized sample of magnetic molecules. These experiments did not appear in a vacuum, but rather in the context of previous experimental efforts to see tunneling in magnetic particles and in the biological macromolecule ferritin (see section 4.3 below). The possibility of doing tunneling relaxation experiments on molecule crystals of large-spin nanomolecules relies essentially on recent advances in the preparation of such molecules by molecular chemists .

There have also been a large number of tunneling relaxation experiments on nanomagnetic particles. I shall make no attempt to review all of these; however some of them are very interesting, specially as they indicate what remarkable possibilities exist for the future.

4.1(a) MAGNETIC NANOMOLECULES- STRUCTURE and EXPERIMENTS

I will only touch the surface here of a huge field, which was for many years the purview of chemists and the occasional molecular physicist. Reviews of some of the chemistry are given by, eg., Gatteschi et al [70]; and I would particularly recommend the book of Kahn [71]. The low-T experiments are too new to have been reviewed.

(i) Spin Structure at Low Energies: The first experiments were done on the "Mn-12" system, first in powder samples and then in large single crystals (refs.[1-6]). Early theoretical studies of this molecule, by the Grenoble-Firenze group, led to the proposal of a "giant spin" Hamiltonians $H_o(\vec{S}) = ({}^{\|}H_o^{(2)} + {}^{\perp}H_o^{(4)})$, where

$$
{}^{\|}H_o^{(2)} = -(D/S)S_z^2 \tag{128}
$$

$$
{}^{\perp}H_o^{(4)} = -(K_4/S^3)\left[S_+^4 + S_-^4\right] \tag{129}
$$

and in which S = 10, D $\sim$ 6-7K, and K_4 is hard to determine (present estimates for K_4/D range from 0.005 to 0.03). Magnetisation and specific heat measurements indicate that, as far as the low-T dynamics is concerned, this is a good approximation for $T \leq 30K$. The Mn-12 molecule is of course very complicated, and early numerical attempts to establish the low-energy H_{eff} did not succeed in predicting the S=10 ground state, partly because no magnetic anisotropy was included. The difficulty of the problem can be seen by considering Fig.18(a) below, which shows the important exchange couplings, acting between the 12 different Mn ions in the molecule. In reality these couplings are of course anisotropic, and there will also be single-ion anisotropy terms (cf eqn (10)). Because the molecule lacks inversion symmetry, one also expects "Dzyaloshinskii-Moriya" terms, of form $V_{DM} = \sum_{<ij>} \vec{D}_{ij} \cdot (\vec{s}_i \times \vec{s}_j)$, acting between spins.

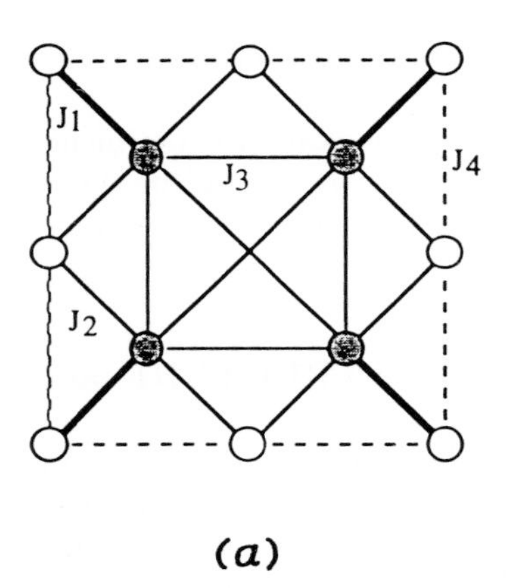

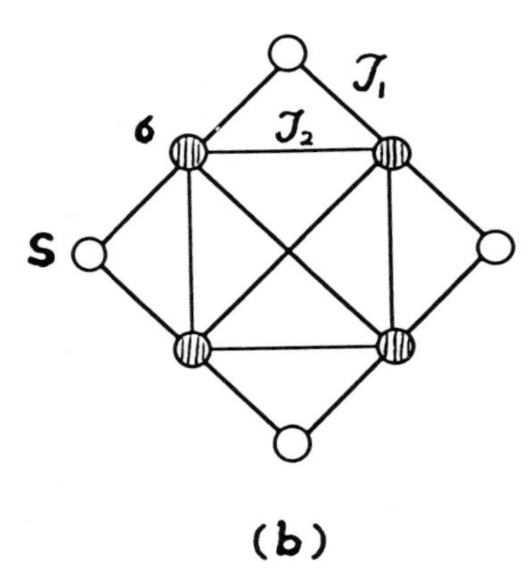

FIG. 18: The structure of the exchange couplings in the Mn_{12}, after Sessoli et al. [73]. In (a) is shown the couplings between the 12 Mn ions; whereas in (b) we see the couplings between the 8 spins in the reduced model described in the text, where the very strongly coupled "diagonal" spin pairs (with estimated AFM coupling $J_1 \sim -215K$ in (a)) are coupled to form single spin-1/2 at low energy. Sessoli et al. also estimate $J_2 \sim J_3 \sim -85K$ and $|J_4| \leq 45K$. The renormalised couplings in (b) are discussed in the text.

Thus any serious attempt to diagonalize the full spin Hamitonian (in a Hilbert space of dimension $5^8 4^4 = 10^8$), with a large number of couplings which are very difficult to determine, is rather pointless. A different tactic was adopted by Tupitsyn et al, [72], following the early work of Sessoli et al. [73], in an effort to explore the region $T \leq 150K$. It was assumed that the "diagonal" exchange couplings in Fig.18(a) are so strong that at low T, the Mn^{+3}/Mn^{+4} ionic pair is locked into a spin-1/2 state. This reduces the problem to an 8-spin one, with only 10^4 degrees of freedom (Fig.18(b)). At this point one writes down an effective Hamiltonian for this 8-spin system, which is consistent with the known molecular symmetry, and which captures both exchange and anisotropy. The simpliest such form, which ignores any time-reversal symmetry breaking (which could come from Dzyaloshiniskii-Moriya interactions) can be written as follows:

$$\hat{H}_o = \mathcal{J}_1 \sum_{k=1}^{4} \vec{S}_k \cdot \vec{\sigma}_k + \mathcal{J}_2 (\sum_{k=1}^{3} \vec{\sigma}_k \cdot \vec{\sigma}_{k+1} + \vec{\sigma}_4 \cdot \vec{\sigma}_1) + K_{\parallel} \sum_{k=1}^{4} S_k^z \cdot \sigma_k^z, \qquad S = 2, \sigma = \frac{1}{2}, \tag{130}$$

where we couple 4 spin-2 spins $\vec{S}_k$ and 4 spin-1/2 spins $\vec{\sigma}_k$ (the latter should not be confused with nuclear spins!)

We notice immediately that we have not *derived* this Hamiltonian. In fact if the reader is getting used to the idea of low-energy effective Hamiltonians, he/she may well be asking - from what starting point *is* one supposed to derive H_o? The answer is of course that the only reasonable starting point is one where we have some way of determing the input parameters experimentally. One might of course attempt the very ambitious task of calculating the various couplings in the 12-spin $Mn_{12}O_{12}$-ac system, starting from an underlying Anderson lattice model for the molecule. From a microscopic viewpoint this certainly makes sense - we are dealing with a hydrated transition metal oxide, to which the Anderson-Mott-Hubbard ideas apply rather directly. But the problem is obvious - with no knowledge of the parameters in such a model, we can do no more than guess what superexchange couplings they might lead to, between the Mn ions. In fact, the values generally assumed for these couplings are no more than educated guesses made by the chemists.

The natural result of this line of reasoning is to treat the above 8-spin Hamiltonian as our starting point and attempt to *deduce* the values of the parameters $\mathcal{J}_1$, $\mathcal{J}_2$, and $K_{\parallel}$ from experiments. The way this was done by Tupitsyn et al. was simple. They first calculated the magnetisation M_z in a longitudinal field H_z, including all 10^4 levels of the 8-spin model. The results for $M_z(H_z, T)$ were compared with a large number of experimental results on a single oriented Mn_{12} - ac crystal, taken at different T and H. The parameters $\mathcal{J}_1, \mathcal{J}_2$, and $K_{\parallel}$ were then adjusted to give the best possible fit to these experiments; it was found that this gave $\mathcal{J}_1$ = -85K, $\mathcal{J}_2$ = 55K, and $K_{\parallel}$ = 7.5K.

To test this model, it was then used to predict the results of a *transverse* magnetisation measurement, ie., the magnetisation $M_x(H_x, T)$ was calculated over a wide range of H_x and T values. These were then compared with transverse susceptibility measurements. This is a sensitive test of the theory, because the transverse field mixes together states which have little to do with each other in a longitudinal field. The results were very good for $T \leq$ 150K, undoubtedly because above this temperature, the "diagonal couplings" are too weak to prevent the break-up of the 4 fake spin-1/2 pairs, into their spin-2 and spin-3/2 constituents. Thus up to 150K this sort of exercise is very useful, whereas the spin-10 "giant spin" Hamiltonian breaks down seriously above 30K. With further thermodynamic measurements, eg., measurements of $C_v(T,H)$ (with the phonon contribution subtracted off) it should be possible to refine these models even more.

I should emphasize, however, that such calculations, although they can in principle give us a good understanding of the gross structure and origin of the eigenstates up to $\sim 100K$, are utterly useless when it comes to understanding the *tunneling*. This is because, as we have already seen in Chapter 2, the tunneling matrix elements are not only *exponentially smaller* than the energy scale relevant to these calculations (ie., composed to anisotropy energies); they also depend exponentially on any change in these energy scales! Thus any uncertainty in our knowledge of the correct effective Hamiltonian (whether it be at the level of the giant spin $H_o(\vec{S})$, or at the more microscopic level just discussed) will be magnified exponentially when it is used to try and determine the tunneling matrix elements. I shall return to this point again below, in discussing the interpretation of the experiments.

(ii) Quantum Relaxation Experiments: Let us now proceed directly to the low-T experimental results on the Mn-12 and Fe-8 systems, (ie., for $T \leq 5K$). We begin with Mn-12, for which many experiments have been done in recent years. The clearest of these are those by Thomas et al. [5] on a single crystal. The essential results found by the various authors were

(i) When $T > 2K$, the relaxation of an initially polarized sample of molecules was found to decay roughly exponentially (although more recent measurements by these authors also find non-exponential relaxation around the resonances at high T). The remarkable thing here was the series of resonant increases in the experimental relaxation rate $\tau^{-1}(H,T)$ around fields nH_1, where $n = 0, \pm 1, \pm 2, \ldots$, and $H_1 \sim 0.44T$. The crucial point is that if we add a longitudinal field bias term $-g\mu_B SH_z$ to (4.1), then the resonant fields correspond precisely to "level crossings", ie., to fields at which eigenstates $|m\rangle$ of (4.1) become degenerate with eigenstates $|-m+n\rangle$ (here we define eigenstates $|m\rangle$ via the equation ${}^{\|}H_o^{(2)}|m\rangle = \varepsilon_m^o|m\rangle$, where $\varepsilon_m^o = -D_m^2/S$). The existence of resonant increases in τ^{-1} at these fields, and a proposed explanation in terms of thermally activated resonant tunneling, was first given by Novak and Sessoli [2] (see also Barbara et al. [3]).

(ii) Below 2K, the relaxation appears to become rather T-dependent in the

Mn-12; moreover, recent unpublished data shows it is not at all exponential, a fact emphasized in refs. [1,4] (but not ref. [5]).

(iii) The resonant peaks in $\tau^{-1}(H,T)$ appear to be Lorentzian around the H_n. Their half-width is rather large (roughly $0.1T$).

The Mn_{12} experiments were followed by some very striking results [7] obtained in the "Fe_8" molecule. In many ways this molecule is very similar to Mn-12; a giant spin is again formed via -oxo bonds, in a typical example of antiferromagnetic superexchange in a transition metal oxide. In this system, all eight Fe ions are in a spin-5/2 valence state, which align again in a spin-10 ground state, which again is described to first approximation by an easy axis term ${}^{\|}H_o^{(2)}$ (eg. (4.1)); in this system $D \sim 2.7K$. The lowest transverse term which breaks rotational symmetry about S_z must take the form ${}^{\perp}H_o^{(2)} = -(E/S)[S_x^2 - S_y^2]$ (a term disallowed in Mn_{12}, which crystallizes in tetragonal form); just as in Mn_{12}, ${}^{\perp}H_o/{}^{\|}H_o \ll 1$.

The relaxation results in Fe_8 molecular crystals share several features in common with those of Mn_{12}. Again, one sees resonantly enhanced relaxation at fields H_n; but now the relaxation rate $\tau^{-1}(H,T)$ increases over 4 orders of magnitude around the zero field resonance (which, incidentally, is shifted from $H = 0$ by several hundred Gauss)! A particularly valuable feature of the Fe_8 system is that one may see reasonably rapid relaxation down to very low T; in fact, Sangregorio et al. [7] followed it down to 70mK. Very direct evidence appears in these experiments of a crossover, below $T \sim 360 - 400mK$, to a regime of quantum relaxation. The relaxation $M(t)$ of the magnetisation M in an initially polarised sample is completely T-independent below 360 mK. It is very noticeably non-exponential; in fact the authors attempted to fit it using a stretched exponential.

How we might understand these results? Before looking at the relaxation itself, I think we need some cautionary remarks concerning the use of the giant spin Hamiltonian in discussing this sort of experiment. These concern the relationship between the values of the tunneling matrix elements Δ_m, between levels $|m\rangle$ and $|-m\rangle$ of the longitudinal part of the giant spin Hamiltonian. In what follows we will treat a matrix element like Δ_{10}, coupling the 2 lowest levels of ${}^{\|}\mathcal{H}_o^{(2)}$ in Mn_{12} and Fe_8, as an *independent parameter.* One can of course calculate Δ_{10} if the weak anisotropy couplings are known. Thus a 2nd-order (in S^+, S^-) anisotropy term ${}^{\perp}\mathcal{H}_o^{(2)} = E(S_x^2 - S_y^2)$ yields

$$\Delta_{10} = \frac{4S^2 E^S (2S-1)!!}{(2D)^{S-1}(2S)!!} \approx \frac{4S^{3/2}}{\sqrt{\pi}} E(E/2D)^{S-1} \,. \tag{131}$$

Using $E = -0.046\ K$ for Fe_8 molecules [7] then gives $\Delta_{10} \sim 10^{-9}\ K$! On the other hand the 4th-order anisotropy term ${}^{\perp}\mathcal{H}_o^{(4)} = B(S_+^4 + S_-^4)$ yields

$$\Delta_{10} = \frac{(BS^2)^{S/2} S^2 (2S)!}{(D)^{S/2-1} 16^{[S/4]} [S!!]^2} = BS^2 (BS^2/D)^{S/2-1} \frac{(2S)!}{S^{S-2} 16^{[S/4]} [S!!]^2} . \quad (132)$$

In the Mn_{12} system the ${}^{\perp}\mathcal{H}_o^{(2)}$ anisotropy term is absent, and the parameter $g_4 = (BS^2)/D$ is rather small (estimates range from $g_4 \approx 0.03$ down to $g_4 \sim 0.005$. For $g_4 = 0.03$ one gets $\Delta_{10} \approx 10^{-10}\ K$.

From these examples we learn that even rather small higher-order anisotropy terms may determine high-order tunneling matrix elements. This is very simply because, as already noted in section 2, a higher-order transverse coupling (eg., a coupling of 8-th order in S^+, S^-) appears raised to a lower power in Δ_{10} than a 2nd- or 4th- order coupling term, and so can play an important role even if it is much smaller than a 2nd- or 4th-order term. Consequently, in, eg., Fe_8, Δ_{10} may be considerably larger than Eq.(131), because of such higher-order terms (up to the twentieth order in S^+, S^-, in fact!). They are almost impossible to obtain experimentally, since their contribution to the transitions studied by EPR spectroscopy is likely to be negligible. All I am doing here, of course, is repeating the warning made earlier about the exponential dependence of tunneling amplitudes on microscopic parameters, in a different way.

Note that as a corollary to this argument, one has to be particularly cautious in predicting the effects of the applied weak transverse field $H_\perp$ on Δ_{10}, since this field has a contribution of order $H_\perp (H_\perp/D)^{2S-1}$, ie., to the $2S$-th power in the field! Essentially this means that its direct effect at low T is very small- thus, the effect of any applied transverse field on Δ_{10} is negligible unless H is of order *Tesla*. Likewise, the effect of *static* internal transverse fields (such as transverse hyperfine or dipolar fields when $kT \ll D$) on Δ_{10} is negligible.

4.1(b) THEORETICAL INTERPRETATIONS

To deal with relaxation we must take proper account of environmental effects. Historically the first attempt was made by Villain and co-workers, even before resonant tunneling was established [53,54]. They considered the coupling between a giant spin, described by the Hamiltonian in equations (4.1) and (4.2), and a bath of phonons. The physics of spin-phonon couplings is well understood at the microscopic level [52], and Politi et al. [54] gave a treatment of the phonon-mediated tunneling relaxation of a giant spin $\vec{S}$, for the particular case of a single isolated Mn_{12} molecule. However this theory disagreed with the experiments. More recently they have extended it, and some other authors [74,75] have given similar treatments.

In what follows I will lay considerable stress on the fact that none of these treatments is capable of explaining the low-T relaxation in the *Quantum* Regime, for the simple reason that at these temperatures the phonons can play no role. It can be then more or less *deduced* from the experiments that the relaxation must be mediated by nuclear spins, which constitute the only dynamic environment remaining in the quantum regime. In this sense the low-T results are a kind of "smoking gun" for the role of the spin bath in the quantum dynamics of the system. However it is clear that at higher T both phonons and nuclear spins must be involved (as well as the rapidly fluctuating magnetic dipolar fields).

(i) <u>Phonon-mediated Relaxation</u>: Let us briefly review this theory and its results. This is useful because despite the fact that the calculations of Villain et al.[53,54] are correct, their results disagree strongly with the experiments, particularly at low T and low H (ie., in the Quantum regime), and it is important to understand why.

Consider first the form of the coupling between $\underset{\sim}{S}$ and phonons (parametrized, as in Chapter 2, by operators $b_{\vec{q}}, b^{+}_{\vec{q}}$, frequencies $\omega_q \sim c_s q$, and a Debye temperature Θ_D). For a strongly easy-axis "Ising-type" anistropy, there is the transverse term already given previously (cf. eqtn. (52)). That the coupling should be proportional to Ω_o, the "bounce frequency" introduced in section 2, is easily shown using instanton methods where applicable [17], but is also fairly obvious from dimensional arguments. For strongly anisotropic "Ising-like" molecules like Mn_{12} or Fe_8, $\Omega_o \rightarrow D$ (recall that Ω_o is roughly the energy separation between the 2 lowest levels of the symmetric van Hemmen/Sutö Hamiltonian, and the next levels). The same result is found using standard spin-phonon theory [52].

There are of course other spin-phonon coupling terms, but before asking why this one is the most important, let us see what it does to the dynamics of $\vec{S}$. Consider 2 possible transitions between eigenstates $\mid m >$ of the longitudinal part ${}^{\parallel}H_o^{(2)}$ of the giant spin Hamiltonian (here $\mid m >=\mid S >, \mid S-1 >, \ldots \mid -S >$). These are mediated simultaneously by the tunneling term ${}^{\perp}H_o(s)$, and by the coupling to the phonons. Now the crucial point here is that a non-diagonal coupling like the one just given allows a transition $\mid m > \rightarrow \mid -(m \pm 1) >$, provided $\mid m >$ and $\mid -m >$ are in resonance (ie., provided $\mid \varepsilon_m - \varepsilon_{-m} \mid \leq \Delta_m$, where Δ_m is the tunneling matrix element between $\mid m >$ and $\mid -m >$, coming from ${}^{\perp}H_o$, and ε_m is defined by ${}^{\parallel}H_o \mid m > = \varepsilon_m \mid m >$). This inelastic transition goes at a rate proportional to ξ^3,where $\xi = (\varepsilon_m - \varepsilon_{-m-1})$ is the relevant energy difference between initial and final states, for Debye phonons. This standard result [52] comes from the available phase space for phonons. A simple calculation[54,17] gives a non-diagonal relaxation rate

$$\tau^{-1}(\varepsilon, T) \sim e^{-\varepsilon_m/kT} \frac{S^2\Delta_m^2}{\Theta_D} \left(\frac{\varepsilon}{\Theta_D}\right)^3 \tag{133}$$

at low T. On the other hand a *diagonal coupling* (ie., one not involving operators S_x, S_y) will have a much smaller available phase space. This is why the non-diagonal coupling dominates.

Suppose now that we are in the low-T limit; this arises precisely when $kT \leq \Omega_o/2\pi$. The crossover occurs experimentally for $T_c \sim 2K$ in Mn_{12}, and $T_c \sim 0.4K$ in Fe_8. Below T_c, only the 2 lowest levels are involved. In this case a more accurate formula for the phonon relaxation rate is[17]:

$$\tau^{-1}(H, T) \sim \frac{S^2\Delta_{10}^2}{\Theta_D} \left(\frac{\xi}{\Theta_D}\right) \coth(\xi/2kT) \tag{134}$$

Now we immediately notice 2 things about this result. First, $\tau(H,T)$ is *extremely long* for low fields (ie., where the bias $\xi = (\varepsilon_{10} - \varepsilon_{-10})$ is generated only by internal bias fields like nuclear hyperfine fields or intermolecular dipolar fields - in this case $\xi \sim 0.1 - 1K$ only). Thus for $S = 10$, and for the values of Δ_{10} estimated for Mn_{12} and Fe_8 (for which Δ_{10} is almost certainly less than $10^{-8}K$, perhaps much smaller), one finds that $\tau(H,T)$ at low T and low H becomes longer than the Hubble time!

The 2nd thing we notice is that around $H = 0$, this sort of spin-phonon theory predicts a rapid increase ($\sim H^3$) of $\tau^{-1}(H)$ as $|H|$ is increased.

Unfortunately these 2 results are flatly contradicted by the experimental results, which show (for Fe_8 and Mn_{12}) a very sharp resonant *maximum* in $\tau^{-1}(H)$ around $H \sim 0$; in the middle of this the relaxation rates can be inverse seconds only (although, as already noted, the short time relaxation is far from exponential in the low T regime).

Thus we see that the low-T results are quite fatal to a theory which only involves a phonon environment. However, this is by no means all that one can deduce from these low-T results.

(ii) Nuclear Hyperfine, and Molecular Dipole fields: Since, at low T, phonons are irrelevant to the relaxation, let us consider the problem solely in the presence of nuclear and dipolar couplings. Below T_c the problem then reduces to an effective Hamiltonian:

$$H = \frac{1}{2}\sum_{ij} V_{ij}^{(d)} \tau_z^{(i)} \tau_z^{(j)} + \sum_i \Delta_{10}\tau_x^{(i)} + \sum_{ik} V^{(N)}(\tau_z^{(i)}, \vec{I}_k) + H^{NN} \tag{135}$$

where the first term describes the *static* dipolar-dipolar interactions between molecules, the second describes tunneling, the third couples magnetic molecules to nuclear spins $\{\vec{I}_k\}$, and the last term describes interactions between the nuclear spins. This effective Hamiltonian operates in the subspace of the two lowest levels of each molecule; we choose the basis set to be $|S_z = \pm S\rangle$; τ_z and τ_x are Pauli matrices, and $\{i\}$, $\{j\}$ label molecular sites. Only the longitudinal dipolar interaction appears; transverse "flip-flop" processes are exponentially small in the parameter $(\varepsilon_9 - \varepsilon_{10})/kT$.

Now it is crucial that if we also treat the hyperfine fields as static, we get a really gross contradiction with experiment. As already mentioned, the bias $\xi = (\varepsilon_{10} - \varepsilon_{-10})$ due to the combined nuclear and dipolar couplings, is not enough for phonons to act; moreover, the transverse fields coming from these couplings are far too small to change Δ_{10}. However the most crucial point is simply this: all but a very tiny fraction of molecules are blocked from any transitions at all, because $\xi \gg \Delta_{10}$ for almost all molecules. From this standpoint it is quite incredible that there is any relaxation at all, in the low-T limit (recall that at the lowest temperatures in the Fe-8 experiments, kT was nearly 500 times less than the barrier height, and some 100 times smaller than the energy $\sim 2\mathcal{D}$ required to excite any but the 2 lowest levels of each molecule!). **Nevertheless one sees rapid relaxation - why?**

I have very recently argued, in a paper with N.V. Prokof'ev [76], that the explanation can in fact be *deduced* from the experiments. The basic point is that no relaxation at all can proceed unless we give the environment some dynamics; as we have just seen, a static Hamiltonian won't work. But at $70mK$ there is only one source of such dynamics, as already seen in Chapters 2 and 3; it comes from the transverse "spin diffusion" fluctuations in the hyperfine field, which have a typical frequency $T_2^{-1} \sim 10 - 100kHz$. Unlike phonons, or intermolecular dipolar flip-flop transitions, or nuclear T_1 transitions (all of which disappear in the low-T limit, and are exponentially small in the ratio $\mathcal{D}/kT$), the nuclear T_2 dynamics persists down to temperatures in the nK range, and is fairly independent of T in the present experimental temperature range.

We have already seen in Chapter 3 what these rapid hyperfine fluctuations can do. For the present problem, the hyperfine coupling is sufficiently weak that no nuclear spins are likely to be flipped when $\vec{S}$ flips. Thus a molecule near resonance (ie., where ξ is small), will relax *incoherently* between states $|\,10>$ and $|\,-10>$ (or vice versa) at a rate given by

$$\tau_N^{-1}(\xi) \approx \tau_0^{-1} e^{-|\xi|/\xi_o} \,. \tag{136}$$

$$\tau_0^{-1} \approx \frac{2\Delta_{10}^2}{\pi^{1/2}\Gamma_2} \ . \tag{137}$$

where we have assumed a fluctuating bias field $\xi(t) = \xi + \delta\xi(t)$, for which the fluctuations $\delta\xi(t)$ extend over a range $\xi_o \sim T_2^{-1}$ (compare eqtn.(127)). The basic physics is simple; not only do the fluctuations greatly widen the "resonance window" width (from Δ_{10} to ξ_o, an increase of several orders of magnitude), they also make the transitions *inelastic* (and therefore cause irreversible relaxation). Thus, at very short times, molecules with a bias within ξ_o of zero are sucked into the resonance window, where they undergo irreversible inelastic relaxation.

At this point the long-range dipolar fields come in. Once a fraction ξ_o/W of molecules has relaxed via this mechanism (where $W \sim 0.1 - 1K$ is the total spread in longitudinal bias fields around the sample), the adjustment of the long-range dipolar fields caused by these transitions is quite enough to bring more molecules into the resonance window. At first glance this seems like a very complex problem; however, we can treat it using standard kinetic theory methods.

We begin by defining a normalised 1-molecule distribution function $P_\alpha(\xi, \vec{r}; t)$, with $\sum_\alpha \int d\xi \int d\vec{r} P_\alpha(\xi, \vec{r}; t) = 1$. It gives the probability of finding a molecule at position $\vec{r}$, with polarisation $\alpha = \pm 1$ (ie., in state $|S_z = \pm S\rangle$), having a bias energy ξ, at time t. Molecules having bias energy ξ undergo transitions between $|S_z = S\rangle$ and $|S_z = -S\rangle$ at a rate given by (136). Flipping these molecules then changes the dipolar fields acting on the whole ensemble, bringing more molecules into near (or away from) resonance, and leading to a self-consistent evolution of $P_\alpha(\xi)$ in time. The general solution of this problem requires a kinetic equation for $P_\alpha(\xi, \vec{r}; t)$.

To derive a kinetic equation for $P_\alpha(\xi, \vec{r}; t)$, we again assume that dipolar and hyperfine fields are frozen (apart from the nuclear T_2 fluctuations just discussed), *unless* a molecule flips. All kinetics then come from these flips, along with the resulting adjustment of the dipolar field. We may then derive a kinetic equation in the usual way, by considering the change in P_α in a time δt, caused by molecular flips, at the rate $\tau_N^{-1}(\xi)$, around the sample. This yields

$$\begin{aligned} \dot{P}_\alpha(\xi, \vec{r}) = \quad & -\tau_N^{-1}(\xi)[P_\alpha(\xi, \vec{r}) - P_{-\alpha}(\xi, \vec{r})] \\ & - \sum_{\alpha'} \int \frac{d\vec{r}\,'}{\Omega_0} \int \frac{d\xi'}{\tau_N(\xi')} \Big[P^{(2)}_{\alpha\alpha'}(\xi, \xi'; \vec{r}, \vec{r}\,') \\ & - P^{(2)}_{\alpha\alpha'}(\xi - \alpha\alpha' V_D(\vec{r} - \vec{r}\,'), \xi'; \vec{r}, \vec{r}\,') \Big] \ , \end{aligned} \tag{138}$$

where $P^{(2)}_{\alpha\alpha'}(\xi, \xi'; \vec{r}, \vec{r}''; t)$ is the two-molecule distribution, giving the normalized joint probability of finding a molecule at site $\vec{r}$, in state $|\alpha\rangle$ and in a bias ξ, whilst another is at $\vec{r}\,'$, in state $|\alpha'\rangle$, and in a bias ξ'. $P^{(2)}$ is linked to higher multi-molecule distributions by a BBGKY-like hierarchy of equations. The first term on the right-hand side of (138) describes the local tunneling relaxation, and the second non-local term (analogous to a collision integral) comes from the change in the dipolar field at $\vec{r}$, caused by a molecular flip at $\vec{r}'$; the dipolar interaction $V_D(\vec{r}) = E_D[1 - 3\cos^2\theta]\Omega_0/r^3$, where Ω_0 is the volume of the unit molecular cell, and $\int d\vec{r}\,'$ integrates over the sample volume.

As discussed in our paper, with this kinetic equation we can make a number of very clear predictions concerning the low-T relaxation in these magnetic molecular crystals. These include (i) The prediction that an initially polarized sample will, at short times, relax according to a universal $\sqrt{t}$ law (where t is the elapsed time); the sample magnetisation will obey

$$M(t) = \left[1 - (t/\tau_{short})^{1/2}\right] M_o \tag{139}$$

where M_o is the saturated magnetisation, provided $1 - (M/M_o) \ll 1$. (ii) The timescale τ_{short} depends radically on the sample shape. For an ellipsoidal sample one finds

$$\tau_{short}^{-1} = \frac{\xi_0}{E_D\tau_0} \frac{32\pi}{3^{5/2}(c^2 + 16\pi^2/3^5)} \,. \tag{140}$$

where the constant c is given by magnetostatic theory; for a prolate spheroid, one finds

$$c = (2\pi/3)[a^4 + a^2 - 3a\sqrt{a^2-1}\ln(a + \sqrt{a^2-1}) - 2]/(a^2-1)^2 \tag{141}$$

where a is the ratio of the longitudinal axis to its perpendicular; analytic formulas can be found for any ellipsoid.

The analytic simplicity of these results arises because in the short-time limit, $P^{(2)}(\xi, \xi'; \vec{r}, \vec{r}'; t)$ is factorizable, and because for an ellipsoid, the initial demagnetisation field is homogenous (a standard result of magnetostatic theory). For a sample of arbitrary shape, we still get $\sqrt{t}$ relaxation, but in this case τ_{short} above is replaced by

$$(\tau_{short}^{(inh)})^{-1} \sim \xi_0 N(0) \tau_{short}^{-1} \,, \tag{142}$$

where $N(0) = \int d\vec{r} \sum_\alpha P_\alpha(\xi = 0, \vec{r}; t = 0)$ is the *initial* "density of states" for the dipolar field distribution, integrated over the whole sample, at bias $\xi = 0$; typically $N(0) \sim 1/E_{Dm}$, where E_{Dm} is the average demagnetization field.

Notice that the prediction for these timescales shows that they depend essentially on both the nuclear fluctuation timescale T_2 *and* on the sample shape. Once the condition $1 - (M/M_o) \ll 1$ is relaxed, we must worry about the non-factorisability of $P^{(2)}$, and analytic calculations are no longer so easy. However it is easy to both verify the above results for short times, and to extend them to longer times, using Monte Carlo (MC) simulations of the relaxation in samples of various shapes [76]. It is also interesting to see how the 1- molecule distribution function $P_\alpha(\xi, \vec{r}; t)$ itself relaxes; this is shown in Fig.19, at short times, for a spherical sample.

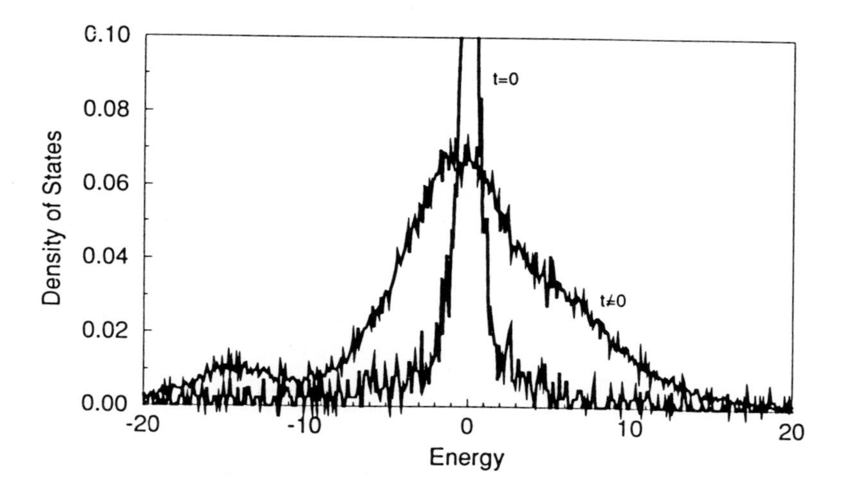

FIG. 19: Monte-Carlo (MC) simulation of the time evolution of the time-dependent density of states $N_\alpha(\xi, t) = \int d\vec{r} P_\alpha(\vec{r}, \xi; t)$ of the unflipped spins in a spherical sample of diameter 50 sites, with spins arranged in a cubic lattice. Energy and density of states use units where $\xi_0 = 1$ and $E_D = 20$, and results are shown for $\alpha = +1$ at time $t = 0$ (for an initially polarized sample) and at $t = 0.1\tau_0$. The Lorentzian shape is distorted by lattice effects at high energy (giving the hump at $E = -15$). The noise, and the finite width of the δ-function at $t = 0$, are finite size effects.

Why is the short-time behaviour so simple (and *universal*)? Essentially because we have a small parameter (the fraction of flipped spins). These spins behave like a set of dilute dipoles, creating a solution to the kinetic equation which is in fact an old friend to anyone who has worked in NMR. Consider first for simplicity the case of an ellipsoidal sample, so that the initial field is homogeneous. Then at short times the solution to our kinetic equation, if the normalised magnetisation at $t = 0$ is unity, just becomes

$$\dot{M}(t) = -M(t)\frac{2}{\tau_0}\int d\xi e^{-|\xi|/\xi_0}\frac{\Gamma_d(t)/\pi}{[\xi - E(t)]^2 + \Gamma_d^2(t)} \ . \tag{143}$$

where the field distribution in an ellipsoidal sample is nothing but a Lorentzian distribution (up to a high energy cut-off E_D) related to that found by Anderson for dilute *static* dipoles:

$$\begin{aligned} P_\alpha(\xi) &= \frac{1+\alpha M(t)}{2}\,\frac{\Gamma_d(t)/\pi}{[\xi - \alpha E(t)]^2 + \Gamma_d^2(t)} \ ; \\ \Gamma_d(t) &= \frac{4\pi^2}{3^{5/2}} E_D(1 - M(t)) \ ; \qquad (144) \\ E(t) &= cE_D(1 - M(t)) \ , \qquad (145) \end{aligned}$$

where c is the sample shape dependent coefficient defined above, and $E(t)$ is the time-dependent internal field (assuming here that the external field is zero).

This explains the universality of the $\sqrt{t}$ behaviour for an ellipsoidal sample (and note the distinctive and strong dependence of τ_{short} on the sample shape- this should be testable experimentally). But what about samples of arbitrary shape? Then the problem becomes essentially inhomogeneous. We thus return to the kinetic equation (138), and notice that if the demagnetisation varies on a length scale much greater than the average distance between flipped spins, then (143) is simply modified to

$$\dot{M}(\vec{r},t) = -M(\vec{r},t)\frac{2}{\tau_0}\int\frac{d\xi}{\pi}\frac{\Gamma_d(\vec{r},t)e^{-|\xi|/\xi_0}}{[\xi - E(\vec{r},t)]^2 + \Gamma_d^2(\vec{r},t)} \ . \tag{146}$$

where $\Gamma_d(\vec{r},t)$ and $E(\vec{r},t)$ are defined in terms of $M(\vec{r},t)$ analogously to (144) and(145); the solution is then identical to (139) except that τ_{short} is modified to $\tau_{short}^{(inh)}$. You can think about this in a geometrical way. In an ellipsoidal sample, with initially uniform field, the molecules are either initially in the "resonance window" (of width $1/\xi_o$) or they are not. If they are, then the rapid decrease in the instantaneous relaxation rate, giving the square root form, comes about because the molecules are being pushed away from resonance as the inhomogeneous Lorentzian fields develop. Only a Lorentzian shape gives a $\sqrt{t}$ relaxation; and when the flipped spins become dense enough, the Lorentzian form must break down (in fact it probably becomes more gaussian, at least for intermediate times). Now turning to a sample of arbitrary shape, one should imagine the surface inside the sample which is the locus of points where the total field (ie., the sum of external field H_o and the internal field $E(\vec{r},t)$) is zero. When the sample starts relaxing, you can imagine that an expanding region around this surface becomes involved in the relaxation, with molecules inside it being either brought into or forced out of resonance. In a sample of complicated shape (particularly with edges and corners) this surface (and the related density of states) will be complicated, with van Hove singularities of various kinds in $N(\xi)$ (in principle classifiable using Morse theory). However, this does not *in any way* affect the way in which the function $[N_+(\xi,0) - N_+(\xi,t)]$, integrated over the sample, develops a Lorentzian shape around $\xi = 0$ at short times- the relaxation digs a "Lorentzian hole" in $N_+(\xi,t)$, as time goes on.

I strongly emphasize that as soon as we have a reasonably large concentration of flipped molecules (roughly 10 to 15 per cent, according to the MC simulations), this analysis must begin to break down, because the bias field distribution, due to the flipped spins, will stop being Lorentzian. Moreover, we also expect the assumption of factorisation of the 2-molecule distribution function to break down, and the problem then apparently becomes intractable- in fact we have a kind of *Quantum Spin Glass* problem. We expect a complicated crossover to a long time behaviour which shows "ageing" behaviour, and whose form seems very difficult to predict. At even longer times $\ll T_1$, exponential behaviour will again set in (see below).

However, one can also make analytical predictions for long times under a slightly different circumstance, viz., if the sample is first depolarized at *high* T, to a value $M/M_o \ll 1$, and the cooled to the low-T quantum regime. In this case we also have factorizability of the 2-particle distribution function; and in fact another analytic solution for the homogeneous (ie., ellipsoidal) case can be found from (138), when $M \ll 1$ and $P^{(2)}_{\alpha\alpha'}(\xi,\xi';\vec{r},\vec{r}\,') = P_\alpha(\xi)P_{\alpha'}(\xi')$. One finds *exponential* relaxation, at a rate

$$\tau_{long}^{-1} \approx \frac{2\xi_0}{E_{max}\tau_0[1+\kappa\ln(E_{max}/\pi\xi_0)]}\ , \tag{147}$$

where $\kappa \sim 1$ is a numerical coefficient, and E_{max} is the spread in dipolar fields in this nearly depolarized limit.

It will be very interesting to see if these low-T predictions are confirmed. I emphasize that the low-T, low-H limit is crucial here - in this case the predictions can more or less be deduced directly, along the lines just described. In my personal opinion, confirmation of the $\sqrt{t}$ law, and the dependence of τ_{short} on T_2 and on sample shape, would give very powerful evidence for the nuclear relaxation mechanism discussed. In any case, we see here a very nice example of the role a nuclear spin bath can play in tunneling dynamics. The essential result which the spin bath theory gave us is the formula for $\tau_N^{-1}(\xi)$ in eqtn (137) (and this result is just a special case of eqtn (126)).

What more theoretical work can be done in this low-T limit? The first obvious thing is to extend this kind of theory to much higher fields (both longitudinal, which will bring level $\mid 10>$ into resonance successively with levels $\mid 9>$, $\mid 8>$, etc; and transverse fields, which will slowly reduce the total barrier height, as well as slowly changing the tunneling matrix elements. The effect of longitudinal fields will not be simple - even in the very low-T limit, the resonant tunneling between $\mid 10>$ and $\mid -9>$ will be quite different from that between $\mid 10>$ and $\mid -10>$, since both spontaneously-emitted phonons and dipolar processes will play a role.

Another unsolved problem of great interest at low T is the effect of an AC field on the nanomolecular dynamics. A number of papers have already been written on this, essentially treating the problem as a simple Landau-Zener problem, involving only the giant spin levels and the AC field. However I should emphasize strongly that the Landau-Zener transitions, when they take place, are doing so in the presence not only of the applied AC field, but also of the much more rapidly fluctuating nuclear fields, even at the very lowest temperatures (and of course at higher T, they have to contend with the large and wildly fluctuating intermolecular dipolar fields!). The experiments have already shown a very rich behaviour (see particularly recent work of Novak et al.,[77]), which depend strongly on H and T.

(iii) Thermally-Activated Regime: I concentrated on the low-T, low-H regime above, in discussing the experiments, simply because their interpretation in this regime was straightforward (and provides clear evidence for nuclear spin effects). What about experiments above T_c?

It is always a good idea, in trying to understand a set of experiments, to see what happens in regimes where the theory is simple - and this often means looking for "limiting cases", in which one or more parameters are very small (or very large), so that some processes dominate and others are irrelevant. We have just seen this in the low-T limit (where all dynamics except transverse nuclear spin fluctuations are frozen out). Is there any analogous simplifying limit at higher T?

At first glance this is not obvious - at higher T many more giant spin levels are involved, as are now the phonons, and quite a number of recent papers have dealt with processes involving phonon-mediated transitions between these levels. For reasons quite obscure to me, most of these papers treat both the nuclear hyperfine and the molecular dipoles fields as *static*, when in fact they are both fluctuating rather violently and rapidly in time. Thus the relaxation, at first, seems very messy.

In fact, however, things are simple for the following reasons:

(i) The timescales involved in the various physical processes are much shorter than the experimental relaxation timescales (and this is exactly what should be expected once T is considerably greater than T_c).

(ii) The competition between the experimentally increasing tunneling matrix elements, and the experimentally decreasing thermal occupations, as one rises up through higher and higher giant spin levels, means that at given values of H and T, one transition will dominate the tunneling relaxation.

There is nothing particularly radical about these circumstances- they are perfectly normal. Let us see how they arise.

We note that first that thermal equilibration between different electronic levels on the *same side* of the energy barrier is fast, over timescales $\sim \mu secs$: this is easily shown assuming spin-phonon coupling strengths. Such processes do *not* involve tunneling. In contrast to the low-T case, dipolar "flip-flop" processes now also occur very rapidly - in fact even faster than spin-phonon transitions. Thus, for small H, transitions like $|9\rangle|-10\rangle \rightarrow |10\rangle|-9\rangle$ (or vice-versa) happen at a frequency exceeding $10^6 Hz$ once we are well out of the quantum regime.

These 2 processes are the only ones which can cause transitions between 2 giant spin levels on the same side of the barrier - they were both frozen at low T, but their rate increases exponentially fast as T rises. The now very rapid dipolar fluctuations also completely change the nuclear spin dynamics; the *longitudinal* nuclear relaxation rate T^{-1} rises with the dipolar flip-flop rate, since the latter drives the former. This is important, because it means that the bias energy $\xi = g\mu_B H S_z$ acting on each molecule is fluctuating very fast (because of the fluctuations of both $\vec{H}_{dip}$ and $\vec{H}_{hyp}$) over a fairly large range of energies (for $Fe_8, \vec{H}_{dip} \gg \vec{H}_{Hyp}$, and ξ fluctuates over a range $\sim 0.5K$, whereas for for Mn_{12}, the hyperfine field is much larger, indeed not much smaller than the dipolar fields, so that the fluctuations in ξ may well be some 50 per cent larger). Notice that although we do not know $T_1(T)$ (it has not been measured for any of these molecules at the relevant temperatures), we do expect it to have roughly the same T-dependence as the dipolar flip-flop rate; and we expect $T_1 \ll \tau_{exp}$ (where τ_{exp} is the experimental relaxation rate), once we are well inside the thermally-activated regime.

Thus the picture that energies in the thermally activated regime is one where molecules cycle between giant spin states on the same side of the barrier, and between all the nuclear multiplets available for a given giant spin state

$|m\rangle$. Inelastic tunneling transitions are much slower. What then controls the tunneling rate in this regime?

The answer is startlingly simple. The transitions on one side of the barrier occur so quickly that they *disappear completely from the physics* - their only role is to keep the molecules on one side of the barrier in a state of quasi-equilibrium, in which the probability of occupying a state of energy ϵ is just $Z^{-1}e^{-\epsilon/kT}$, where Z is the total partition function. Under these circumstances we must expect *exponential relaxation*!

There is nothing in this conclusion that will surprise many - in fact a number of papers have already simply *assumed* the existence of quasi-equilibrium thermal populations on each side of the barrier, in attempting to calculate a relaxation rate . However this assumption obviously needs to be justified, particularly as we know that the low-T relaxation is predicted to be *non-exponential*! More generally, as emphasized some time ago[17], the crossover to exponential relaxation will occur on a timescale roughly equal to T_1, and so in the region of temperature where one sees crossover from quantum to thermally activated behaviour (ie., for $T \sim \mathcal{D}/2\pi$), one may even expect to see $\sqrt{t}$ relaxation for $t \ll T_1$, with a crossover to exponential relaxation for $t \gg T_1$. Experimental measurements of T_1 would be very useful here!

However a cautionary note. It is not entirely obvious that one will always expect to see exponential relaxation in the experiments, even at high T. The reason for this is very simple. In a typical relaxation experiment, as we have already seen at low T, the total field (internal field $E(\vec{r}, t)$ plus external applied field $\vec{H}_o$) changes in time as the system relaxes; moreover, so does the distribution function $P_\alpha(s, \vec{r}, ; t)$, and hence the density of states $N(\xi; t)$ available for transitions. Consequently, in the course of the relaxation, the relaxation rate $\tau^{-1}(H)$ must become a function of time - essentially we are sweeping through regions of different relaxation rate. Thus the high-T relaxation is probably going to be rather complex (and indeed the various experiments do not obviously agree in this regime).

4.2 Macroscopic Quantum Coherence?

Ever since the well-known early discussions of Leggett et al. [48,41,78], on the possibility of superpositions of "macroscopically distinguishable" quantum states in SQUIDs, there has been intense interest in its experimental realisation. Most discussions of this "Macroscopic Quantum Coherence" have been in the context of the spin-boson model, following Leggett et al. (see his chapter in this book).

Despite several attempts to find Macroscopic Quantum Coherence (MQC) in superconducting systems, no results have yet been reported. In section 3.4(b) above, I already discussed why I think the observation of MQC in superconductors is going to be very difficult, because of the nuclear bath effects. However in the different field of magnetism, a rather dramatic claim for the discovery of MQC has been made by the group of Awschalom et al.[8,9], in a series of papers going back to 1992. There has also been strong opposition to these claims from various quarters, most notably in the work of A. Garg[79].

I don't think this issue is yet settled, but it is clearly of interest; the purpose of the present sub-section is simply to present the results, and to explain the role of the nuclear spin bath in the problem.

4.2(a) EXPERIMENTAL RESULTS on FERRITIN MACROMOLECULES

The series of experiments reported by Awschalom et al. actually began with results [80] on arrays of Iron pentacarbonyl ($FeCo_5$) ellipsoidal grains, of rather large size. The AC susceptibility absorption was sharply peaked in these experiments, at very low frequencies (ranging from $\omega_o \sim 60$ Hz for grains of size 150 Å× 700 Å, up to $\omega_o \sim 400$ Hz for grains of size 150 Å× 380 Å). Curiously, it was found that ω_o varied roughly exponentially with the volume of the particles (decreasing with larger volume) in a way that would be expected if the particle magnetisation was tunneling coherently between 2 orientations, or else being driven by the AC field between 2 eigenstates, which themselves are superpositions of 2 "semiclassical" states $\mid \Uparrow$ and $\mid \Downarrow$ (the particles are uniaxial).

Nevertheless, as recognized by these authors, this interpretation is untenable, principally because the energy scale ω_o is so small (a frequency 400 Hz is equivalent to a temperature $\sim 8 \times 10^{-8} K$). For all the grains to show such a coherent response would require that any external bias field, acting on the grains, would be so small that it would split the degenerate states by a bias energy ξ considerably less than ω_o. This is of course impossible; apart from anything else the dipolar magnetic interaction between the grains is at least 5 orders of magnitude larger than ω_o, for even the smallest grain size! In fact, to this date I am aware of no explanation offered for these results.

However in 1992 experiments done on frozen solutions of ferritin molecules in the protein apoferritin were reported[8]. Ferritin is a very interesting molecule, found naturally in all eukaryotic cells; it is of roughly spherical shape, and diameter $\sim 80 \mathring{A}$. The core contains some 4500 Fe^{3+} ions (spin 5/2), and is protected from the outside world by a "cage" of apoferritin protein molecules.

In bulk the ferrihydrite would order antiferromagnetically (giving a Néel vector for this system of magnitude $N = |\vec{N}| \sim 18,000\mu_B$). In reality of course the molecules possess a surface moment, which must certainly vary from one molecule to another - in their first paper Awschalom et al. [8] estimated this moment by various means, yielding values ranging from 217 μ_B to 640 μ_B. The main result of this experiment, like the earlier $FeCo_5$ one, was a sharp resonance in $\chi''(\omega)$ (the AC absorption). However this time the frequency was $\sim 940kHz$ (ie., an energy $\sim 45\mu K$). This was only seen in strongly diluted (1000:1) solutions.

More recently this group has done similar experiments on artificially engineered samples of ferritin molecules with smaller core sizes, and again they saw ω_o rise as a roughly experimental function of the inverse particle size. One should also notice that no resonance was found in experiments on an apoferritin sample containing no ferritin, and that increased dilution of the sample tended to sharpen the absorption lines.

In contrast to the $FeCo_5$ results, Awschalom et al. believe that the ferritin results provide evidence for the resonant absorption of EM waves at a frequency ω_o corresponding to the tunneling matrix element Δ_o, between Néel states $| \uparrow\rangle$ and $| \downarrow\rangle$. This belief is partly based on the estimated value for the anisotropy field, which is not incompatible with a tunneling matrix element of this size.

A large number of objections have been raised to this straightforward interpretation. These include (i) the existence of randomly varying dipolar couplings between the molecules (the molecules are not oriented)[37] (ii) the field dependence of the resonance frequency and linewidth [79] (iii) the power absorption in the experiments; [79] (iv) a claimed contradiction with high-T blocking [81]; and (v) nuclear spin effects [14,16,40]. Many of these queries will probably only be definitively settled if and when other experimental groups carry out an independent check on the experiments. I will make no attempt to address all of the issues, and refer the reader to the literature. Instead, and in the context of this chapter, we will look at the single question of the the effect of nuclear spins on the coherence.

4.2(b) NUCLEAR SPIN EFFECTS on MQC in FERRITIN

To investigate the effects of the nuclear bath (and other environmental spins) on the dynamics of the ferritin molecules, we need a realistic effective Hamiltonian. The problem is analogous to the SQUID problem we already looked at, and it is pedagogically useful because it deals with an antiferromagnetic system. We begin with a single ferritin molecule, for which the biaxial Hamiltonian (9) is usually used (with the understanding that $\vec{S}$ now represents the Neél vector, and that the extra surface moment will in reality complicate things- the following is not meant to be a complete study). Awschalom et al. [8,9] give a value 1.72 Tesla (in field units) for $(K_{\|}K_{\perp})^{1/2}$, so we will take a small oscillation frequency $\Omega_o \sim 40GHz$ as a rough estimate. If we also suppose that the resonance at frequency ω_o represents resonant tunneling, then $\Delta_o \sim 1MHz$ for the sample of naturally occurring ferritin.

From these numbers we immediately see that topological decoherence effects will be very small; with roughly 100 Fe^{57} nuclei in the sample (some 2 per cent of all the nuclei have a spin), and a hyperfine coupling $\omega_k \sim 50-60MHz$, we get $\alpha_k \sim 2\times10^{-3}$, and so the mean number of nuclear spins co-flipping with the ferritin is $\lambda \sim 4\times10^{-5}$. However as noted in Prokof'ev and Stamp[14,16], and discussed at some length later by Garg[40], the degeneracy blocking effects are very severe; the spread E_o in the nuclear multiplet around each ferritin level is roughly $500-600MHz$, and so the mean bias is some 600 times greater than Δ (the contrary claim of Levine and Howard[82] was later retracted[83]). There will also be an effect of orthogonality blocking, whose effect is very hard to quantify; it will come from the surface moments on the ferritin, and even more from the surface interactions with the apoferritin (particularly between surface moments and any paramagnetic spins in the apoferritin; this was described as an effect of "loose spins" in the early papers[14]). The ensuing random fields acting on the ferritin moment mean that the initial and final states of the Neél vector will not be exactly antiparallel.

If we ignore degeneracy blocking, and also ignore the nuclear spin dynamics, then the situation does not in fact look too bad for the experiments, at least at first glance. This is because the lineshape in a situation of pure degeneracy blocking (already shown in Fig.XXX), is rahter sharply peaked, and so resembles the experiments. There is certainly a problem of power absorption, but as noted by Awschalom et al., this is very hard to quantify[79,81]. I ignore here the other difficulties mentioned above (points (i)-(iv)).

However there is also the problem of the nuclear spin dynamics at low T; moreover, as we have seen already in the context of the magnetic molecules, this will drive a time-dependent change in the dipolar field distribution in the sample (which presumably starts off being pretty much Lorentzian in the sample of dilute and randomly-ordered molecules). However there is one big difference here from the case of Mn_{12} and Fe_8 discusssed earlier. This is that in all of the experiments done on ferritin so far, Δ_o is *bigger* than the inverse timescale T_2^{-1} of the nuclear fluctuations. This changes things a great deal- now the bath fluctuations are slow and thus have a much smaller effect on coherent tunneling.

One could go on and discuss this problem in quantitative detail. However I will not for the following simple reason. This is that *none* of the calculations described here (or those appearing in the literature) have yet addressed what is the key theoretical problem in the interpretation of the ferritin experiments, which is the AC response of a central spin, coupled to a spin bath, when the driving field is *fast*. This is nothing but the **Landau-Zener** problem for a system coupled to a spin bath, and it is not yet solved! I would stress that the solution to this problem is not at al obvious- indeed with a *finite* spin bath it is clear that under many circumstances a response function (in the strict sense of linear response) does not even exist! Thus, since whereof one cannot speak..

4.3 Acknowledgements

:

Much of the work of my own described herein was done with Drs. M. Dube' and N.V. Prokof'ev, whom I thank for innumerable discussions. At the Seattle workshop I also enjoyed very useful discussions with Drs. P. Ao, G. Bertsch, O. Bohigas, A. Bulgac, H. Grabert, P. Hanggi, P. Leboeuf, A.J. Leggett, A. Lopez-Martens, C. Sa de Melo, B. Spivak, M. Stone, D.J. Thouless, S. Tomsovic, and J. Treiner, although not all of the subjects we discussed appear here! I thank the Institute for Nuclear Theory (Seattle), the Canadian Institute for Advanced Research, NSERC Canada, and the Laboratoire de Champs Magnetiques Intenses (Grenoble), for support during the time this chapter was written.

1. C. Paulsen and J.G. Park, in *"Quantum Tunneling of Magnetisation-QTM'94"* (ed. L. Gunther and B. Barbara), Kluwer publishing, pp. 189-207 (1995).
2. M. Novak and R. Sessoli, pp. 171-188 in ref. 1.
3. B. Barbara *et al.*, J. Mag. Magn. Mat. **140-144**, 1825 (1995).
4. J.R. Friedman *et al.*, Phys. Rev. Lett., **76**, 3830-3833 (1996).
5. L. Thomas *et al.*, Nature **383**, 145-147 (1996).
6. J.M. Hernandez *et al.*, Europhys. Lett., **35**, 301-306 (1996).
7. C. Sangregorio, T. Ohm, C. Paulsen, R. Sessoli, and D. Gatteschi, Phys. Rev. Lett., **78**, 4645 (1997).
8. D.D. Awschalom et al., Science **258**, 414 (1992), and refs. therein.
9. S. Gider et al., Science **268**, 77 (1995)
10. K. Hong, N. Giordano, Europhys. Lett. **36**, 147 (1996), and refs. therein.
11. W. Wernsdorfer *et al.*, Phys. Rev. Lett. **78**, 1791 (1997); and May 1997 preprint.
12. R.P. Feynman, F.L. Vernon, Ann. Phys. **24**, 118 (1963)
13. A.O. Caldeira, A.J. Leggett, Ann. Phys. **149**, 374 (1983)
14. N.V. Prokof'ev, P.C.E. Stamp, J. Phys. CM **5**, L663 (1993)
15. P.C.E. Stamp, Physica B **197**, 133 (1994) [Proc. LT-20, Aug. 1993]. See also P.C.E. Stamp, UBC preprint (Sept. 1992, unpublished), and Nature **359**, 365 (1992); and ref. [22] below.
16. N.V. Prokof'ev, P.C.E. Stamp, pp. 347-371 in ref. 1
17. N.V. Prokof'ev, P.C.E. Stamp, J. Low Temp. Phys. **104**, 143 (1996)
18. N.V. Prokof'ev, P.C.E. Stamp, Rep. Prog. Phys. (to be published)
19. A. Shimshoni, Y. Gefen, Ann. Phys. **210**, 16 (1991)
20. A.O. Caldeira, A.H.Castro-Neto, T. Oliveira de Carvalho, Phys. Rev. **B48**, 13974 (1993).
21. S. Sachdev, Physics World **7**, 25 (1994), and refs. therein; R.N. Bhatt, S. Sachdev, J. Appl. Phys. ; C.T. Muruyama, W.G. Clark, J. Sanny, Phys. Rev. **B29**, 6063 (1984).
22. P.C.E. Stamp, Phys. Rev. Lett. **61**, 2905 (1988)
23. N. Nagaosa, A. Furusaki, M. Sigrist, H. Fukuyama, J. Phys. Soc. Jap. **65**, 3724 (1996)
24. M. Yamanaka, N. Nagaosa, P.C.E. Stamp, to be published.
25. D.L. Hill, J.A. Wheeler, Phys. Rev. **89**, 1102 (1953)
26. A. Bulgac, G. Do Dang, D. Kusnezov, Phys. Rev. **E54**, 3468 (1996), and Ann. Phys. **242**, 191 (1995); and refs. therein.
27. K.G. Wilson, Rev. Mod. Phys. **47**, 773 (1975).
28. M. Dube, P.C.E. Stamp, J. Low. Temp. Phys. (accepted Aug. 1997); /cond-mat 9708191

69. N.V. Prokof'ev, P.C.E. Stamp, /cond-mat 9511011 (unpublished).
70. D. Gatteschi et al., Science **265**, 1054 (1994)
71. O. Kahn, "Molecular magnetism", VCH publishers (1993).
72. I. Tupitsyn et al., (preprint)
73. R. Sessoli et al. J. Am. Chem. Soc. **115**, 1804 (1993); and R. Sessoli et al., Nature **365**, 141 (1993)
74. A. Garunin, E.M. Chudnovsky (preprint)
75. A. Bartolome et al (preprint)
76. N.V. Prokof'ev, P.C.E. Stamp, submitted to Phys. Rev. Letts, 2 Oct 1997 (/cond-mat 9710246).
77. M. Novak et al. preprint
78. A.J. Leggett, Prog. Th. Phys. Supp. **69**, 80 (1980)
79. See A. Garg, Phys. Rev Lett. **70**, C2198 (1993), and reply of D. D. Awschalom et al., *ibid.*, C2199 (1993); or A. Garg, Phys. Rev. Lett. **71**, 4241 (1993), and the associated Comment of D. D. Awschalom et al., *ibid.*, C4276 (1993); or A. Garg, Science **272**, 425 (1996), with the reply of D. D. Awschalom et al., *ibid.*, 425 (1996)
80. D.D. Awschalom et al., Phys. Rev. Lett. **65**, 783 (1990)
81. See J. Tejada, Science **272**, 424 (1996), and the reply by D. D. Awschalom et al, *ibid.*, 425 (1996)).
82. G. Levine, J. Howard, Phys. Rev. Let. **75**, 4142 (1995)
83. G. Levine, J. Howard, Phys. Rev. Lett. **76**, 3241 (1996)

COHERENT MAGNETIC MOMENT REVERSAL IN SMALL PARTICLES WITH NUCLEAR SPINS

ANUPAM GARG
Department of Physics and Astronomy
Northwestern University
2145 Sheridan Road
Evanston, IL 60208, USA
E-mail: agarg@nwu.edu

A discrete WKB method is developed for calculating tunnel splittings in spin problems. The method is then applied to the issue of how nuclear spins affect the macroscopic quantum coherence of the total magnetic moment in small magnetic particles. The results are compared with numerical work, and with previous instanton based analytic approaches.

1 Motivation for this Work

Small magnetic particles have now been investigated as good candidates for experimental observation of macroscopic quantum tunneling (MQT) and coherence (MQC).[1,2,3] This is because at first sight the main critera for a system to be a good candidate for seeing MQP (P for phenomena) are met. One of these is that the energy barrier through which the system must tunnel be microscopic, even though the tunneling variable itself is macroscopic. This demand is met for the physical reason that the anisotropy energy barrier originates in spin-orbit or spin-spin interactions at the microscopic level, both of which are relativistic effects, and hence small. A second criterion is that there be a well defined macrovariable, whose dynamics can, to a first approximation, be isolated from those of other microscopic degrees of freedom. This criterion can be met by working at temperatures sufficiently below the equivalent anisotropy energy gap, as spin wave excitations are then frozen out, and we may focus on the net magnetic moment of the particle as the macrovariable.

There are also extremely good reasons to believe, however, that MQP are very hard to observe in general. Chief among these reasons is that the couplings of the macrovariable to the microscopic degrees of freedom give rise to decoherence. This effect can be especially severe in the case of macroscopic quantum coherence, even when the coupling is so small that the semiclassical dynamics of the macrovarible are significantly underdamped. The best studied example which provides a detailed illustration of this point is that of the spin-boson problem.[4,5,6] Further support for this point is provided by the *scantiness of observational evidence* for quantum coherence even in systems that one would regard as microscopic. Thus, we know of only a few dozen or so "flexible" molecules like NH_3 which display coherent flip-flop between different nuclear configurations. [7] The frequency of flip-flop is generally in the 0.1–100 GHz range. Given the vast range of molecular structures, and bonding strengths, there must surely be many naturally occurring molecules whose energy barriers and attempt frequencies are such as to put the flip-flop frerquency at anout 1 Hz. Yet such flip-flop has never been seen. The reason almost certainly is that such molecules are never naturally encountered in isolation by themselves, and collisions and other environmental interactions are very effective in wiping out the quantum coherence.

It is thus important to identify mechanisms for decoherence in the magnetic particle system, and several have been put forth (phonons, magnons, Stoner excitations). The most critical, however, is the spin of the nuclei in the particle. The hyperfine coupling between the nuclear and electronic spins in magnetic solids is of order 100 MHz or more (in frequency units) per nucleus, which is rather high on the scale of the expected MQC frequencies. At the same time this frequency is rather low on the scale of the attempt frequencies associated with the electronic moments. Nuclear spins are therefore likely to be extremely efficient decoherers or "observers" of the direction of the moment of a small particle. This expectation is confirmed by theoretical calculations for both MQT, [8] and MQC. [9] It is the latter that I wish to focus on and revisit in this article, for several reasons. First is that this decoherence mechanism falls outside the scope of the harmonic oscillator bath, [10,11] so one cannot rely on previous results. The calculations in Ref. 9 are done using an instanton technique with many physically motivated approximations about the trajectories likely to give the dominant contribution to some path integral. It is not obvious even to me that this calculation is done in strict adherence with the *Rheinheitsgebot.* On the occasions that I have given seminars on the subject, the quizzical looks on the faces of my audience make it clear to me that it does not fully believe or understand my approach. It is therefore desirable to study this problem using different methods, and we shall do so in this article using the discrete WKB method. Although this method itself is quite old (see Braun's review [12] for references), it does not appear to have been used in spin tunneling problems except for some work by van Hemmen and Sütő. [13] These authors have not fully exploited the power of this method, however. We shall see that calculations which have traditionally been done by instanton methods can be done much more simply to the same accuracy using this method. The opportunity to present the technical aspects of this method in a tutorial volume devoted to tunneling in complex systems is greatly welcome, and provides me with another reason for writing this article.

It should be noted that the same subject has also been studied by Prokof'ev and Stamp in several papers. The first of these [14] correctly notes that the tunneling amplitude is suppressed by nuclear spins, but fails to recognize the implications of this fact for the nature of the tunneling spectrum. See footnote *d* for more on this. Subsequent papers [15] puport to make detailed calculations of the tunneling spectrum including lineshapes, and also to include a host of other physical effects, such as spin diffusion, the Suhl-Nakamura interaction, and others with less familiar names such as "topological decoherence", "orthogonality blocking", and "degeneracy blocking". I cannot comment on the later work, simply because I do not understand much of it, especially some of the more mathematically specific and detailed conclusions, about lineshapes, for example. My goal in this article will be much more modest. It was argued in Refs. 3 and 9 that in the presence of nuclear spins the tunneling spectrum is broken into several resonance lines. A physical interpretation was attached to this broken-up spectrum, and formulas were presented for their frequencies and spectral weights. This article corroborates these claims. As in the previous papers, I have not attempted to model or calculate the details of the relaxation, i.e., the lineshapes. To this extent, I am only prepared to claim a qualitative understanding for the very low frequency part of the tunneling spectrum. Fortunately, it is the high frequency end that is most likely to be experimentally relevant if at all, and here the situation is much better.

The plan of this article is as follows. In Sec. 2, I will give a brief introduction to the physical problem and the models studied in this paper. Sec. 3 contains a general discussion of the discrete WKB method. This is used to calculate the bare tunnel splitting, i.e., without nuclear spins, in Sec. 4. Nuclear spins are added to the problem in Sec. 5. This section has the bulk of the new results in this article. The results for the tunnel splitting(s) obtained via the discrete WKB method are checked against those from exact numerical diagonalization of model Hamiltonians. I also compare them with those from the instanton approach, [9] and identify which features of the latter appear to be quantitatively robust, and which are only qualitatively correct. I conclude in Sec. 6 with a summary of the effects of nuclear spins, and some general remarks on the observability of MQC in magnetic particles.

2 Introduction to Physical Problem and Models

2.1 Physical System

The physical system is a small insulating magnetic particle, about 50 Å in diameter, at millikelvin temperatures. The anisotropy energy gap is much larger than this, so spin wave excitations are frozen out, and the individual atomic spins are orientationally locked together. Furthermore, at this size the particle typically contains only one magnetic domain. Thus the magnitude of the total spin or magnetic moment of the particle is essentially fixed proportional to the number of atomic spins (assuming one magnetic species for simplicity), and the only relevant dynamical variable is its direction. (It may be useful to think of the system in terms of a Heisenberg-like model, with additional single-ion anisotropy terms.) The existence of anisotropy implies that some directions are energetically favored over others, and we wish to investigate whether the spin orientation can display quantum mechanical behavior, in particular tunneling and/or coherent oscillation between different energy minima.

A model Hamiltonian which incorporates the above features is: [1]

$$\mathcal{H}_0 = -k_1' S_z^2 + k_2' S_x^2 \tag{1}$$

Here S_x, S_y, and S_z are the components of our large spin, and k_i' are phenomenological anisotropy coefficents. This particular Hamiltonian is time-reversal invariant, which should be the case if our particle is not subject to any external magnetic fields. We take $k_1' > 0$, $k_2' > 0$, so that in classical language, $\pm\hat{\mathbf{z}}$ are easy directions, and $\pm\hat{\mathbf{x}}$ are hard directions. Quantum mechanically, the two classical ground states $\pm\hat{\mathbf{z}}$ will be split by tunneling. A generally valid approximate expression for the tunnel splitting can be obtained by writing

$$\Delta_0 \simeq \omega_e \exp(-Su/\omega_e). \tag{2}$$

[A more exact expression is given in Eq. (24) below.] In this formula,

$$\omega_e = 2(k_1 k_{12})^{1/2} \tag{3}$$

is the oscillation (or precession) frequency for small deviations of the spin orientation from the classical equilibrium directions $\pm\hat{\mathbf{z}}$,[a] and we have also defined

$$k_1 = Sk_1', \quad k_2 = Sk_2', \quad k_{12} = k_1 + k_2. \tag{4}$$

Further, Su is a quantity proportional to the energy barrier, of order Sk_{12}. We work throughout in units such that $\hbar = 1$.

We next wish to consider the influence of nuclear spins. To model this let us suppose that N_n of the atoms have nuclei with spins $\mathbf{I}_i$, and take all of these to be of magnitude 1/2. Let us further simplify the problem and assume that the hyperfine interaction for each of these atoms is of identical strength and of the form $\mathbf{s}_i \cdot \mathbf{I}_i$, where $\mathbf{s}_i$ is the electronic spin on atom i. All the assumptions

[a] This result may be derived by writing down the Heisenberg equation of motion for $\mathbf{S}$, and linearizing it in small deviations from $\mathbf{S} = S\hat{\mathbf{z}}$.

except that of identical interaction strength are immaterial, and even this is a rather good approximation. The effects of relaxing it will be briefly discussed in Sec. 5.3. The total hyperfine contribution to the Hamiltonian can then be written as

$$\mathcal{H}_{\rm hf} = -\frac{\omega_n}{s} \sum_{i=1}^{N_n} \mathbf{s}_i \cdot \mathbf{I}_i = -\frac{\omega_n}{S} \mathbf{S} \cdot \mathbf{I}_{\rm tot}, \tag{5}$$

where we have expressed the coupling constant in terms of ω_n, the nuclear Larmor frequency that would be obtained if the electronic spin orientation were fixed. The first expression in Eq. (5) is just the sum of the interactions for the individual atoms, and the second follows from assuming that the atomic spins are all parallel to one another, which permits one to write $\mathbf{s}_i = (s/S)\mathbf{S}$ for all i. The remaining sum equals $\sum_i \mathbf{I}_i$, which we call $\mathbf{I}_{\rm tot}$, the total nuclear spin. For any given value of N_n, $I_{\rm tot}$ can take on values ranging in integer steps from $N_n/2$ to either 0 (if N_n is even) or $1/2$ (if N_n is odd), with multiplicities that can easily be found. (See below.) Since $\mathcal{H}_{\rm hf}$ commutes with $\mathbf{I}_{\rm tot}^2$, we can consider the problem for one value of $I_{\rm tot}$ at a time. Writing $\mathbf{I}$ instead of $\mathbf{I}_{\rm tot}$ henceforth, we arrive at a model Hamiltonian obtained by adding Eqs. (1) and (5), i.e.,

$$\mathcal{H}_I = \frac{1}{S}\left(-k_1 S_z^2 + k_2 S_x^2 - \omega_n \mathbf{S} \cdot \mathbf{I}\right). \tag{6}$$

[For completeness, we give here the formula for the multiplicity, i.e., the number of times a given value of $I_{\rm tot}$ appears when N_n spins of magnitude $1/2$ are added together. For $I_{\rm tot} = (N_n/2) - k$, the multiplicity is given by

$$\binom{N_n}{k} - \binom{N_n}{k-1}. \tag{7}$$

As an example, if $N_n = 6$, multiplets with $I_{\rm tot} = 3$, 2, 1, and 0, occur 1, 5, 9, and 5 times respectively.]

It should be noted at this point that for a particle of diameter 50 Å, which corresponds to $S = \mathcal{O}(10^4)$, and typical material or anisotropy parameters, the tunnel splitting Δ_0 as given by Eq. (2) is unobservably small. Of course the splitting goes up if S is decreased, but particles much smaller than 50 Å seem difficult to attain controllably and reproducibly with present day technology.[b]

[b]The last few years have seen some very interesting work[16,17,18] on systems with much smaller values of S, of order 10. These systems are based on magnetic molecules, where the value of S is highly reproducible, but it is my belief that the interesting questions here are quite different, and have little to do with MQP. Interestingly, semicalssical methods, such as those developed in this paper, are likely to be quite valuable in analyzing these systems. For particles with intermediate values of S, say about 100, on the other hand, it will probably be necessary to devise new ways of treating the environment because the law of large numbers will no longer be applicable.

MQC is a more likely proposition in *antiferromagnetic* particles[19,20,21] where the two states involved differ in the orientation of the Néel vector, or the spins on one of the sublattices.[c] The essential aspects of the earlier papers, [3,9] in which the influence of nuclear spins was studied using instantons, hold equally for ferro- and antiferro-magnetic particles. My goal in this article is to try and verify the predictions of the instanton approach as quantitatively as possible. It is simpler to do this using a ferromagnetic model, and I shall therefore limit myself to that in this article.

I also do not wish to discuss at length the experimental observability of the reversal phenomenon. In addition to the values of the physical parameters mentioned above, this hinges on a number of other issues such as the temperature, the signal size, stray magnetic fields, the nature of the substrate, etc. We refer readers to previous papers[3,9,22,23] for detailed discussions of these points. For the purposes of this article, we will only note that the ratio ω_n/ω_e is of order 10^{-3} to 10^{-2} in antiferomagnets, and is about one order of magnitude higher in ferromagnets. Throughout this article therefore, we shall assume that $\omega_n/\omega_e \ll 1$, and work to leading order in this ratio. We shall further assume that $\Delta_0/\omega_n \ll 1$, which, by virtue of the exponentially small WKB or Gamow factor, is almost certain to be the case for values of S of interest to us.

2.2 What to Calculate; Preliminary Arguments

Having settled on a theoretical model, let us ask what physical quantity we should calculate. One way the dynamical behaviour of the moment can be described is in terms of a time dependent probability to find it along some direction given an initial state in which it was prepared.[4,5] This description best applies to experiments on a single system. Another way is to find the appropriate frequency-dependent dynamical susceptibility $\chi(\omega)$. This description better applies to an assembly of identical or nearly identical systems, such as is obtained in an NMR experiment. We will adopt the second approach.

Let us suppose then that we wish to calculate $\chi''(\omega)$ for our magnetic particle. Since the Hamiltonian (6) describes a closed system, χ'' can only consist of a set of delta functions. The real system is of course not closed, and has means of energy relaxation such as phonons, with which the spins can interact in a variety of ways. These will lead to line widths and broadening in the usual way. Now in a setup such as NMR, the resonance frequencies are few in number and usually known quite accurately to begin with. Interest

[c]The reason for this is quite simple. The tunnel splitting can still be written in the form (2). The energy barrier Su is of the same order of magnitude, but the electronic spin attempt frequency, ω_e, is about 100 times higher for antiferromagnets than ferromagnets, being given by the geometric mean of an exchange energy and an anisotropy energy in contrast to Eq. (3).

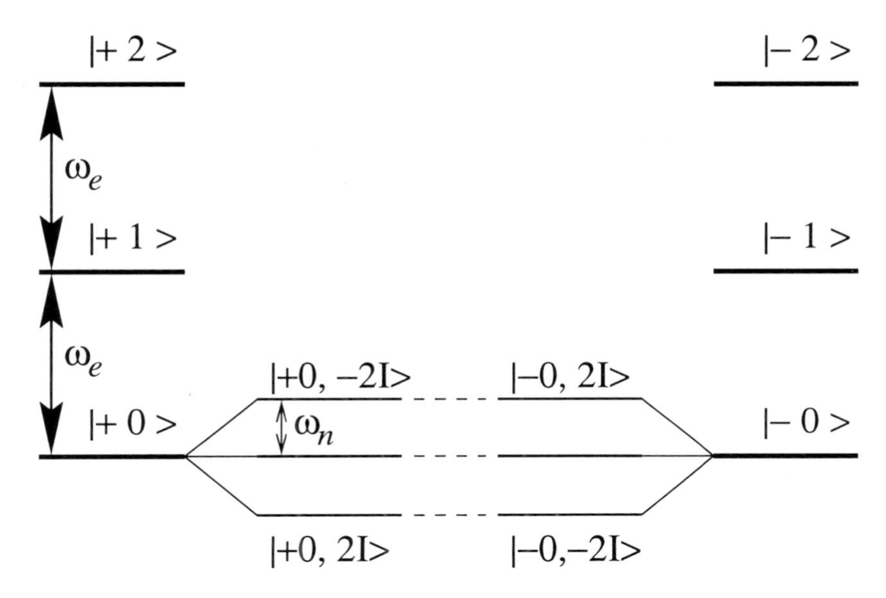

Figure 1: Structure of low lying energy levels of coupled electron and nuclear spin systems. The states $|\pm 0\rangle$ are electronic spin ground states with $\mathbf{S}$ localized along $\pm\hat{\mathbf{z}}$ in the absence of the nuclear spins. The corresponding first two excited states are $|\pm 1\rangle$ and $|\pm 2\rangle$. The central part of the figure shows how the ground electronic levels are modified by the nuclear spins, still ignoring tunneling. The ratio $\omega_e/\omega_n \gg 1$ and is not accurately represented in the figure. If $\omega_n \gg \Delta_0$, only states joined by dashed lines mix with each other to any appreciable extent once tunneling is turned on.

then attaches to the calculation of line shifts and widths and shapes and so on. In our problem, on the other hand, the presence of nuclear spins alters the resonance frequencies themselves drastically. To see this, let us denote by $|\pm 0\rangle$ the approximate eigenstates of the bare Hamiltonian $\mathcal{H}_0$ that correspond to the spin having a mean value $\langle S_z\rangle \approx \pm S$, i.e., to states in which the spin is localized in the ground state in one of the two energy minima centerd at $\pm\hat{\mathbf{z}}$. Tunneling mixes these states and splits them by Δ_0. In addition to these states, let us consider the next few excited states in each well, and denote them by $|\pm 1\rangle$, $|\pm 2\rangle$, and so on. Since the small oscillations (or more accurately precessions) around the $\pm\hat{\mathbf{z}}$ directions are approximately harmonic, these states will form an approximately equally spaced ladder with spacing ω_e, i.e., the states $|\pm n\rangle$ will lie an energy ω_e above $|\pm(n-1)\rangle$ for low values of n. (See Fig. 1.) Now let us bring in the nuclear spins. Since $\langle \pm 0|\mathbf{S}|\pm 0\rangle = \pm S\hat{\mathbf{z}}$ to very good accuracy, the combined elctronic-nuclear spin states will be eigenstates of I_z, which we label by $|+0,p\rangle$, and $|-0,p\rangle$, where $p=2I_z$, will be split into $2I+1$ Zeeman levels with a spacing ω_n. (The higher states $|\pm 1\rangle$, $|\pm 2\rangle,\ldots$, will also be similarly split, but we do not consider that for the moment.) As long as I is small enough so that $2I\omega_n \ll \omega_e$, our basic assumption that $\omega_n \gg \Delta_0$ combined with the physically obvious but important fact that significant resonance is only possible between states that are degenerate to within the matrix element connecting them implies that the state $|+0,p\rangle$ will resonantly tunnel only to the state $|-0,-p\rangle$. Since this tunneling now involves a change in the nuclear spin state in addition to that of the electronic spin, the splitting should in general be different from Δ_0. We denote the magnitude of this splitting by $\Delta(I,p)$, i.e.,

$$\Delta(I,p) = \pm(E_{g(0,p)} - E_{u(0,p)}), \tag{8}$$

where $E_{g(0,p)}$ and $E_{u(0,p)}$ denote the energies of the antisymmetric and symmetric linear combinations $(|0,p\rangle \mp |0,p\rangle)/\sqrt{2}$. Note that the antisymmetric state need not be the one with higher energy.

To orient further discussion, let us summarize the results of the instanton approach.[3,9] In this approach too the starting point is that the nuclear spins spoil the degeneracy between the $|\pm 0\rangle$ states. An instanton connecting degenerate states must involve the flipping over of a certain number of nuclear spins along with that of $\mathbf{S}$. Since the nuclear spins can only respond on a time scale ω_n^{-1} to any perturbation, and the instanton has a temporal width of order ω_e^{-1}, perturbation theory shows that each nuclear spin coflip reduces the tunneling amplitude by a factor $\sim (\omega_n/\omega_e)$.[d] The amplitude for p nuclear spin coflips was found[3,9] (in the case of an antiferromagnetic particle) to be

$$\Delta_p = (\pi\omega_n/2\omega_e)^{|p|}\Delta_0. \tag{9}$$

In other words, the tunneling amplitude decreases geometrically with the number of units by which I_z must change. If this is so (and we will find here that by and large it is), then clearly the more interesting problem is find the tunneling frequencies $\Delta(I,p)$ themselves, and the associated spectral weight to be assigned to each frequency. The issue of linewidths and relaxation becomes secondary. As stated in the previous section, however, the calculation on which these conclusions are based is not totally satisfactory. It is with this viewpoint that we focus in this article on calculating the splittings as carefully as possible.

---

[d]This reduction of tunneling amplitudes due to nuclear spin coflips was also found by Prokof'ev and Stamp.[14] The highly chopped-up nature of the χ'' spectrum was missed by them, however, and the mechanism for decoherence initially studied by them is rather different from mine.[3,9]

It should be clarified that the amplitudes Δ_p are not to be identified with the tunnel splittings themselves. To see this, consider a multiinstanton trajectory that starts and ends at a specific state with $S_z = S$ and $I_z = p/2$. The intermediate states with a long residence time in this trajectory are those with $S_z = S$, $I_z = p/2$, and $S_z = -S$, $I_z = -p/2$. Since p spins can be chosen from N_n in many ways, there can be several such intermediate states, and the amplitudes for all the corresponding trajectories must be added together. This leads to interference effects. When the combinatoric factors are added together, it is found that the *splittings* are given by [9]

$$\binom{n}{|p|}\Delta_p, \quad n = |p|, |p|+1, \ldots, (N_n + |p|)/2. \tag{10}$$

In this article we will try and see how correct these results are. The main flaw in the previous work is the conclusion that $\Delta_p = \Delta_{-p}$, i.e., that the $|0,p\rangle \leftrightarrow |-0,-p\rangle$ and $|0,-p\rangle \leftrightarrow |-0,p\rangle$ splittings are the same. In yet more words, the splittings for p coflips are independent of whether the coflipping spins are parallel or antiparallel to the large electronic spin $\mathbf{S}$. We will see that this is no longer true. Further, the geometric dependence $(\omega_n/\omega_e)^{|p|}$ seems to be only approximately correct. Finally, we will show in Sec. 5 that the binomial factor in Eq. (10) follows from a simple effective tunneling Hamiltonian connecting the states $|0,p\rangle$ and $|-0,-p\rangle$, which holds as long as $\omega_n/\omega_e \ll 1$.

One trivial point should be noted and disposed of once and for all. Since the Hamiltonian (6) is time-reversal invariant, the bare tunnel splitting vanishes unless S is an integer, and all the splittigns $\Delta(I,p)$ vanish unless $S+I$ is an integer. For the purposes of interpreting the relation (9) (or a more accurate replacement) it is useful to *define* Δ_0 for all S by Eq. (2) or its more accurate version, Eq. (24) below.

3 The Discrete WKB Method

Like its continuous counterpart, the discrete WKB method is applicable to a wide variety of settings. (See the review by Braun [12] for examples.) We will only give an overview and physically motivated discussion of this method using the Hamiltonian (1) as an illustrative example.

Let us write a general eigenfunction of $\mathcal{H}_0$ as $\sum_m a_m|m\rangle$, where as usual $S_z|m\rangle = m|m\rangle$. Schrödinger's equation then becoems a three-term recursion relation for the coefficients a_m:

$$w_m a_m + t_{m,m+2}\, a_{m+2} + t_{m,m-2}\, a_{m-2} = E a_m, \tag{11}$$

where $w_m = \langle m|\mathcal{H}_0|m\rangle$, and $t_{m,m\pm 2} = \langle m|\mathcal{H}_0|m \pm 2\rangle$. Now for large S, the differences $w_{m+2} - w_m$ and $t_{m,m+2} - t_{m-2,m}$, etc., are of order $1/S$ relative

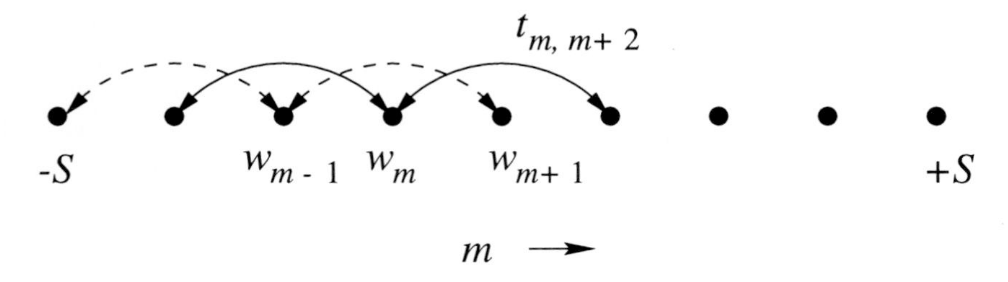

Figure 2: Mapping of spin problem onto an electron hopping on a lattice. The diagram is drawn for a case such as Eq. (11) with only second neighbour hopping. It is evident that the sites connected by dashed and solid lines then belong to two disjoint subspaces of the Hamiltonian.

to w_m and $t_{m,m+2}$ themselves. It is thus extremely useful to view Eq. (11) as arising from a tight-binding model for an electron on a one-dimensional lattice with sites labeled by m, on-site energies w_m, and hopping energies $t_{m,m\pm2}$, that *vary slowly with position*.[e] (See Fig. 2.) This viewpoint immediately suggests the approximation of semiclassical electron dynamics, and indeed this approximation is identical to discrete WKB.

To apply semiclassical dynamics, we define local m-dependent functions $w(m)$ and $t(m)$ by

$$\begin{aligned} w(m) &= w_m, \\ t(m) &= (t_{m,m+2} + t_{m,m-2})/2, \end{aligned} \tag{12}$$

which we extend to continuous values of m by demanding that they be smooth, and that dw/dm and dt/dm be of relative order S^{-1} with m formally regarded as a quantity of order 1. A particle of energy E can be assigned a local, m-dependent, wavevector $q(m)$ in complete analogy with the continuous WKB approach. The only difference is that the kinetic energy is given by $2t(m)\cos q(m)$ instead of $q^2(m)/2\mu$ (μ being the mass). We thus obtain

$$q(m) = \cos^{-1}\left(\frac{E - w(m)}{2t(m)}\right). \tag{13}$$

The general solution to (11) is given by linear combinations of

$$a_m \sim \frac{1}{\sqrt{v(m)}} \exp\left(\pm i \int^m q(m')\frac{dm'}{2}\right). \tag{14}$$

[e]For the Hamiltonian (1), the hopping connects sites differeing by $\Delta m = 2$, so that the eigenvalue problem divides into two subspaces depending on whether $S - m$ is even or odd. This point is of no consequence for application of the discrete WKB method itself, as the problem can be recast as one of nearest-neighbor hopping in each subspace.

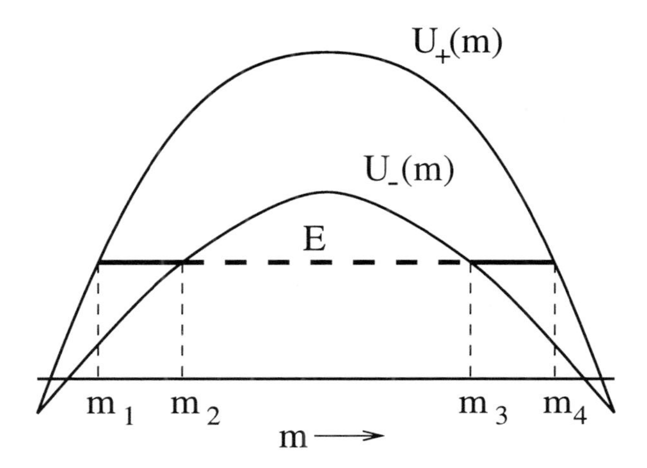

Figure 3: Discrete WKB method potential energy diagram for a symmetric double-well problem. The functions $U_{\pm}(m) = w(m) \pm 2t(m)$ are the local band edges. For a particle with energy E as shown, the regions $m_1 \leq m \leq m_2$, and $m_3 \leq m \leq m_4$ are classically accessible, and represent the two wells. The particle must tunnel across the classically forbidden region between m_3 and m_4.

In this equation,

$$v(m) = -2t(m) \sin q(m) \tag{15}$$

is the local particle velocity (equal to $\partial E/\partial q$), and the factor $[v(m)]^{-1/2}$ fixes the normalization so that the probability current is conserved. Also, the factor of $1/2$ in the integrand reflects the step length $\Delta m = 2$ in Eq. (11).

We conclude here our general discussion of the discrete WKB method, and continue the illustration with the Hamiltonian (1) in the next section. Braun's review [12] contains a proper proof of the above aprroximations, as well as connection formulas at turning points, Bohr-Sommerfeld quantization rules, etc. Most of these are physically apparent, and the chief novelty arises from the fact that for particle in a periodic potential (or a tight-binding model), the allowed energies lie in a *band*, and are bounded both above and below. This gives rise to turning points when the energy equals the local *upper* band edge $w(m)+2|t(m)|$, in addition to those that occur when the energy equals $w(m)-2|t(m)|$. These features are illustrated in Fig. 3. Braun refers to the first kind of turning point as "unusual", but the associated formulas are easily derived from those for a "usual" turning point by making a gauge transformation which changes the sign of every other coefficient a_m. The associated bookkeeping can be fairly cumbersome, however, and it is easy to make mistakes.

4 Tunnel Splitting for the Hamiltonian (1)

We now turn to applying the discrete WKB method to finding the ground state wavefunction and tunnel splitting Δ_0 for the Hamiltonian (1). (We can also find the higher splttings and wavefunctions.[24]) There are three steps in finding the wavefunction itself: (i) find it in the classically allowed region, (ii) find it in the classically forbidden region $|m| \ll S$, (iii) match the two parts of the wavefunction using the connection formulas or otherwise. The splitting is then obtainable by a textbook formula.[f]

To find the wavefunction in the classically allowed region near $m = -S$, e.g., it is advantageous to write $C_n = a_{-S+n}$, with $n = 0, 1, \ldots$. It is generally necessary to obtain the functions $w(m)$ and $t(m)$ to an accuracy such that the first two terms in an expansion in powers of S^{-1} are correctly given; if only the leading term is kept, then retention of the $\left(v(m)\right)^{-1/2}$ factor in Eq. (14) can not be justified. In carrying out this exercise near the ends of the chain, as is the case now, it is necessary to treat $n = S + m$ rather than m as a quantity of order S^0. It is also useful to add a constant $k_1' S(S+1)$ to the Hamiltonian. The Schrödinger equation then reads

$$(2k_1+k_2)\left(n+\frac{1}{2}\right)C_n+\frac{1}{2}k_2\sqrt{n(n-1)}C_{n-2}+\frac{1}{2}k_2\sqrt{(n+1)(n+2)}C_{n+2} = EC_n. \tag{16}$$

This equation, however, can be solved exactly. It is that for the harmonic oscillator Hamiltonian

$$\mathcal{H}_{\rm ho} = (2k_1 + k_2)a^\dagger a + \frac{1}{2}k_2(a^2 + a^{\dagger 2}), \tag{17}$$

where $a^\dagger$ and a are raising and lowering operators for the number eigenstates $|n\rangle$ obeying $a^\dagger a|n\rangle = n|n\rangle$, and $C_n = \langle n|\psi\rangle$ for an energy eigenfunction $|\psi\rangle$. We can diagonalize $\mathcal{H}_{\rm ho}$ by writing $a = (x+ip)/2^{1/2}$, $a^\dagger = (x-ip)/2^{1/2}$, where x and p are canonical position and momentum operators obeying $[x,p] = i$ as usual. This yields

$$\mathcal{H}_{\rm ho} = k_{12}x^2 + k_1 p^2. \tag{18}$$

---

[f] See Eq. (23) below. This formula is anologous to one that appears in the solution to Problem 3, Sec. 50, in the famous text by Landau and Lifshitz.[25] However, it is *not* the final formula in that solution! We have found it preferable to use the unnumbered intermediate equation in which the splitting is given as a product of the wavefunction and its derivative at the symmetry point of the potential. The three dimensional analog of the latter formula is often named after C. Herring who first used it find the splitting of the electron terms in the H_2^+ ion. This problem is also discussed by Landau and Lifshitz in the solution to the problem accompanying Sec. 81.

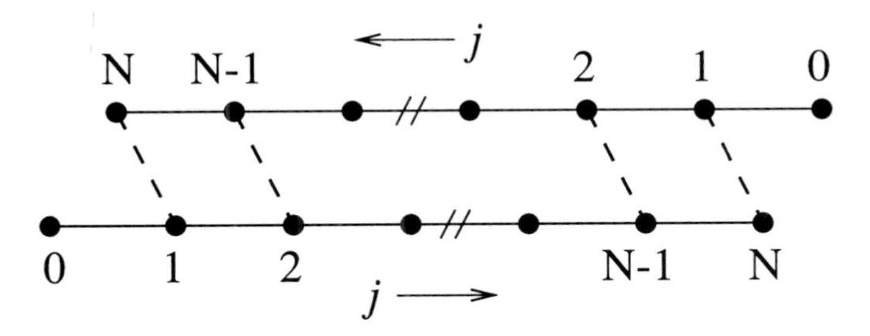

Figure 6: Same as Fig. 5 for $\Delta(1/2, 1)$. Only sites in the subspace $S_z = S$, $I_z = 1/2$ are shown this time.

The dominant contribution to the sum arises from the middle of the chains, i.e., from values of j such that $(N-j), j \gg 1$. Using Eq. (22) and neglecting terms of order unity in comparison with j and $N-j$, we obtain

$$V = -(-1)^N 2\,\omega_n (S/\pi)^{1/2} \operatorname{sech}\theta \tanh^N \theta \sum_j \left[(j+1/2)(N-j+1/2)\right]^{-1/2}. \tag{28}$$

For $N \gg 1$ the sum may be approximated by an integral, which is easily shown to equal π. Making use of Eq. (24), and recalling that $S = N + 1/2$, we can write the tunnel splitting (which equals $2|V|$) as

$$\Delta_{-1} = \Delta(1/2, -1) = \frac{\pi}{2}\frac{\omega_n}{\omega_e}(\tanh\theta)^{-1/2}\Delta_0. \tag{29}$$

The splitting $\Delta(1/2, 1)$ $(= \Delta_1)$ can be found in the same way. The appropriate lattice diagram is drawn in Fig. 6. This time the two low energy states are $|+0, p=1\rangle$ and $|-0, p=-1\rangle$, and the wavefunctions are given by the same C'_j as above with the sites labeled as shown. The key quantity is again the off-diagonal element. It is obvious from the figure that this is again given by a sum as in Eq. (27). The only change required is that we replace the product of wave functions in that summand by $C'_j C'_{N+1-j}$. The sum is again dominated by the central region, and since $C'_{j+1} \approx -\tanh\theta C'_j$ in this region, the change in the summand has the effect of multiplying the each term in the sum by $-\tanh\theta$. The final result for the splitting is, therefore,

$$\Delta_1 = \Delta(1/2, 1) = \frac{\pi}{2}\frac{\omega_n}{\omega_e}(\tanh\theta)^{1/2}\Delta_0. \tag{30}$$

We thus see that apart from the factors of $(\tanh\theta)^{\pm 1/2}$, the results (29) and (30) agree with the instanton answer, Eq. (9). This agreement shows that

Table 2: Comparison between numerical and analytical [Eqs. (29) and (30)] results for one-coflip tunnel splitting. The parameters are $k_1 = 5.0$, $k_2 = 20.0$, and $\omega_n = 0.1$. For each value of S, the entry in the upper row is Δ_1, and that in the lower row Δ_{-1}.

S	$\Delta_{\pm 1}$ (numerical)	$\Delta_{\pm 1}$ (analytic)	Error(%)
$10\frac{1}{2}$	2.576×10^{-5}	2.681×10^{-5}	4.1
	6.661×10^{-5}	7.018×10^{-5}	5.4
$11\frac{1}{2}$	1.034×10^{-5}	1.072×10^{-5}	3.6
	2.674×10^{-5}	2.805×10^{-5}	4.9
$12\frac{1}{2}$	4.134×10^{-6}	4.267×10^{-6}	3.2
	1.069×10^{-5}	1.117×10^{-5}	4.5
$13\frac{1}{2}$	1.646×10^{-6}	1.694×10^{-6}	2.9
	4.257×10^{-6}	4.435×10^{-6}	4.2
$14\frac{1}{2}$	6.534×10^{-7}	6.705×10^{-7}	2.6
	1.689×10^{-6}	1.755×10^{-6}	3.9
$15\frac{1}{2}$	2.587×10^{-7}	2.648×10^{-7}	2.4
	6.688×10^{-7}	6.933×10^{-7}	3.7
$16\frac{1}{2}$	1.022×10^{-7}	1.044×10^{-7}	2.2
	2.641×10^{-7}	2.732×10^{-7}	3.5
$17\frac{1}{2}$	4.024×10^{-8}	4.105×10^{-8}	2.0
	1.041×10^{-7}	1.075×10^{-7}	3.3
$18\frac{1}{2}$	1.583×10^{-8}	1.612×10^{-8}	1.9
	4.092×10^{-8}	4.221×10^{-8}	3.2
$19\frac{1}{2}$	6.217×10^{-9}	6.322×10^{-9}	1.7
	1.608×10^{-8}	1.655×10^{-8}	3.0
$20\frac{1}{2}$	2.438×10^{-9}	2.476×10^{-9}	1.6
	6.303×10^{-9}	6.482×10^{-9}	2.9

the instanton method is basically sound, and although it can not in its simplest version give the prefactors in Δ_0 correctly, the overall picture it provides is correct. In fact, the $(\tanh\theta)^{\pm 1/2}$ factors can also be found from the instanton approach by exercising a little more care. These origin of these factors is actually very easy to understand. They reflect the fact that the state with antiparallel alignment of the electronic and nuclear spins has greater quantum fluctuations or zero point motion than the state with parallel alignment. Another way to say this is that the state $|S_z = S, I_z = I\rangle$ is an eigenstate of the hyperfine coupling term in $\mathcal{H}_I$, while $|S_z = S, I_z = -I\rangle$ is not. This difference in zero point fluctuations shows up as a higher tunneling rate for coflips in which the nuclear spins are oppositely aligned to the electronic ones.

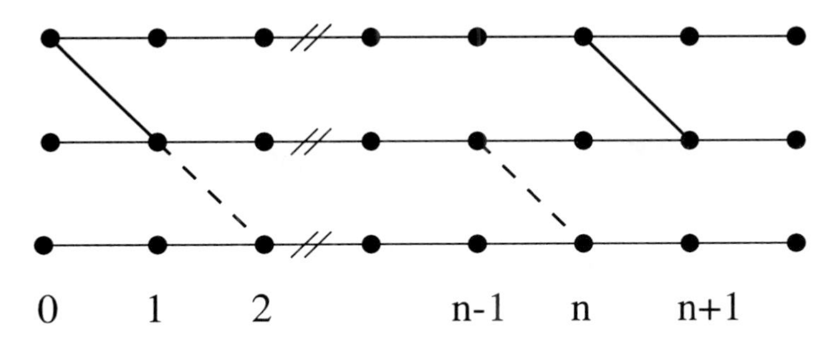

Figure 7: Lattice diagram for the case $I = 1$. The bonds entering into the calculation of the transition matrix element V relevant to $\Delta(1,2)$ and $\Delta(1,-2)$ are shown as dashed and solid diagonal lines respectively.

We also remind readers that in Eqs. (29) and (30), Δ_0 is not the tunnel splitting for the bare problem without nuclear spins (which vanishes because S is half-integral), but rather the pseudo-splitting formally given by Eq. (24).

In Table 2, we show the comparison between the above formulas and exact answers from numerical diagonalization of $\mathcal{H}_I$ with $I = 1/2$. We choose $k_1 = 5.0$ and $k_2 = 20.0$ as before, and $\omega_n = 0.1$. Since $\omega_e = 22.36$, the condition $\omega_n \ll \omega_e$ is well obeyed. Again the agreement is very satisfactory, and improves with increasing S. In particular, the inclusion of the $(\tanh\theta)^{\pm 1/2}$ factors seems to be required.

5.2 Two-Coflip Splittings

Let us proceed to calculate the two-coflip splittings, $\Delta(1,\pm 2)$ $(= \Delta_{\pm 2})$, by studying the Hamiltonian (6) for $I = 1$ and integer S. As before, we carry out the calculation by including the $S_z I_z$ term in the unperturbed Hamiltonian along with $\mathcal{H}_0$, and treating the $S_+ I_-$ and $S_- I_+$ terms as a perturbation.

Suppose we wish to find $\Delta(1,2)$. The appropriate lattice diagram is shown in Fig. 7. This tunneling process requires making a transition from the state $|-0, p = -2\rangle \equiv |\mathrm{I}\rangle$ to the state $|0, p = 2\rangle \equiv |\mathrm{IV}\rangle$. However, there is no direct matrix element of $\mathcal{H}_I$ between these states, and we must go through intermediate states. The lowest energy intermediate states are $|-1, p = 0\rangle \equiv |\mathrm{II}\rangle$ and $|1, p = 0\rangle \equiv |\mathrm{III}\rangle$. Keeping only these we obtain a four-state effective Hamiltonian

$$\mathcal{H}_{\text{eff}} = \begin{pmatrix} E_0 & V & 0 & 0 \\ V & E_1 & \Delta^{(1)}/2 & 0 \\ 0 & \Delta^{(1)}/2 & E_1 & V \\ 0 & 0 & V & E_0 \end{pmatrix}, \tag{31}$$

where the equality of various elements is assured by symmetry. Here, $E_0 = \langle \mathrm{I}|\mathcal{H}_I|\mathrm{I}\rangle = \langle \mathrm{IV}|\mathcal{H}_I|\mathrm{IV}\rangle$, $E_1 = \langle \mathrm{II}|\mathcal{H}_I|\mathrm{II}\rangle = \langle \mathrm{III}|\mathcal{H}_I|\mathrm{III}\rangle$, $V = \langle \mathrm{II}|\mathcal{H}_I|\mathrm{I}\rangle$, and $\Delta^{(1)}$ is the tunnel splitting between the states $|\mathrm{II}\rangle$ and $|\mathrm{III}\rangle$. This splitting is identical to the splitting between the first excited states of the Hamiltonian $\mathcal{H}_0$ without nuclear spins, and is given by [24]

$$\Delta^{(1)} = \frac{2S}{\sinh\theta \cosh\theta}\Delta_0. \tag{32}$$

Secondly, the energy difference $E_1 - E_0 = \omega_e + \mathcal{O}(\omega_n)$.

The key quantity in calculating Δ_2 is thus the matrix element V as before. This is given by a sum involving the lattice site wavefunctions of the states $|\mathrm{I}\rangle$ and $|\mathrm{II}\rangle$ and a $I_- S_+$ matrix element. Refering to Fig. 7, consider the contribution of the bond connecting site n in the lower row to site $n-1$ in the middle row. The matrix element for this bond equals

$$-\frac{\omega_n}{2S}\langle -S+n-1, I_z = 0|S_+ I_-| -S+n, I_z = -1\rangle = -\frac{\omega_n}{\sqrt{2S}}\left(n(2S-n+1)\right)^{1/2}. \tag{33}$$

The wavefunctions at the two ends of this bond, on the other hand, are given by $\langle n|\psi_0\rangle$ and $\langle n-1|\psi_1\rangle$, where $|\psi_0\rangle$ and $|\psi_1\rangle$ are the ground and first excited state of the harmonic oscillator Hamiltonian $\mathcal{H}_{\mathrm{ho}}$, Eq. (17), and $|n\rangle$ are the eigenstates of the number operator $a^\dagger a$. Thus,

$$\begin{aligned} V &= -\frac{\omega_n}{\sqrt{2}S}\sum_{n=0}^{\infty}\left(n(2S-n+1)\right)^{1/2}\langle\psi_1|n-1\rangle\langle n|\psi_0\rangle \\ &= -\frac{\omega_n}{\sqrt{S}}\sum_{n=0}^{\infty} n^{1/2}\langle\psi_1|n-1\rangle\langle n|\psi_0\rangle \\ &= -\frac{\omega_n}{\sqrt{S}}\sum_{n=0}^{\infty}\langle\psi_1|a|n\rangle\langle n|\psi_0\rangle \\ &= -\frac{\omega_n}{\sqrt{S}}\langle\psi_1|a|\psi_0\rangle \\ &= \frac{\omega_n}{\sqrt{S}}\sinh\theta. \end{aligned} \tag{34}$$

The second equality in this chain follows from noting that the sum is dominated by values of n of order unity, which allows us to neglect n in comparison to S. The final equality follows if one makes use of the Bogoliubov transformation (21). Then, $\langle\psi_1|a|\psi_0\rangle = -\sinh\theta\langle\psi_1|b^\dagger|\psi\rangle_0 = -\sinh\theta$.

Table 3: Comparison between numerical and analytical results for two-coflip tunnel splittings, for $k_1 = 5.0$, $k_2 = 20.0$, $\omega_n = 0.5$, $S = 15$, and $I = 1$. In the column labeled 'ratio' we give the ratio of the numerically obtained splittings to those given by the instanton method, i.e., to Eqs. (36) and (39), but with a numerical factor of $\pi^2/4$ instead of 2.

	Numerical	Ratio
Δ_2	2.559×10^{-8}	0.56
Δ_{-2}	3.488×10^{-7}	1.11

To leading order in ω_n, the energy splitting of the lowest two states of the effective Hamiltonian (31) is easily shown to be given by

$$\frac{V^2}{E_1 - E_0 - \Delta^{(1)}/2} - \frac{V^2}{E_1 - E_0 + \Delta^{(1)}/2} \approx \frac{V^2}{(E_1 - E_0)^2}\Delta^{(1)}. \tag{35}$$

Combining Eqs. (32), (34), and using the fact that $E_1 - E_0 = \omega_e + \mathcal{O}(\omega_n)$, we finally obtain

$$\Delta_2 = \Delta(1,2) \approx 2\frac{\omega_n^2}{\omega_e^2} \tanh\theta \Delta_0. \tag{36}$$

The calculation of $\Delta(1,-2)$ proceeds very similarly. The effective Hamiltonian has the same structure as Eq. (31), except that this time the states $|\mathrm{I}\rangle$ and $|\mathrm{IV}\rangle$ must be taken as $|-0, p = 2\rangle$ and $|+0, p = -2\rangle$ respectively, while the states $|\mathrm{III}\rangle$ and $|\mathrm{IV}\rangle$ are unchanged. This time the bonds connect site n in the upper row with site $n+1$ in the lower row, and the matrix element is

$$-\frac{\omega_n}{2S}\langle -S+n+1, I_z = 0|S_- I_+| - S + n, I_z = 1\rangle \approx -\frac{\omega_n}{\sqrt{S}}(n+1)^{1/2} \tag{37}$$

instead of Eq. (33). Using the same technique as before, the transition matrix element V is found to be

$$V = -\frac{\omega_n}{\sqrt{S}}\sum_{n=0}^{\infty}(n+1)^{1/2}\langle\psi_1|n+1\rangle\langle n|\psi_0\rangle = -S^{-1/2}\omega_n\cosh\theta. \tag{38}$$

Substituting this in Eq. (35), we obtain the splitting as

$$\Delta_{-2} = \Delta(1,2) \approx 2\frac{\omega_n^2}{\omega_e^2}(\tanh\theta)^{-1}\Delta_0. \tag{39}$$

The results (36) and (39) are very close to but not exactly what one would expect from the instanton method, even after the fluctuational factors of $\tanh\theta$

have been included. If we define $\eta_\pm = (\pi\omega_n/2\omega_e)(\tanh\theta)^{\pm 1/2}$, then an instanton argument would lead us to expect

$$\Delta_{\pm 2} = \eta_\pm^2 \Delta_0. \tag{40}$$

This would require the numerical factors in Eqs. (36) and (39) to be $\pi^2/4$ instead of 2. Could this really be the case, and if so, what is the source of the discrepancy in our present calculation? One possible answer is that we have ignored the contribution of higher energy intermediate states in the calculation of $\Delta_{\pm 2}$. In fact, it can be shown that these states' contribution is also formally of order $(\omega_n/\omega_e)^2\Delta_0$, but with different numerical prefactors, which decrease rapidly with increasing intermediate state energy. The accurate calculation of these numerical factors is difficult. Similarly, we neglected the terms of higher order in S^{-1} in expanding the matrix elelment (33). Terms of such higher order are also present in the contributions from the higher intermediate states, and it is not clear that the *sum* of all these terms will continue to be of higher order in S^{-1}. When these approximations are taken into account, we cannot exclude the possibility that Eq. (40) holds exactly.

We show a comparison between our theoretical and numerical results in Table 3. The quality of agreement is still fairly good, though not as impressive as for the one-coflip splittings. We have not carried out this numerical work as extensively as in the previous cases, and so can not comment on the behavior with increasing S.

5.3 Higher Coflip Processes; Effective Hamiltonian

It is evident that the discrete WKB method is increasingly ill suited to the calculation of higher coflip processes. Systematic numerical investigation of this problem runs into the difficulty that the splittings necessarily decrease with increasing coflip number p, requiring the numerical diagonalization to be carried out to increasing precision. My own numerical studies are very limited, and while it is undoubtedly possible to improve on them, that would require substantially greater time and effort than I am prepared to commit!

There is, however, one aspect of the higher coflip problem, which can be understood quite simply. [We have alluded to this earlier; see the discussion immediately preceding Eq. (10).] This is that splittings $\Delta(I,p)$ for different I but equal p can be related to each other. To see this, let us consider the problem in terms of N_n nuclear spins each with spin 1/2 rather than the model (6). There are then a large number of states in the $|+0,p\rangle$ group that are degenerate with one another, and with an equally large number of mutually degenerate states in the $|-0,-p\rangle$ group. (Recall that $p = 2I_z^{\text{tot}}$.) A p-coflip process can take us from any of the states in the first group to several states in the second group. A convenient algebraic method for keeping track of which states are connected is as follows. To avoid cluttering up the formulas, it is best to treat the cases $p > 0$ and $p < 0$ separately. Let us do the $p > 0$ case first. Let $\pi_{\pm p}$ be projection operators onto the set of states with $I_z^{\text{tot}} = \pm p/2$, irrespective of the value of I. Further, let

$$\begin{aligned} \sigma_+ &= |+0\rangle\langle -0|, \\ \sigma_- &= |-0\rangle\langle +0|, \end{aligned} \tag{41}$$

be transition operators for the large electronic spin between the two states involved in the MQC process. Finally, let Δ_p be the amplitude for the transition between any two of the states in the groups with opposite I_z. Then, an effective Hamiltonian that describes all the coflip processes of order p is

$$\mathcal{H}_p^{\text{cf}} = \Delta_p\Big(\sigma_+\pi_p Q_p^+ \pi_{-p} + \text{h.c.}\Big); \tag{42}$$

$$Q_p^+ = \sum_{j_1,\, j_2,\ldots,j_p}\sum\cdots\sum I_{j_1}^+ I_{j_2}^+ \cdots I_{j_p}^+, \tag{43}$$

where $I^{\pm} = I_x \pm iI_y$ as usual, and the sum in Eq. (43) is over all distinct p-tuplets of indices chosen from the indices $1, 2, \ldots, N_n$.

Let us check that Eq. (42) does what it is supposed to. Consider the first term on the right, and let it operate on a ket. The projection operator π_{-p} ensures that this term is only relevant when we operate on a state with $I_z = -p/2$. The operator Q_p^+ then raises (or flips) p nuclear spins leading to a state with $I_z = p/2$. The projection operator π_p in this term is added to ensure that the result is sensible when we operate on a bra. The second term in (42) ensures hermiticity and describes processes in which I_z is lowered.

It is now very easy, however, to find the I dependence of the matrix elements of $\mathcal{H}_p^{\text{cf}}$. We first note that because $(I_j^+)^2 = 0$ for any j, we can write

$$Q_p^+ = \frac{1}{p!}\Big(I_1^+ + I_2^+ + \cdots + I_{N_n}^+\Big)^p = \frac{1}{p!}(I^+)^p. \tag{44}$$

The only non-zero matrix element of $\mathcal{H}_p^{\text{cf}}$ between states of given I is thus equal to

$$\frac{\Delta_p}{p!}\left\langle I, m = \frac{1}{2}p\middle|(I^+)^p\middle|I, m = -\frac{1}{2}p\right\rangle = \Delta_p\binom{I + \frac{1}{2}p}{p}. \tag{45}$$

The case $p < 0$ is now easily treated by minor changes of notation. We simply change Q_p^+ in the first term in Eq. (42) to its hermitean adjoint Q_p^-,

Table 4: Numerical results for higher coflip tunnel splittings. The parameters are $k_1 = 5.0$, $k_2 = 20.0$, $\omega_n = 0.5$, $S = 15$, and $I = 0, 1, 2, 3$. The table shows $\Delta(I,p)$ divided by the combinatoric factor in Eq. (46). This combinatoric number is shown next to the result in parentheses.

p	$I=0$	$I=1$	$I=2$	$I=3$
0	$9.412 \times 10^{-5}(1)$	$9.435 \times 10^{-5}(1)$	$9.479 \times 10^{-5}(1)$	$9.546 \times 10^{-5}(1)$
-2		$3.488 \times 10^{-7}(1)$	$3.497 \times 10^{-7}(3)$	$3.507 \times 10^{-7}(6)$
2		$2.559 \times 10^{-8}(1)$	$2.586 \times 10^{-8}(1)$	$2.628 \times 10^{-8}(6)$
-4			$2.134 \times 10^{-9}(1)$	$2.138 \times 10^{-9}(5)$
4			$2.655 \times 10^{-11}(1)$	$2.580 \times 10^{-11}(5)$
-6				$1.735 \times 10^{-11}(1)$
6				$3.624 \times 10^{-13}(1)$

with the sum in the analog of Eq. (43) being over all $|p|$-tuplets. Thus $Q_p^- = (I^-)^{|p|}/|p|!$ for $p < 0$, and the matrix element analogous to (45) is identically evaluated. The upshot is that we can write

$$\Delta(I,p) = \binom{I + \frac{1}{2}|p|}{|p|} \Delta_p \tag{46}$$

for all p, positive or negative.

We show in Tables 4 and 5 some numerical results for tunnel splittings involving up to 6 nuclear spin coflips. Instead of tabulating $\Delta(I,p)$ itself, we have divided it by the combinatorial factor in Eq. (46). We expect that the resultant quantities will be independent of I, and we can see from the tables that indeed they are. In all the cases, the spread in the values of Δ_p so deduced is less than 2%.

Table 5: Same as Table 4 for odd numbers of coflips. Now $S = 31/2$, and $I = 1/2, 3/2, 5/2$. The other parameters are unchanged.

p	$I=1/2$	$I=3/2$	$I=5/2$
-1	$3.265 \times 10^{-6}(1)$	$3.272 \times 10^{-6}(2)$	$3.284 \times 10^{-6}(3)$
1	$1.327 \times 10^{-6}(1)$	$1.326 \times 10^{-6}(2)$	$1.323 \times 10^{-6}(3)$
-3		$1.619 \times 10^{-8}(1)$	$1.623 \times 10^{-8}(4)$
3		$2.305 \times 10^{-9}(1)$	$2.306 \times 10^{-9}(4)$
-5			$1.136 \times 10^{-10}(1)$
5			$6.457 \times 10^{-12}(1)$

The fact that $\Delta(I,p)$ grows with I may make one wonder if this might not be a way to boost the resonance frequency for some of the tunneling processes. In fact, the result (46) only applies to the ideal case where the hyperfine couplings are identical for all the nuclei. This limits the relevance of this result for actual systems to small values of p or I. Real particles typically possess some small spread in these couplings. This spread spoils the degeneracy of all the states in the $|0,p\rangle$ group (and likewise for the states $|-0,-p\rangle$). When the spread starts to equal Δ_p, then the constructive interferences which give rise to the large combinatoric factors in Eq. (46) are no longer possible. The spectral weight in χ'' is then shifted from the higher frequency peaks in a given coflip group to lower frequencies. Secondly, we have neglected incoherent processes involving the nuclear spins. MQC is destroyed by a single such process, and since the likelihood that at leat one such process will occur in a given time grows linearly with N_n, the number of nuclear spins, it is evident that it does not pay to increase this number.

6 Conclusions

The main purpose of this article has been to verify previous calculations [3,9] of the coflip tunneling frequencies without using instanton methods. We find that by and large, the instanton calculations lead to correct answers.

Our discussion up to this point has treated the system as closed, with no means for energy to flow in and out. If this were strictly true, this would imply that χ'' consisted of a sum of delta functions at the frequencies $\Delta(I,p)$. This is of course an idealization. To complete the discussion of the problem, we must also consider relaxation. If we put in broadening by hand, the qualitative picture of the χ'' spectrum is as shown in Fig. 8. A proper investigation of the physical mechanisms behind this would take us into very different territory. Two obvious mechanisms which may be mentioned in passing are phonons and dipolar magnetic fields due to other nuclei such as protons which may be present in the material. It does not seem worthwhile to develop a quantitative theory of the relaxation due to even these processes in the absence of experimental impetus, but certain broadly valid comments can nevertheless be made. By appealing to the general principles laid down in Refs. 11 and 29, it is quite plausible that the relaxation can be treated by coupling the resonating variable (which we would regard as different for each coflip line) to a phenomenological harmonic oscillator bath (which would also be different for each coflip line). Broadly speaking, such a bath has two effects. First, it leads to a pulling down or downward renormalization of the bare coherence frequency Δ_b to $\tilde{\Delta}_b$. At the simplest level this renormalization can be understood in terms

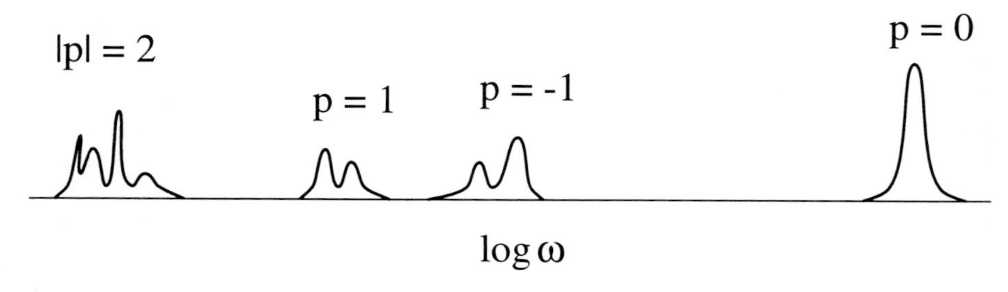

Figure 8: Sketch of the expected tunneling susceptibility spectrum $\chi''(\omega)$ in the presence of nuclear spins, showing the first few polarization blocks. Note that the frequency scale is logarithmic. We have tried to indicate the fine structure within the higher polarization blocks, but we have made no attempt to represent the broadening of the lines correctly.

of a multiplicative Franck-Condon factor. Second, it gives rise to true damping of the resonance, i.e., an intrinsic broadening or linewidth. (There may of course be additional inhomogeneous broadening arising from a spread in $\tilde{\Delta}_b$.) The general point which is noteworthy is that as the bare frequency Δ_b of the coherence phenomenon under study decreases, the environment suppresses it even further. This is especially so if the environment has any subohmic or ohmic component. Environmental degrees of freedom that might not have been relevant if Δ_b had been higher by a factor of 10, say, gang up on the system, as it were, and do start to matter. Further, degeneracy is broken by weaker and weaker stray fields and drifts, effectively eliminating the resonance altogether.

All this implies that the higher coflip resonances in our problem are highly likely to be overdamped and unobservable. This point can be seen even in the simplistic model introduced in Ref. 9, and briefly touched upon at the end of the previous section. If a single nuclear spin undergoes an incoherent transition from $I_z = 1/2$ to $I_z = -1/2$ or vice versa at a rate $1/\tau$, then any MQC in the particle as a whole is damped at a rate N_n/τ. All the lines in $\chi''(\omega)$ thus acquire a width of order N_n/τ irrespective of the number of coflips involved, and the higher coflip lines with $\Delta(I,p) < N_n/\tau$ essentially merge into a mushy continuous background with no clear feature that could be identified as a resonance. This in turn means that while an accurate calculation of $\Delta(I,p)$ for large $|p|$ would still have theoretical interest, the unlikelihood of its relevance to experiment reduces the urgency of this calculation considerably. The most interesting remaining problem at this stage appears to be that of broadening of the very low order coflip lines.

Let us therefore summarize our conclusions about how MQC in small magnetic particles is affected by nuclear spins. The effect is severe, and the spectrum of χ'' is broken up into a large number of lines which can be grouped by the number of nuclear spin coflips required to maintain degeneracy, or more conveniently by the nuclear spin polarization

$$p = \frac{2}{s} \sum_i s_{iz} I_{iz} , \tag{47}$$

where s_i and I_i are the electronic and nuclear spins on atom i. The frequency of a line decreases quasi-geometrically with the number of coflips $|p|$, and there is a further distinction between $p > 0$ and $p < 0$, with the latter having a higher frequency. Within each polarization group, there is a fine structure to the tunneling spectrum controlled by interference between the different ways in which p nuclear spins can coflip. The total spectral weight in a given polarization block is therefore given by f_p, the probability that a particle will have polarization p. This probability can be controlled by thermostatstical factors. For example, if we assume that the nuclear spins are in equilibrium with the elctronic spin, then the Boltzmann weights for $\mathbf{I}_i$ parallel and antiparallel to $\mathbf{s}_i$ are in the ratio $1 : \exp(-\beta\omega_n)$, where $\beta = 1/k_B T$. This leads to

$$f_p = (2\pi\sigma_p^2)^{-1/2} e^{-(p-\bar{p})^2/2\sigma_p^2} , \tag{48}$$

where $\bar{p} = N_n \tanh(\beta\omega_n/2)$, and $\sigma_p = N_n^{1/2} \mathrm{sech}\,(\beta\omega_n/2)$. The dominant resonance line is thus that with zero polarization, and has a weight f_0.

Finally, we note that the effect of nuclear spins on magnetic particle MQC is completely different from that of an oscillator bath. [4,5] While unlike the latter, the present problem does not seem to lend itself to the study of elegant mathematical questions such as renormalization group connections with the Kondo problem, [30,31] it provides a novel and interestingly different example of how interaction with the environment suppresess MQC.

Acknowledgments

This research is supported by the National Science Foundation via Grant number DMR-9616749. I would also like to thank the members and the staff of the Institute for Nuclear Theory, University of Washington, Seattle, for their warm hospitality during the workshop on Tunneling in Complex Systems in the spring of 1997. While the specific research reported herein was not carried out at the Institute, my discussions with its members and various workshop participants have all helped to increase my understanding of this subject.

References

1. E. M. Chudnovsky and L. Gunther, *Phys. Rev. Lett.* **60**, 661 (1988).
2. See A. J. Leggett in *Chance and Matter*, edited by J. Souletie, J. Vannimenus, and R. Stora (North-Holland, Amsterdam, 1987) for a general review of MQT and MQC.
3. A. Garg, *J. Appl. Phys.* **76**, 6168 (1994).
4. S. Chakravarty and A. J. Leggett, *Phys. Rev. Lett.* **52**, 5 (1984).
5. A. J. Leggett et al., *Rev. Mod. Phys.* **59**, 1 (1987); *ibid* **67**, 725 (E) (1995).
6. U. Weiss, *Quantum Dissipative Systems* (World Scientific, Singapore, 1993). See especially Part IV.
7. H. W. Kroto, *Molecular Vibration Spectra* (London, Wiley, 1975); Ch. 9.
8. A. Garg, *Phys. Rev. Lett.* **70**, 1541 (1993).
9. A. Garg, *Phys. Rev. Lett.* **74**, 1458 (1995).
10. R. P. Feynman and F. L. Vernon, *Ann. Phys. (NY)* **24**, 118 (1963).
11. A. O. Caldeira and A. J. Leggett, *Ann. Phys. (NY)* **149**, 374 (1983).
12. P. A. Braun, *Rev. Mod. Phys.* **65**, 115 (1993).
13. J. L. van Hemmen and A. Sütő, *Europhys. Lett.* **1**, 481 (1986); Physica **141 B**, 37 (1986).
14. N. Prokof'ev and P. C. E. Stamp, *J. Phys.: Condens. Matter* **5**, L663 (1993).
15. N. Prokof'ev and P. C. E. Stamp, cond-mat/9511009; cond-mat/9511011; cond-mat/9511015; *J. Low Temp. Phys.* **104**, 143 (1996); cond-mat/9708199.
16. See the articles by M. A. Novak and R. Sessoli, and by C. Paulsen and J. G. Park in *Quantum Tunneling of Magnetization— QTM '94*, edited by L. Gunther abd B. Barbara (Kluwer, Dordrecht, 1995).
17. J. Friedman et al., *Phys. Rev. Lett.* **76**, 3820 (1996).
18. L. Thomas, et al., *Nature* **383**, 145 (1996).
19. B. Barbara and E. M. Chudnovsky, *Phys. Lett. A* **145**, 205 (1990).
20. I. V. Krive and O. B. Zaslavskii, *J. Phys.: Condens. Matter* **2**, 9457 (1990).
21. J. M. Duan and A. Garg, *J. Phys.: Condens. Matter* **7**, 2171 (1995).
22. A. Garg, *Phys. Rev. Lett.* **71**, 4249 (1993).
23. A. Garg, *Science* **272**, 476 (1996).
24. A. Garg, to be submitted to *J. Math. Phys.*
25. L. D. Landau and E. M. Lifshitz, *Quantum Mechanics*, 3rd edition (Pergamon, Oxford, 1977).
26. M. Enz and R. Schilling, *J. Phys. C: Solid Stae Phys.* **19**, 1765 (1986).

27. V. I. Belinicher, C. Providencia, and J. da Providencia, *J. Phys. A: Math. Gen.* **30**, 5633 (1997).
28. A. Garg and G. H. Kim, *Phys. Rev.* B **45**, 12921 (1992).
29. A. J. Leggett, *Phys. Rev.* B **30**, 1208 (1984).
30. S. Chakravarty, *Phys. Rev. Lett.* **49**, 681 (1982).
31. A. J. Bray and M. A. Moore, *Phys. Rev. Lett.* **49**, 1546 (1982).

TUNNELING FROM SUPER- TO NORMAL- DEFORMED MINIMA IN NUCLEI

TENG LEK KHOO

Physics Division, Argonne National Laboratory, 9700 S. Cass Avenue Argonne, IL60439, USA
E-mail: khoo@anl.gov

An excited minimum, or false vacuum, gives rise to a highly elongated superdeformed (SD) nucleus. A brief review of superdeformation is given, with emphasis on the tunneling from the false to the true vacuum, which occurs in the feeding and decay of SD bands. During the feeding process the tunneling is between hot states, while in the decay it is from a cold to a hot state. The γ spectra connecting SD and normal-deformed (ND) states provide information on several physics issues: the decay mechanism; the spin/parity quantum numbers, energies and microscopic structures of SD bands; the origin of identical SD bands; the quenching of pairing with excitation energy; and the chaoticity of excited ND states at 2.5 - 5 MeV. Other examples of tunneling in nuclei, which are briefly described, include the possible role of tunneling in ΔI=4 bifurcation in SD bands, sub-barrier fusion and proton emitters.

1 Introduction

The study of superdeformed nuclei provides new insights into the structure of the nucleus. It also illustrates that the nucleus is a laboratory for investigating more general physics phenomena such as tunneling, chaos and phase transitions in mesoscopic systems. Relating different fields of physics and identifying the applications of common principles is not only interesting but also beneficial. On one hand, the nucleus can provide interesting data for the study of these general phenomena, e.g. for understanding tunneling in complex systems. On the other hand, new insights into the nucleus can also be obtained from this approach. Finally, perspectives and techniques can be profitably shared among the different areas of physics.

These lecture notes will provide a description of superdeformation in nuclei and will emphasize the role of tunneling from the superdeformed to normal-deformed potential wells. As an experimentalist, I shall highlight what has and can be measured, with the hope that this will provide some stimulation to theoreticians.

2 Superdeformation in nuclei

2.1 General Properties

The nucleus, with a finite number of nucleons, is a mesoscopic object. The mesoscopic nature is illustrated by the fact that its energy can be given in terms of a superposition of a macroscopic liquid-drop term and a microscopic quantal shell-correction term. In a major advance in nuclear theory, Strutinsky[1] developed the method to compute the shell-correction energy. The shell energy has a profound impact on nuclear behavior and leads to a number of striking consequences. One is the increased binding (by up to 7 MeV) in the heaviest elements, with atomic number larger than 104, resulting in the creation of a fission barrier, where classically none would exist at all. The existence of these elements is possible only because of the extra binding from the shell-correction energy. Hence, the production and discovery[2] of the heaviest elements with Z up to 112 constitute some of the more exciting achievements in nuclear physics.

Another striking example (illustrated in Fig. 1) is the creation of a secondary minimum, or false vacuum, in the potential energy surface, at a de-

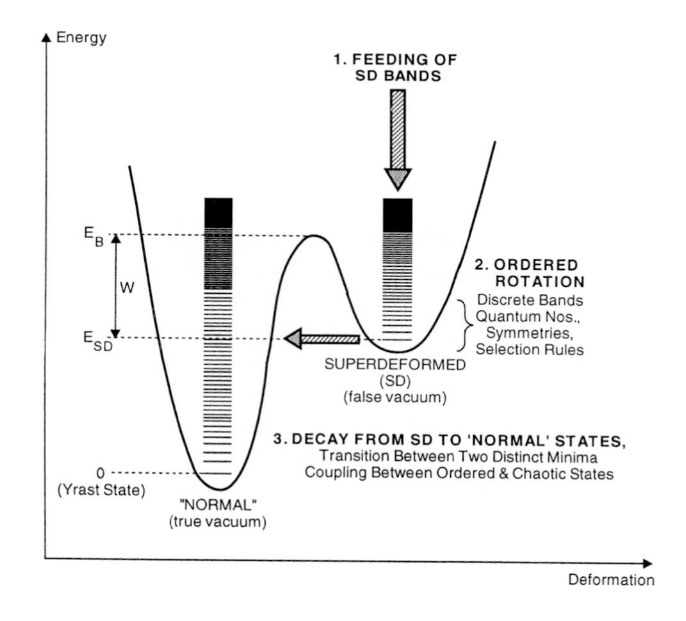

Figure 1: Sketch of potential energy *vs.* deformation, showing two minima created by the shell-correction energy. The three stages in the "life" of a superdeformed band are indicated, together with some characteristic properties.

formation larger than the one corresponding to the lowest minimum (true vacuum). At the superdeformed minimum, the ratio of the long to short axes is around two. The first manifestation of a secondary minimum was the occurrence of fission isomers (reviewed in Ref. [3]). A later manifestation was the existence of superdeformed (SD) bands, which exhibit impressive series of equi- spaced (picket-fence) γ transitions (reviewed in Refs. [4,5]). An example of a γ spectrum from a SD band is shown in Fig. 2, taken from Ref. [6].

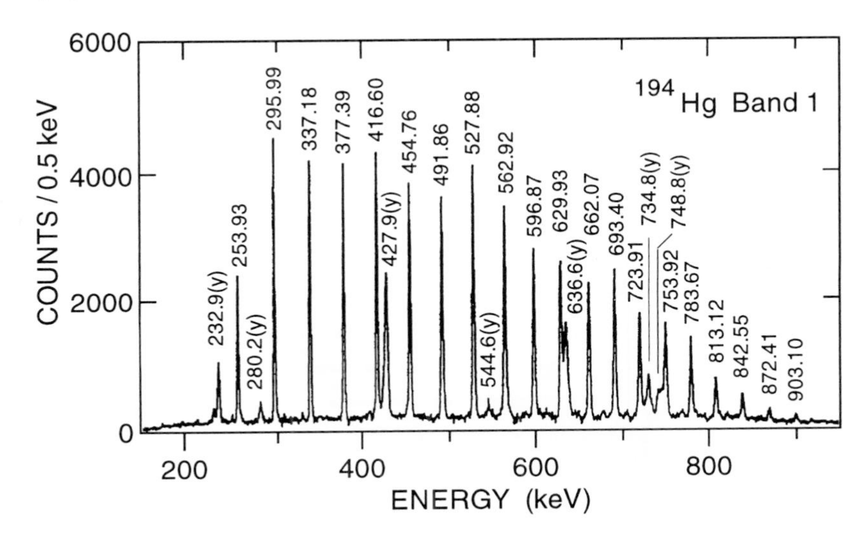

Figure 2: A typical spectrum [6] of a SD band – from ^{194}Hg – showing a characteristic picket-fence spectrum with lines of nearly equal spacing. Transitions labeled (y) are normal-deformed yrast transitions, which are fed after decay from the SD band.

Although the lowest states in the SD false vacuum lie at high excitation energies above the ground or yrast states, they are isolated by a barrier separating SD and normal-deformed (ND) states. (A yrast state is defined as the lowest-lying level at each spin.) As a consequence, the lowest SD states are cold states, characterized by good quantum numbers and symmetries, and their decays are governed by selection rules (Fig. 1). In contrast, ND states that are at the same energies may be described as chaotic compound states, where quantum numbers (apart from spin and parity) are largely lost.

The occurrence of superdeformation owes it origin to large shell gaps. In a harmonic oscillator potential, energy gaps occur when the ratio of the long- to short- axes is a rational ratio [7], e.g. 2/1, corresponding to closed classical orbits. The existence of a spin-orbit interaction in nuclei modifies the pure harmonic oscillator potential, but the occurrence of shell gaps at large deformation persists. When the proton and neutron numbers are both favorable for the occurrence of shell gaps, superdeformation is found. This leads to SD nuclei, which congregate in local regions in the chart of nuclides, with mass numbers around 80, 130, 150, 190 and 240 (Fig. 3, from Ref. [8]).

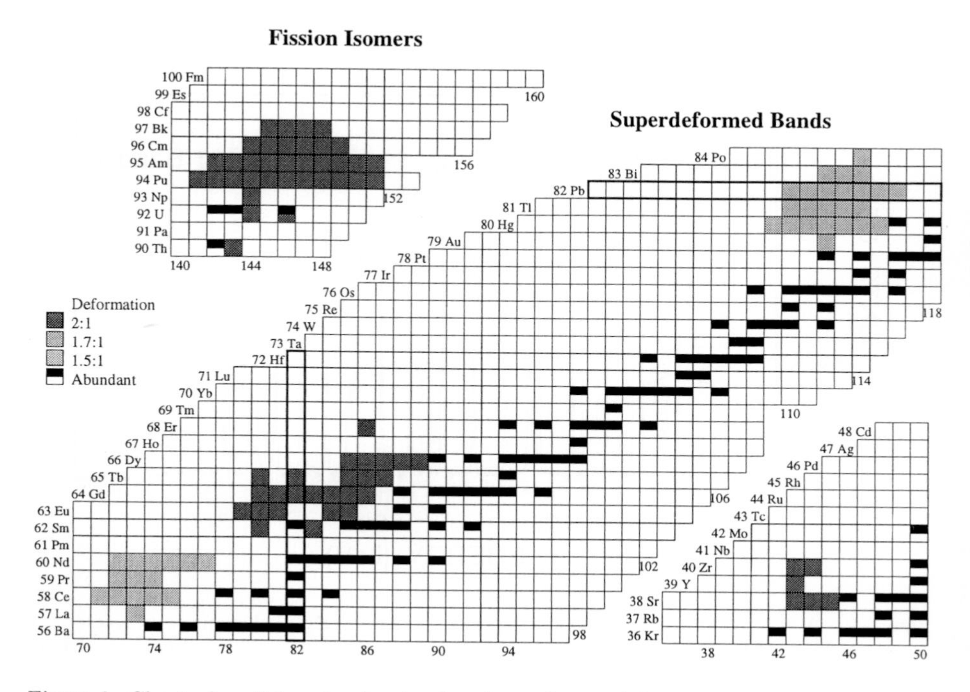

Figure 3: Chart of nuclides, showing local regions of superdeformation in the A $\sim$ 80, 130, 150, 190 and 240 regions; from Ref. [8].

A fascinating feature of almost all SD bands is the sudden drop in intraband transition intensity at low spin, after a string of up to 20 consecutive transitions (Fig. 4). In other words, after a long sequence of transitions within the false vacuum, there is a sudden decay to the true vacuum.

A γ cascade which flows through a SD minimum has three stages (Fig. 1): (a) feeding and trapping into the SD well, (b) intraband transitions within SD bands, and (c) decay from SD to ND states. In stage (a), hot compound nuclear states cool via γ emission. This stage involves the coupling of hot SD and ND states, which includes tunneling between hot states on either side of the barrier. A small fraction (typically around 1 %) of the cascades becomes trapped in the

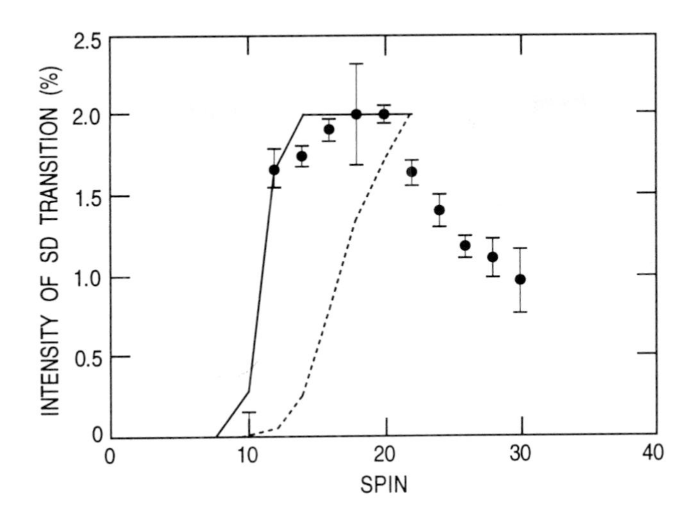

Figure 4: Intensities of SD-band transitions in ^{192}Hg as a function of initial spin, showing a sudden drop around spin 12. The solid and dashed lines are results of calculations using the Vigezzi model [20], with the SD well depth W(I) assumed to either increase with spin (solid line) or remain constant (dashed line).

SD minimum, leading to stage (b), where equally-spaced transitions connect cold SD-band members. Here, selection rules imposed by quantum numbers govern the γ decay, confining the decay to states of the same K (projection of total angular momentum on the symmetry axis), i.e. within a band. With decreasing spin, the excitation energy of the trapped SD state above the ND yrast line rises. It then becomes embedded in a sea of ND states with increasing level density. Finally, when the SD state becomes unavoidably close to a ND level, a sudden tunneling from the cold SD state to the hot ND states occurs – stage (c).

2.2 Tunneling from SD States

Several interesting aspects of the tunneling process are worth noting. The initial SD state is a cold ordered state, characterized by good quantum numbers, whereas the final hot states are (probably) not. At high excitation energy, say near the neutron separation energy, hot ND states are described as chaotic states[9]. The hot ND states (with ~4 MeV excitation energies) to which the SD states tunnel appear to be chaotic (although it still remains to be established whether full chaos is reached).

By analogy with a classical system, where chaos implies no constants of motion, in a quantum system chaos implies a loss of certain quantum numbers. At high excitation energy the residual nucleon interactions mix states over an energy interval. When the spreading width Γ_Q associated with a quantum number Q becomes much larger than the average level spacing D, i.e. $\Gamma_Q/D \gg 1$, chaos is attained. [10] An interesting question is how the good quantum numbers dissolve in the tunneling process? One might also ask if the formalism developed for chaos-assisted tunneling [11] could be usefully applied to this nuclear problem, or whether tunneling from SD states can provide an interesting example of chaos- assisted tunneling. Finally, it should be noted that the tunneling occurs in a multidimensional space, as the nuclear shape is characterized by at least two parameters, β and γ (which are defined later).

2.3 Experiments

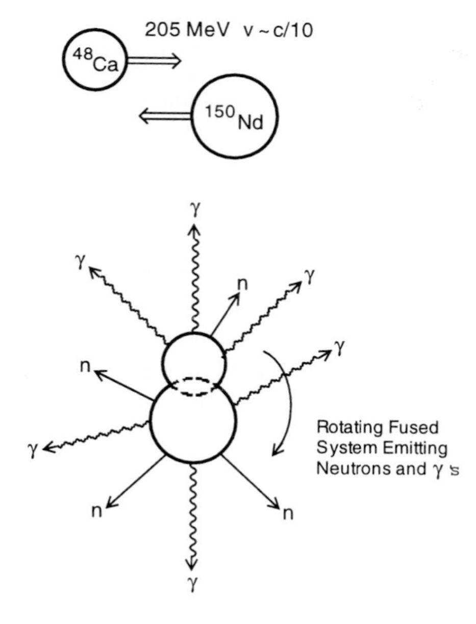

Figure 5: Cartoon of two nuclei fusing to form the compound nucleus198Hg, followed by emission of 4 neutrons and 20- 25 γ rays.

Rapidly spinning nuclei are usually formed in heavy-ion fusion reactions, where a heavy-ion projectile fuses with a target nucleus (Fig. 5). The hot, rapidly-rotating, compound nucleus cools by evaporating neutrons, then γ rays. Most of what we have learned about superdeformed bands has been deduced from the γ rays. A rapid growth of knowledge has occurred with the operation of large, powerful γ-ray detector arrays of Compton-suppressed Ge detectors, such as Gammasphere [12] (which has been located in Berkeley and is now in Argonne) or Euroball (which is now in Legnaro). In fact, superdeformation provided one of the original motivations for constructing these arrays. Fig. 6 shows a photograph of Gammasphere.

Figure 6: Photograph of Gammasphere [12].

The shape of a hot compound nucleus is often not well defined because thermal fluctuations, present in a mesoscopic system, lead to a superposition of many shapes. However, as the nucleus cools towards the yrast line, a variety of ND and SD shapes become defined. These are illustrated in Fig. 7, which also shows the $\beta - \gamma$ values characterizing these shapes. Nuclear shapes are usually described by two deformation parameters: β, a measure of elongation, and γ, a measure of non-axiality. Typically, a SD shape has $\beta \sim 0.5$, $\gamma \sim 0$, while hot ND states have a considerable spread in β and γ because of thermal shape fluctuations.

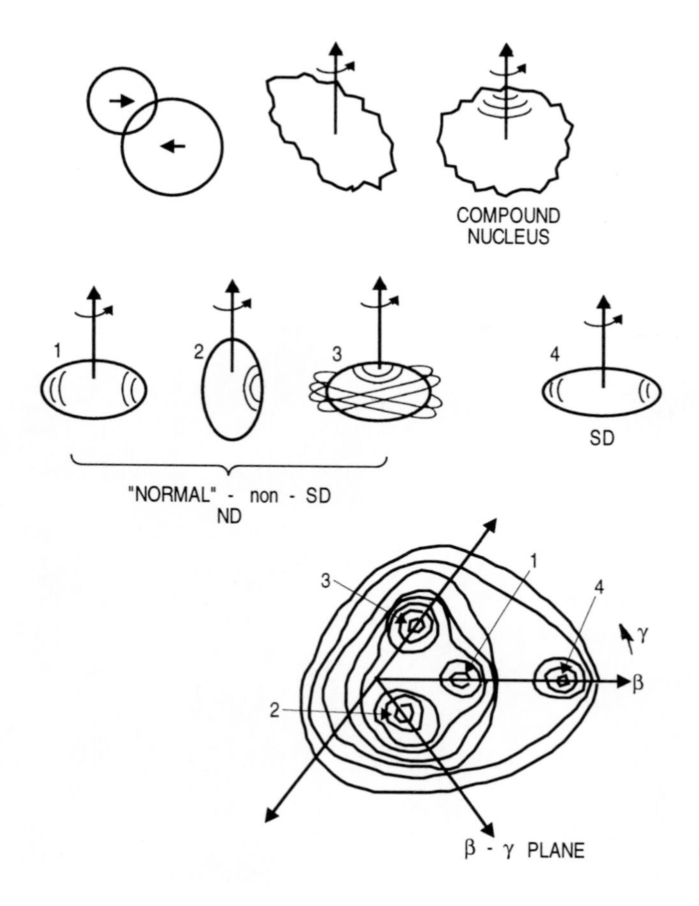

Figure 7: Sequence of nuclear shapes formed after fusion. The shapes in the hot excited phase are largely undefined because of thermal fluctuations, which are present due to the mesoscopic nature of the nucleus. Upon cooling to the vicinity of the yrast line, a variety of defined shapes (labeled 1-4) may be encountered, which are described in terms of the indicated β- and γ- shape parameters.

2.4 Feeding of SD bands

The population of SD bands can be understood in terms of a statistical process (i.e. one governed by level densities), provided tunneling between hot SD and ND states is taken into account. [13,14] As the nucleus cools by emission of so-called statistical γ rays, a small fraction ($\sim$1%) are trapped within the SD minimum. Once they are in the "ground" state in this false vacuum, no further cooling is possible and the decays simply remove angular momentum in intraband cascades. Calculations [13], based on Monte Carlo simulations of the statistical process, are able to reproduce all the observables connected with the feeding process, namely the SD-band intensity, intensity as a function of spin, the entry distributions (starting points for γ deexcitation after neutron emission) and the spectra of γ rays feeding the SD bands.

Fig. 8a from Ref. [13] shows the measured entry distribution leading to all states of the nucleus ^{192}Hg, while Fig. 8b shows that the entry distribution for populating a SD band originates from the high-angular momentum portion

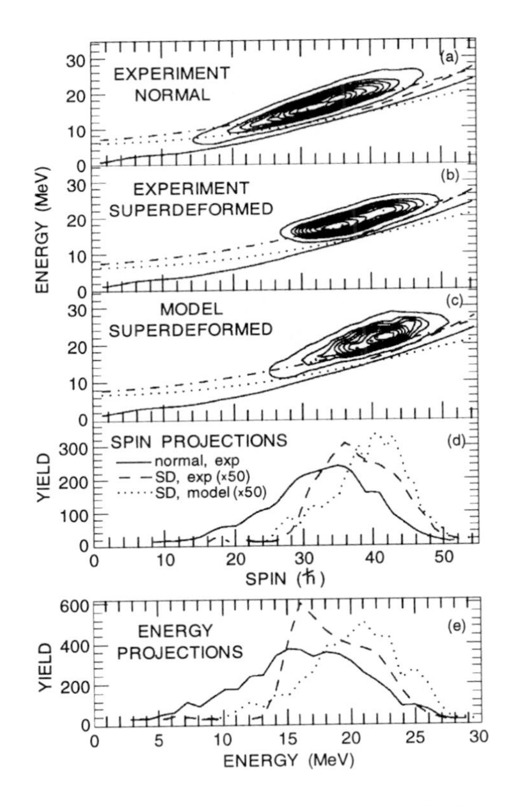

Figure 8: (a-c). Entry distributions in the spin - energy plane leading to the formation of all (mainly normal-deformed) and SD shapes in ^{192}Hg. The distributions in (a, b) are measured with the Argonne-Notre Dame BGO array; that in (c) is calculated. (d, e). Projections of the entry distributions in (a-c) on the spin and energy axes. The solid, dotted and dot-dashed lines represent the ND yrast line, SD yrast line and barrier position used in the calculation. From Ref. [13].

of the total entry distribution. It is only at large spin that SD states have sufficiently large level density to receive significant population. The entry distribution from SD states is reasonably well reproduced by Monte Carlo simulations (Fig. 8c). The projections of the entry distribution on the spin and energy axes are shown in Fig. 8 d and e.

A sample of γ cascades from events that result in trapping in the SD well is shown in Fig. 9, taken from the results [13] of the Monte Carlo simulation. Note that successful trapping in the SD well is favored by rapid cooling via

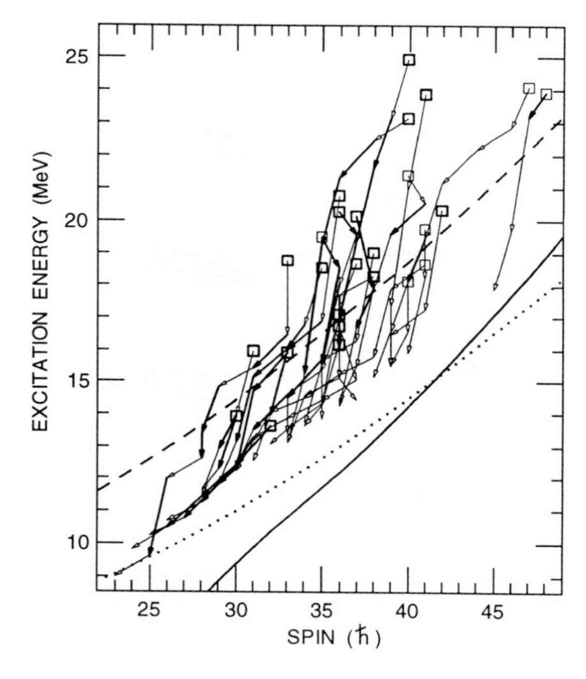

Figure 9: Sample of γ cascades that result in trapping within the SD false vacuum, obtained from a Monte Carlo simulation which reproduces all observed feeding properties. From Ref. [13].

statistical transitions, which remove energy with little loss in angular momentum. Nuclei that cool rapidly are more likely to survive within the SD well in the competition with tunneling to hot ND states. Collective E2 transitions, which remove 2 units of spin but little internal excitation energy, are emitted in a zone about 1.5 MeV below the barrier and 2 MeV above the SD yrast line. In other words, they originate from excited states in the SD false vacuum, thereby providing a probe of the properties and tunneling of these states.

Two of the parameters in the calculations are the well depth W(I) and the excitation energy E^* of the SD band above the yrast line. W(I) is assumed to increase linearly from 1 MeV at spin 0 to a value W(40) at spin I = 40. Values of 3.5-4.5 MeV for W(40) satisfactorily reproduce the data, suggesting that the well is quite deep at high spin. The SD band intensity is sensitively dependent on the energy E^* of the SD states; values of E^* between 3.3 and 4.3 MeV at the point of decay satisfactorily reproduced the band intensities. This provided the first indication that the SD bands lie rather high above the ND yrast line in the Hg region and is consistent with more accurate excitation energies that have been obtained later [15].

Two important open problems to be investigated are the structure of excited SD bands and the tunneling between hot SD and ND states. Information on these problems comes from the spreading width Γ_{rot} of the rotational strength [16] of excited SD bands [17]. The spreading width arises from residual nuclear interactions, which couple not only SD states, but also SD and ND states via tunneling. This is an example of tunneling between hot states. Experimental data on the spreading width can be derived from the widths of so-called valleys and ridges in $E_\gamma - E_\gamma$ correlation plots (see, e.g., Ref. [16]). Unfortunately, hitherto there has been little work on excited SD structures or on the tunneling between hot SD and ND states.

2.5 *Cold cascades within SD rotational bands*

In contrast, there is a large body of data on SD bands, since the predominance of effort has been devoted to the relatively more straightforward identification of these bands. When trapping in the SD well occurs, the γ cascade cools into or near the bottom of the false vacuum. Isolated by the barrier, the nucleus cools no further and loses angular momentum through cold cascades, with highly-collective E2 transitions (~2000 Weisskopf Units) connecting members of rotational bands. This phase gives rise to the beautiful spectrum of nearly equal spacings (see Fig. 2), characteristic of collective, ordered rotation. The spacings between transitions yield values of the dynamical moment of inertia $\mathcal{J}^{(2)}$, which have been compared to theoretical values to infer the microscopic structure of the bands. About 175 SD bands have been identified in the mass 150 and 190 regions. However, in all but three cases, the connections of the excited SD bands to the ND yrast states have proven elusive, so that the excitation energies, spins and parities are not known. Lifetime measurements, on the other hand, have been feasible, showing unambiguously that the bands have large deformation (see, e.g., Ref. [18]). In the mass 190 region, many SD bands have identical quadrupole moments [18] and most have very close $\mathcal{J}^{(2)}(\omega)$

vs. ω curves [5], which slope up with ω. (ω is defined as $E_\gamma/2$, where E_γ is the γ-ray energy.) The upsloping $\mathcal{J}^{(2)}(\omega)$ curves have been attributed to the alignment of high-j single-particle orbitals and to the presence of pair correlations within the SD well [5]. One surprising discovery has been the phenomenon of identical bands (reviewed in Ref. [19]), where many SD bands in neighboring nuclei have either equal transition energies (within 1 keV) or equal $\mathcal{J}^{(2)}$. Some of the topics mentioned here will be discussed again later.

3 Decay from SD bands: tunneling out of the false vacuum

3.1 Physics from Decay Out of SD Bands

This section constitutes the main topic of these lecture notes. From investigating the decay out of SD bands we have been able to obtain interesting information on several physics topics. These include: (i) the mechanism for the depopulation of SD bands; (ii) the spin/parity quantum numbers and excitation energies of SD bands; (iii) the microscopic structure of SD bands; (iv) the origin of identical bands; (v) the quenching of pairing with increasing temperature; and (vi) the onset of chaos (in ND states) with increasing excitation energy above the yrast line.

3.2 Decay mechanism

In this section, we discuss the decay from SD to ND states. Both theoretical and experimental aspects of the decay mechanism are examined.

SD bands exhibit a highly unusual property with decreasing spin: the excitation energy above the ND yrast line increases (Fig. 10), due to their larger moment of inertia. As a consequence, SD states become embedded in a sea of hot ND states with increasing level density. Although SD states are cold and protected from hot ND states by a barrier, the latter squeeze in around the SD state until an inevitable coupling (albeit small) occurs, precipitating a sudden decay.

An important development in understanding the decay mechanism was the work of Vigezzi et al [20], which laid the theoretical foundation for treating the coupling of isolated SD states with ND states of high level density. The SD state couples, with a spreading width Γ to compound ND states, with average level separation D_n. A SD state can decay via a collective E2 intraband branch (with width Γ_{SD}) to the lower rotational member or, via an admixed compound ND wave function, to lower-lying ND states. A compound state, a complicated level with very many components in its wave function, can decay to all states

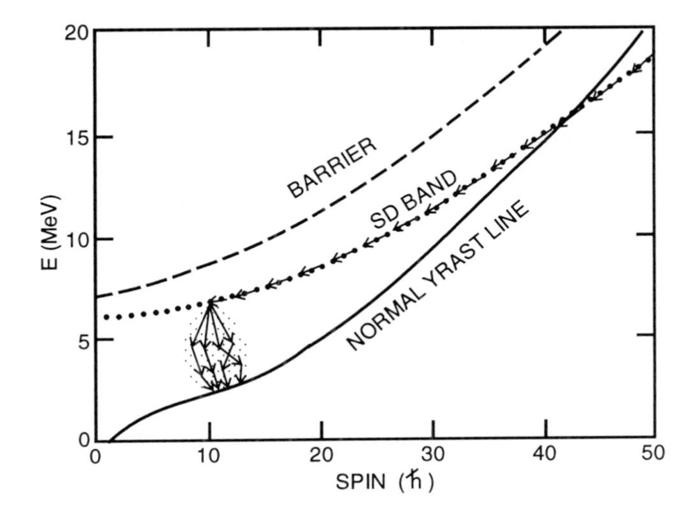

Figure 10: As a SD band loses angular momentum, its excitation energy above the ND yrast line rises. The γ cascade is confined in a single path along the SD band, but is very fragmented in the sudden decay from SD to ND states. The approximate barrier separating SD and ND states is shown, which illustrates that the barrier height (separation between barrier and SD band) for tunneling decreases with lower spin.

by a statistical process. Its decay rate Γ_{CN} can be computed using a γ strength function and level densities.

In the Vigezzi model, [20], the decay rate out of the SD band is governed by one parameter, namely the ratio Γ/D_n. We have applied this model to the decay from SD bands in the Hg region, and are able to reproduce the intensities of SD transitions at the point of decay. A value of $\Gamma/D_n \sim 0.03$ reproduces the spin at which the decay occurs (Fig. 4). Hence, the spreading width is very small and the SD state couples with the nearest one or two excited ND states. At an excitation energy above yrast of $\sim$ 4.3 MeV, $D_n \sim 0.4keV$ giving $\Gamma \sim 13eV$, a very small value indeed. A typical nuclear matrix element is $\sim 100keV$, four orders of magnitude larger. The SD eigenstate is given as

$$|SD> +\alpha|ND^*>, \quad (1)$$

with $\alpha \sim 0.08$ - a very small value.

The decay out of a SD band occurs through the hot ND component, $|ND^*>$. Although the coupling between SD and ND states is very weak, the decay occurs because $\Gamma_{CN}/\Gamma_{SD} \sim 200$ is large. After a long series of up to 20 transitions, this tiny coupling precipitates the rapid decay out of the SD band, a feature which is nearly universal for SD bands. Four factors contribute to the sudden decay. As the spin becomes smaller, Γ_{SD} decreases quickly because of its E_γ^5 dependence. At the same time, the excitation energy above the ND yrast line grows. This leads to an increase in Γ_{CN}, as well as a decrease in D_n, so that the average separation from a ND state becomes smaller. A fourth, and *essential*, ingredient is an exponential increase of Γ with decreasing spin [20]. If Γ were to remain constant, only a gradual decrease in SD intensity would be obtained (dashed line in Fig. 4).

When viewed as a tunneling process, Γ is associated with a tunneling width. Hence, an exponential increase in Γ implies a barrier height W that decreases linearly with diminishing spin. (This trend can be understood from the macroscopic liquid drop term.)

In $A \sim 190$ nuclei, the sudden depopulation of SD bands occurs around spin 8 - 10, whereas in the $A \sim 150$ region it happens around spin 25. Within a given mass region, the level densities, SD excitation energies and the well depths W(I) are roughly the same, so that the decay can happen around the same spin. The fluctuations in decay-out spin (expected because of the stochastic nature of the process) are small since the decrease in W(I) triggers the decay around the same spin values in each mass region.

Shimizu et al [21] have determined Γ or, more specifically, the tunneling action $A(I)$ by computing the well depth and inertial mass as a function of spin. The role of pairing, which increases as the spin becomes smaller, was taken into account in the spirit of the Bertsch hopping model [22], i.e. by counting the number of level crossings. From our application of the Vigezzi model we can also deduce the W(I) that reproduce the sudden reduction in SD transition intensities [23]. We can deduce [23] the action $A(I) = \pi W(I)/\hbar\omega_B$, using an assumed [20] barrier frequency of $\hbar\omega_B = 0.6$ MeV. The action inferred from experiment is compared to that calculated by Shimizu in Fig. 11, taken from Ref. [23]. The actions at the point of decay are in reasonable agreement, although the high-spin values and $dA(I)/dI$ are not. Hence, it still remains a challenge to theoretically describe decay out of SD bands.

3.3 Spectra from Decay out SD States

It is possible to experimentally confirm the postulate of the Vigezzi model, [20] that the decay occurs via a small admixed component of a hot ND state ($|ND^* >$ in Eq. (1)). In this model the decay spectrum should be just that from a hot compound state, which is schematically illustrated in Fig. 12. A hot state can deexcite by primary (i.e. first-step) transitions to near-yrast states, resulting in sharp γ lines. For decay to higher-lying states, with rapidly

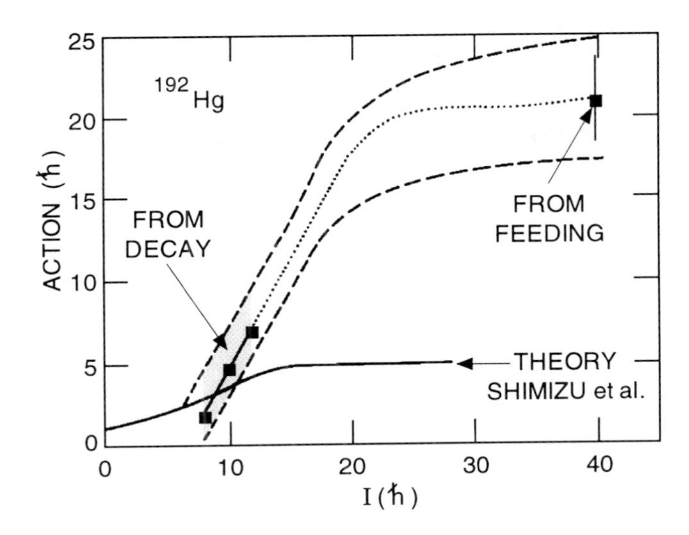

Figure 11: Action $A(I)$ for tunneling between SD and ND states, given as a function of spin. Results from theory [21] (solid line) and those deduced from experiment (filled squares) are shown. The "experimental" actions around spin 10 are deduced by comparing the SD intensities at the decay point with predictions using the Vigezzi model [20]; that at spin 40 is deduced from the comparison of SD-feeding observables (e.g. that shown in Fig. 7) with model predictions [13]. The deduced actions depend on the assumed level density of ND states; the dashed lines are obtained by multiplying the nominal level density by factors of 10 and 0.1. From Ref. [23].

increasing level density, the lines overlap, form structures, and then smear into a quasicontinuum. Primary transitions to highly excited states of high density, together with the subsequent secondary transitions, form an unresolved background. In addition, transitions between near-yrast states will appear as sharp lines at low energy. These are indeed the features observed in the γ spectrum following thermal neutron capture (Fig. 13, from Ref. [24]). Hence, if the spectrum from SD decay has these features, it would constitute strong evidence of the postulated decay mechanism.

This spectrum was first obtained in Ref. [25], where the procedures are described. First, an extremely clean spectrum coincident with a SD band is required and is obtained by setting pairwise coincidence gates on SD lines. A correct background subtraction procedure, such as described in Ref. [26] is essential. Extraneous events from neutron interactions, coincident γ summing and Compton scattering are then removed by the procedures described in Refs. [27,28]. Corrections for the full-energy peak efficiency and angular distribution

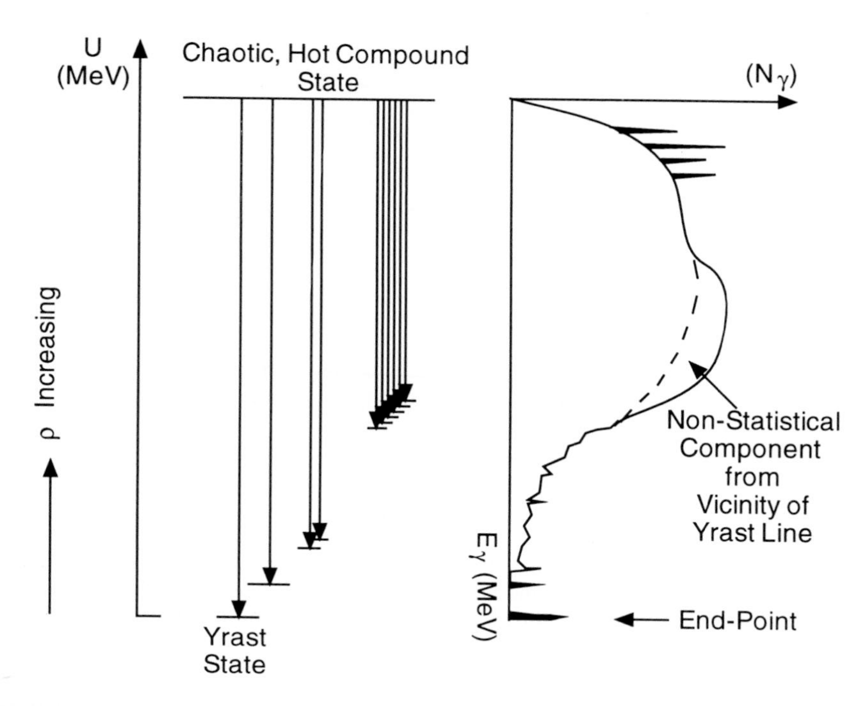

Figure 12: Decay from a highly excited, chaotic, compound state. Some primary transitions to final states at different energies are indicated, together with the characteristic features they yield in the decay spectrum. The sketch of the decay spectrum has a largely statistical form, with some structure coming from secondary transitions occurring near the yrast line. Low-energy peaks are from yrast transitions.

effects are then made. Finally, the spectra are put on an "absolute" scale, by normalizing the spectrum so that the ground-state transition (e.g. $2^+ - 0^+$ transition), has unit area (after correction for conversion electrons). The integral of the spectrum (after correction for coincidence gates) immediately gives the multiplicity, i.e. the average number of steps in the cascade, and the product of multiplicity and average energy gives the total γ energy removed.

The spectrum [25] coincident with the lowest SD band in ^{192}Hg is shown in Fig. 14, together with the spectrum for all states in ^{192}Hg, obtained with a coincident gate on the $2^+ - 0^+$ transition. Differences between the two spectra are immediately apparent, suggesting where the decay-out strength lies in the spectrum. The smooth spectrum in Fig. 14 is a calculation of the feeding statistical spectrum, obtained from a model [13] that reproduces all observables in the feeding of SD bands. Subtraction of this statistical spectrum and of

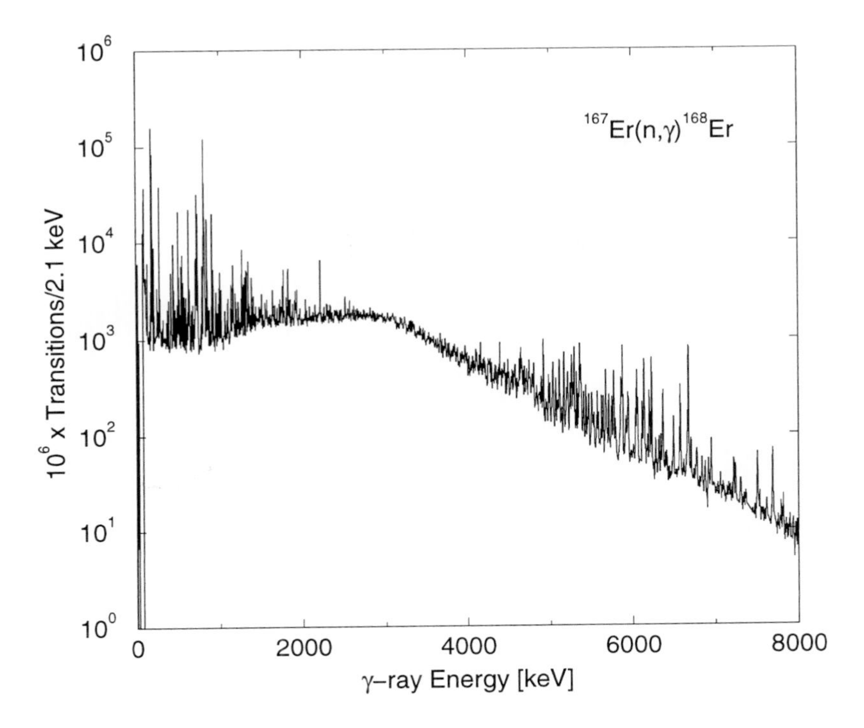

Figure 13: Gamma spectrum following thermal-neutron capture in ^{167}Er; from Argonne-Brookhaven-Manchester- collaboration [24].

the discrete lines yields the spectrum [25] in Fig. 15. Two feeding components at low energy remain: a large E2 peak (A), arising from excited SD bands, and a M1/E2 peak (B), from transitions immediately preceding feeding of the SD bands. The remaining broad, smooth component represents the γ rays connecting SD and ND states; its portion below $\sim 800keV$ cannot be unambiguously extracted. The decay spectrum indeed has the features of the statistical spectrum sketched in Fig. 12. The statistical nature of the decay is further demonstrated by its fragmentation over about 1000 primary pathways. [29]

A similar decay spectrum [30], this time from SD band 1 in ^{194}Hg, is superimposed on a spectrum [24] obtained from thermal neutron capture in ^{167}Er in Fig. 16. The end- points in the two spectra are 4.3 and 7.5 MeV, respectively, but the dispersions of the spectra are adjusted so that the end-points occur at the same channel in Fig. 16. There is a remarkable similarity in shape and

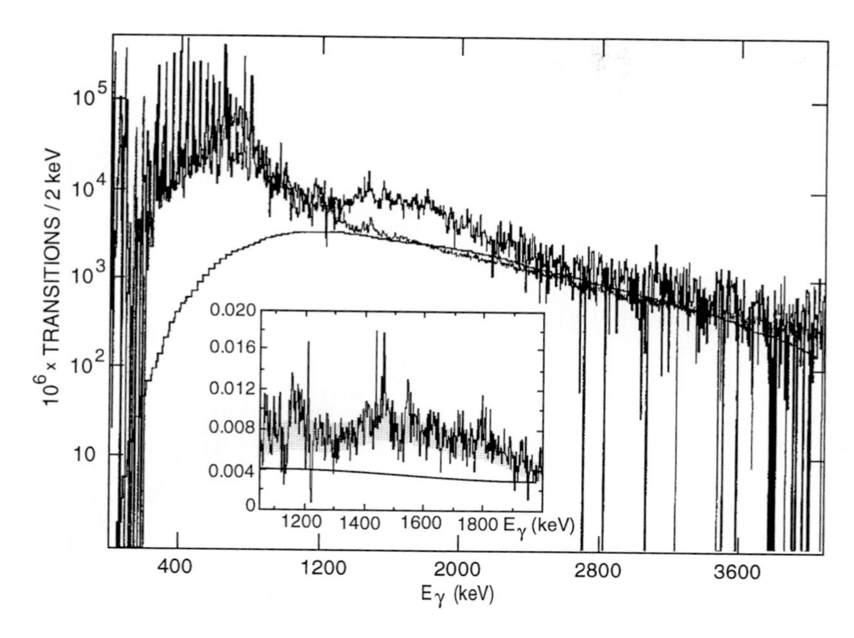

Figure 14: Spectra[25] obtained from pairwise coincidence gates on SD lines (dark histogram) and from a gate on the 2+ - 0+ transition in ^{192}Hg (light histogram). The excess counts around 1.6 MeV suggest significant strength in this region from the decay between SD and ND states. The smooth curve is a calculation of the statistical spectrum feeding the SD band, and is very close to the statistical spectrum (light histogram) feeding ND states. The inset shows a portion of the decay spectrum, revealing that it is composed of lines, overlapping structures and an unresolved component. The integral of each spectrum is its multiplicity.

magnitude between the spectra. (The integral of each spectrum represents the multiplicity.) This remarkable similarity provides the best evidence that decay from SD states is mediated by an admixed component of a compound ND state, and confirms the mechanism suggested by Vigezzi et al. [20]

Sharp peaks at high energy from direct one-step transitions to the yrast line are also expected –see Fig. 12 and 13. Indeed, discrete lines are observed at very high energy ($\sim 4MeV$) in the decay spectrum from ^{194}Hg SD band 1 – Fig. 17a. These lines, finally observed[15] after a decade-long search, complete the parallel between spectra from SD and neutron-capture states.

The decay mechanism is now understood as the coupling of a cold SD state to a hot compound ND state. However, the understanding is not complete since quantitatively correct results for the tunneling process still remain a challenge for theory.

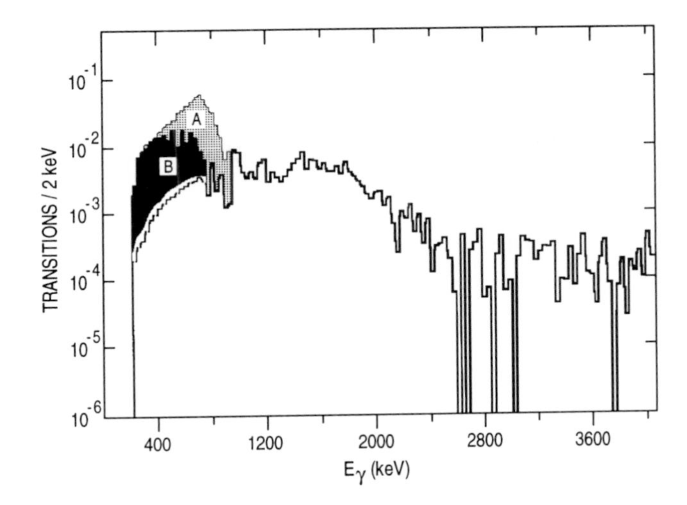

Figure 15: Spectrum from Fig. 14, after subtraction of discrete lines and the calculated statistical spectrum shown there. Components A and B at low energy are from γ rays that feed the SD band. The remaining broad component, which has a statistical-like form, represents the decay spectrum connecting SD and ND states. From Ref. [25].

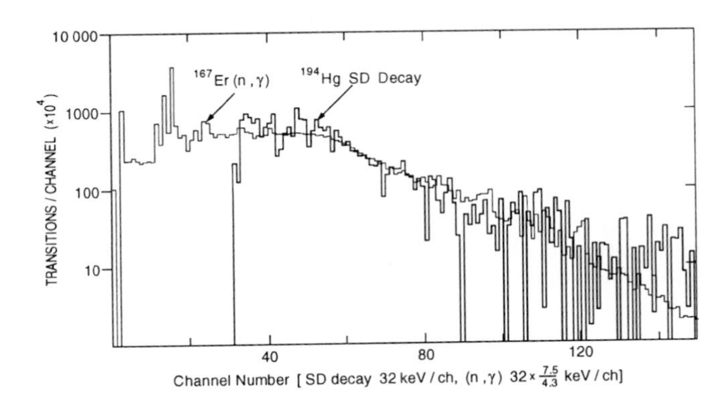

Figure 16: Overlay of spectra from the decay out the ^{194}Hg SD band 1 (thick line) and from thermal-neutron capture (thin line). The spectra have their gains adjusted so that the end-points at 4.3 and 7.5 MeV, occur at the same channel. In each case, the integral of the spectrum gives the multiplicity. Note the remarkable similarity in shape and magnitude of the two spectra, showing that the SD decay occurs via a chaotic process, identical to that from thermal-neutron capture.

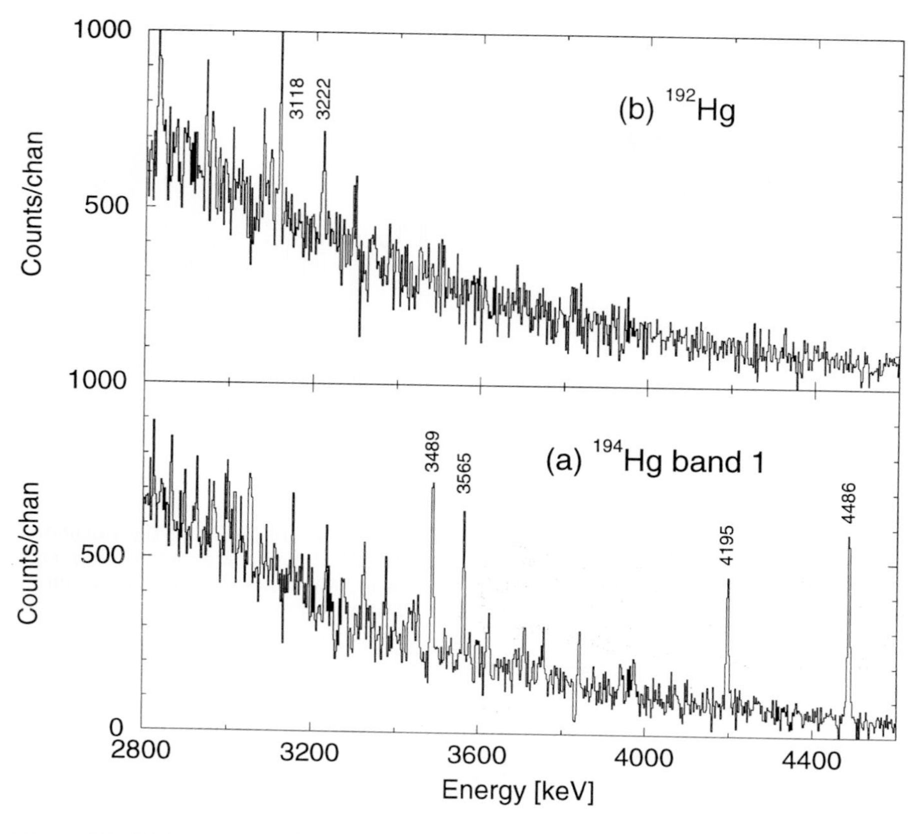

Figure 17: High-energy portions of spectra from (a) ^{194}Hg SD band 1 and (b) ^{192}Hg SD band. One-step transitions directly connecting SD and ND states are prominent in (a), but are not seen in (b). The primary lines in (b) represent the first steps of multi-step decays.

3.4 Spin/parity Quantum Numbers, Energies and Decay Schemes of SD Bands

After a decade of research in superdeformation and the detection of more than 175 SD bands, the essential quantum numbers and excitation energies are still largely unknown. Some early attempts at finding the decay schemes of SD bands were based on trying to establish coincidences among the highly fragmented decay lines. However, with the realization of the analogy between decays from SD and neutron-capture states, it became clear that the simplest method was to identify the primary lines, which directly connect, in one step, SD and ND yrast states. This led to the discovery [15] of high-energy, one-step transitions in ^{194}Hg - see Fig. 17a. It was a thrilling culmination of a long search!

In addition, pairwise coincidence gates set on a one-step decay line and a SD line made it possible to determine [15] the exact connections between the lowest SD states (band 1) in ^{194}Hg and the ND yrast states. One-step transitions from an excited SD band in ^{194}Hg have even been found[31] - see Fig. 18. This figure also displays coincidence spectra that allow definite assignments

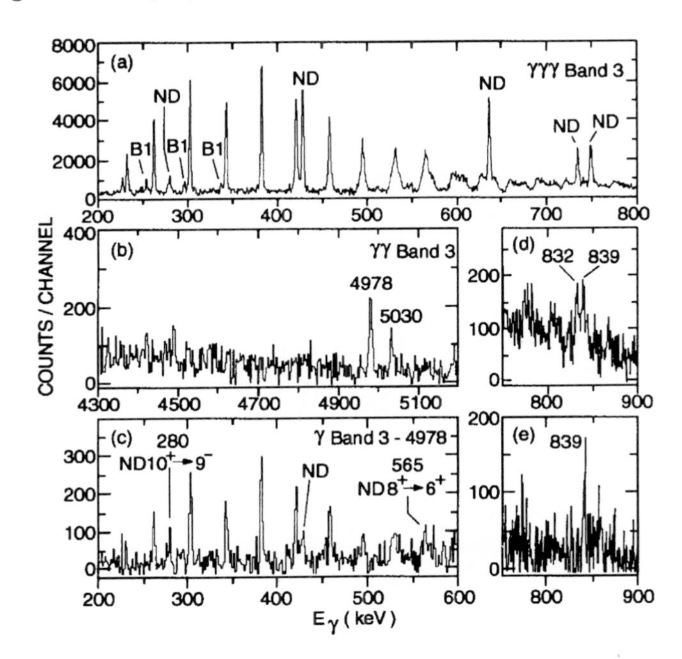

Figure 18: a) Low- and (b) high-energy portions of the spectra coincident with (a) triplets or (b) pairs of transitions from ^{194}Hg SD band 3. (c) Pairwise coincidence spectrum obtained with one gate on the 4978-keV line and another on band-3 transitions, showing transitions from the 10^+ and 8^+ ND levels. (d) and (e) Portions of triple- coincidence spectra, showing the 832 and 839-keV interband lines; (d) from a gate on the 343-keV band-3 line and double gates on higher- lying band-3 transitions; and (e) from a gate on the 296-keV band-1 line and double gates on band-3 transitions.

in a decay scheme of the ~5-MeV one-step lines. A level scheme (Fig. 19)[15,31] giving the energies, spins and likely parity of two SD bands in ^{194}Hg has been constructed, defining for the first time these quantities for SD bands in the mass 150 or 190 regions. At the point of decay, the excitation energies above the ND yrast states are ~ 4.3 and 5.0 MeV for the two bands in ^{194}Hg. The

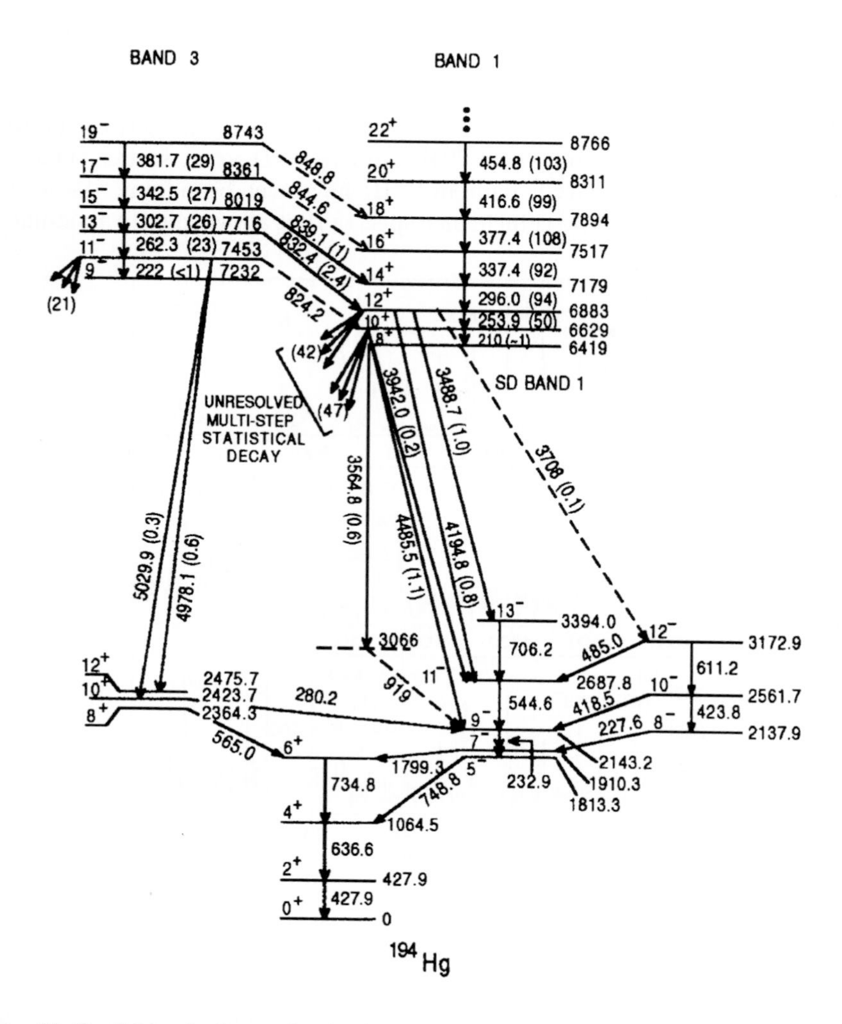

Figure 19: Partial level schemes for the decay of SD bands 1 and 3 in ^{194}Hg. Note the 1-step decays from band 3 to both the SD and ND yrast lines. Dashed lines indicate tentative assignments. The transition intensities (in brackets) are normalized to 100 and 30 for the full- strength transitions in bands 1 and 3, respectively.

SD band in ^{194}Pb represents the only other case where these quantities are known [32,33]. In this case the excitation energy at the point of decay is lower, ~2.7 MeV.

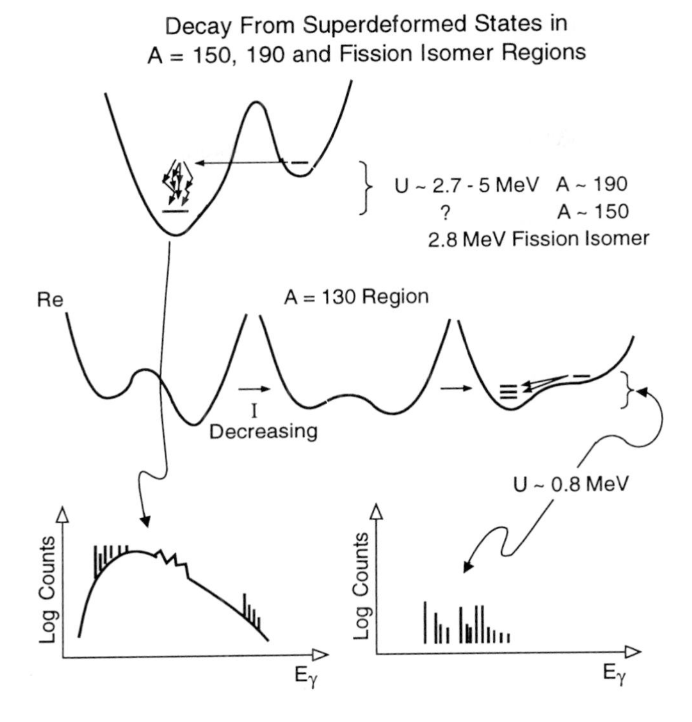

Figure 20: Comparison of decays from SD bands in regions with A=130, 150, 190 and 240, with sketches of the potential energy surfaces, energies of SD states (above ND yrast line) at the points of decay, and cartoons of decay spectra.

We shall now compare the properties of SD states in various mass regions - see Fig. 20. In ^{194}Hg, which has an excitation energy above yrast of ~4.3 MeV, about 10% of the decay strength is found in discrete primary lines; the rest of the decay proceeds through unresolved γ rays which form a quasicontinuum spectrum. In ^{194}Pb, where the excitation energy (~2.7 MeV) is lower, more of the decay is expected as primary lines and, indeed, ~21% is observed [33]. For the zero-spin fission isomers, which lie around 2.8 MeV, the bulk of the decay has been observed as discrete lines in ^{236}U [34], probably due to the lower level density for low-spin states. The highly-deformed bands that have been located (in spin and energy) [35] in the mass 130 region lie about 0.8 MeV above the less-deformed states. A substantial fraction (~50%) of the decay is in the form of discrete lines in these cases. Here, unlike the other cases discussed so far, the potential energy pocket is much shallower and there is probably no barrier separating the states of different deformation at the point of decay. (This may not be the case for bands from which no discrete-line decays have been observed.) In the mass 150 region there is still no SD band whose energy and quantum numbers are definitely known, although the first SD band was observed in ^{152}Dy.

The detection and placement of one-step transitions from SD bands has proven to be quite difficult, despite the power of the modern detector arrays. To understand this, one notes that that the decay is a statistical process governed by level densities and a γ strength function. Although the one-step lines benefit from their larger energies, they are disfavored by a very small weight in level density. Hence, the bulk of the decay occurs via unresolved γ rays from decay to excited states, with high level density, followed by secondary transitions, which together form a quasicontinuum. Therefore, the one-step lines can be favored when the branching ratio for the unresolved component is reduced. This can be achieved by making the phase space for the quasicontinuum component smaller, e.g. by choosing a case with low SD excitation energy (such as ^{194}Pb) or a nucleus with a pair gap. The latter, which represents a region above the yrast line with a paucity of levels, is more pronounced in even- even nuclei and when the spin is low. (A further discussion of the role of the pair gap is given later.) These considerations explain the decay properties of SD state in the various regions discussed above. In particular, they account for the increase in the branching ratio for sharp primary lines with smaller SD excitation energy in ^{194}Pb. They also account for the absence of one-step decay lines in the mass 150 region, where the decay occurs around spin 25, where a pair gap is not expected. The statistical nature of the decay has one further ramification: the intensities of the primary lines are subject to fluctuations[36]. With luck, larger-than-average intensities can be obtained; that probably happens in ^{194}Hg and ^{194}Pb. Without luck, the one-step strengths may drop below the detection limit. For example, the high-energy spectrum (Fig. 17b) from the ^{192}Hg SD band, which has similar statistics as that from ^{194}Hg band 1, does not reveal one-step lines, although other primary transitions, representing the first of several decay steps to the ND yrast line, are visible.

As discussed above, of the ~175 SD bands which have been found in the mass 150 and 190 regions, there are only three which have finally been precisely located in spin and energy. Figure 18 shows the decay schemes for two SD bands in ^{194}Hg. Bands 1 and 3 are the two lowest bands in the SD false vacuum. Band 1, the yrast SD band, has even spin and parity. No signature partner, with odd spin, and equal population intensity has been detected. Hence, the yrast band is just like a $K^{\pi} = 0^{+}$ "ground" band, but in a false vacuum. The same applies for the yrast SD band[32,33] in ^{194}Pb. Hence,

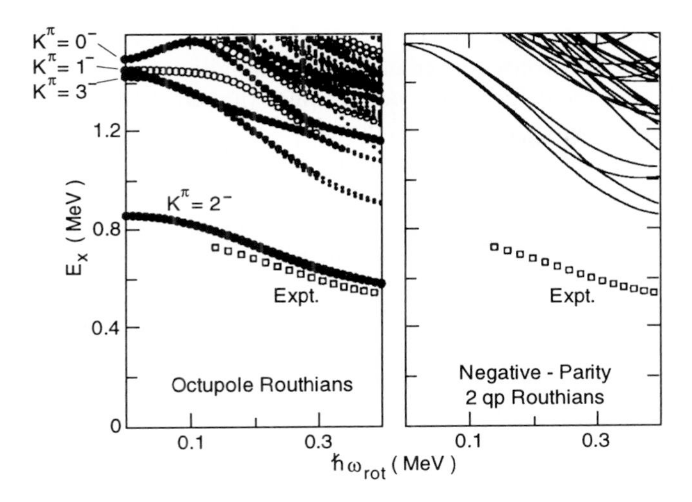

Figure 21: Routhians (orbital energies in intrinsic frame) for theoretical octupole vibrations (open and filled circles) and negative-parity 2-quasiparticle states (lines), compared with experimental data (open squares) for ^{194}Hg SD band 3.

although the SD states lie at high excitation energy, they are characterized by good quantum numbers K and signature σ. K, the projection of spin on the symmetry axis, is the quantum number characterizing rotational invariance about this axis. Good signature is a consequence of invariance with respect to 180° rotation about an axis perpendicular to the symmetry axis. Therefore, the SD states near the bottom of the false vacuum are indeed ordered states, despite their large absolute excitation energies. In contrast, ND states at the same excitation energy probably do not have good K and σ (see discussion below).

The excited SD band 3 in ^{194}Hg has odd spin and parity and has been interpreted [31] as an octupole vibrational band on the basis of its low energy, odd spin and negative parity. Its energy as a function of frequency (Routhian) is very close to that predicted [37,31] for a $K = 2^-$ octupole vibrational band (see Fig. 21), and is considerably lower than the calculated energies for two-quasiparticle states. Another excited band, band 2 in Ref. [6], does not have observable one-step decay transitions. However, its transition energies, which are midway between those of band 3, and its equal population intensity suggest that it is the even spin signature partner of band 3. Low-lying excited SD bands in ^{190}Hg and ^{196}Pb have also been assigned as octupole vibrational bands. [38,39] The present body of data supports the predictions of Ref. [37] that octupole vibrations constitute the lowest excited SD bands in even-even mass 190 nuclei. This is a consequence of the intermingling of positive and negative parity levels at large deformation, and implies that the SD shape is reasonably soft with respect to octupole distortion.

^{192}Hg is believed to have closed proton and neutron shells at large deformation. Therefore, the energies of the ND ground and SD states in $^{192,194}Hg$ and ^{194}Pb constitute an important triad, which gives the 2-proton and 2-neutron separation energies in both the ND and SD wells. All relevant ND and SD masses are known, with the exception of that for the ^{192}Hg SD band, where there is only a tentative value. The separation energies and SD excitation energies provide an important test of effective nuclear forces, e.g. of different Skyrme interactions for Hartree-Fock-Bogolyubov calculations. Calculations of both separation and excitation energies have been performed [40] and agree with experimental energies within about 1 MeV. It remains a challenge for theory to make more accurate predictions and also for experiment to obtain a definite value for the energy of the SD band in ^{192}Hg, for which there is now only a tentative value.

3.5 Identical Bands

One of the most exciting discoveries about SD bands is the phenomenon of identical bands, where bands in neighboring nuclei have energies and moments of inertia which are equal for a large span of transition energies. There have been many attempts at explaining the origin of identical SD bands (reviewed in Ref. [19]), but the phenomenon is still not understood. There could be a "heroic" explanation, based on the existence of a symmetry (which has yet to be identified). On the other hand, the explanation may be rather "unheroic", based on the accidental cancellations of several effects. Progress in understanding will be made only when there is knowledge of the spins, parities, microscopic structures and quadrupole moments of identical bands. A crucial question to be addressed is whether transitions of equal energies originate from states with identical spins (in even-even nuclei), parities and quadrupole moments.

Band 3 in ^{194}Hg and the vacuum band SD band in ^{192}Hg constitute a pair with transition energies that are equal within 1 keV for a large range of energies (382 to 854 keV). The quadrupole moments of the identical bands in $^{192,194}Hg$ have been measured to be equal [18]. Unfortunately, the spins and parity of the ^{192}Hg SD band have not been measured. However, it is reasonable to expect that they will both be even, as observed for the vacuum bands in ^{194}Hg and ^{194}Pb. In addition, by using a model-dependent method [41,31] even spins are also obtained. Hence, there is reasonable confidence that the transitions of equal energy are emitted from states with spins differing by 1 $\hbar$ and with opposite parity. The difference in the quantum numbers of states emitting identical-energy transitions may pose a problem for the "heroic" class of explanations. If a symmetry is responsible for identical bands, then the simplest picture would have them share the same quantum numbers.

Insight about the origin of identical bands can be obtained by examining the so-called dynamical and kinetic moments of inertia, $\mathcal{J}^{(1)}$ and $\mathcal{J}^{(2)}$, for the identical bands. $\mathcal{J}^{(1)} = (4I-2)/E_\gamma\hbar^2$, $\mathcal{J}^{(2)} = 4/\Delta E_\gamma\hbar^2$, where ΔE_γ is the energy difference for successive intraband transitions. The $\mathcal{J}^{(1)}$ and $\mathcal{J}^{(2)}$ moments of inertia for ^{194}Hg bands 1, 3 and for the ^{192}Hg SD band are shown in Fig. 22 (from Ref. [31]). It is significant that $\mathcal{J}^{(1)}$ and $\mathcal{J}^{(2)}$ converge at zero frequency for each of bands 1 and 3 in ^{194}Hg . This can be understood from the familiar equations based on the Harris[42] expansion (with $\hbar = 1$) of :

$$\mathcal{J}^{(2)} = dI_x/d\omega = \mathcal{J}_0 + 3\mathcal{J}_1\omega^2, \tag{2}$$

$$I_x = \mathcal{J}_0\omega + \mathcal{J}_1\omega^3 + i, \tag{3}$$

$$\mathcal{J}^{(1)} = I_x/\omega = \mathcal{J}_0 + \mathcal{J}_1\omega^2 + i/\omega. \tag{4}$$

The known spins yield alignment $i = 0$, so that $\mathcal{J}^{(1)} = \mathcal{J}^{(2)} = \mathcal{J}_0$ at $\omega = 0$, explaining the convergence of $\mathcal{J}^{(1)}$ and $\mathcal{J}^{(2)}$. Since $\mathcal{J}^{(2)}$ and, hence, $dI_x/d\omega$ for ^{194}Hg band 3 is larger than that for ^{192}Hg at low ω (see Fig. 22a), the difference in $I_x(\omega)$ for the two bands grows with ω at low ω (Fig. 22b). The larger $\mathcal{J}^{(2)}$ for ^{194}Hg band 3 is largely due to a decrease in pairing [43,44,31] in the excited SD band, which, as an octupole vibration, is composed of a coherent superposition of 2-quasiparticle excitations. The growth in ΔI_x does not continue, but saturates at 1 because the rate of increase in $\mathcal{J}^{(2)}$ for ^{192}Hg starts to become larger when $\omega > 0.15$ MeV. The latter is due to a larger alignment of particle spin [5] from high-N orbitals (N=7 neutrons, N=6 protons) for ^{192}Hg when $\omega > 0.15$ MeV. In other words the unit difference in spin and identical transition energies for the two identical bands result from *an accidental cancellation between the effects of pairing and particle alignment.* Hence, an unheroic explanation is indicated for this pair of identical bands, for which we have the most complete information. However, a heroic explanation cannot be ruled out since some underlying symmetry may be responsible for the cancellation. However, neither the symmetry nor its cause have been identified, so that the case for a heroic explanation, alas, needs bolstering.

3.6 Quenching of Pairing with Excitation Energy

The γ decay out of SD bands is a statistical process, which samples and probes all of the lower-lying ND states. Therefore, the decay spectrum is sensitively

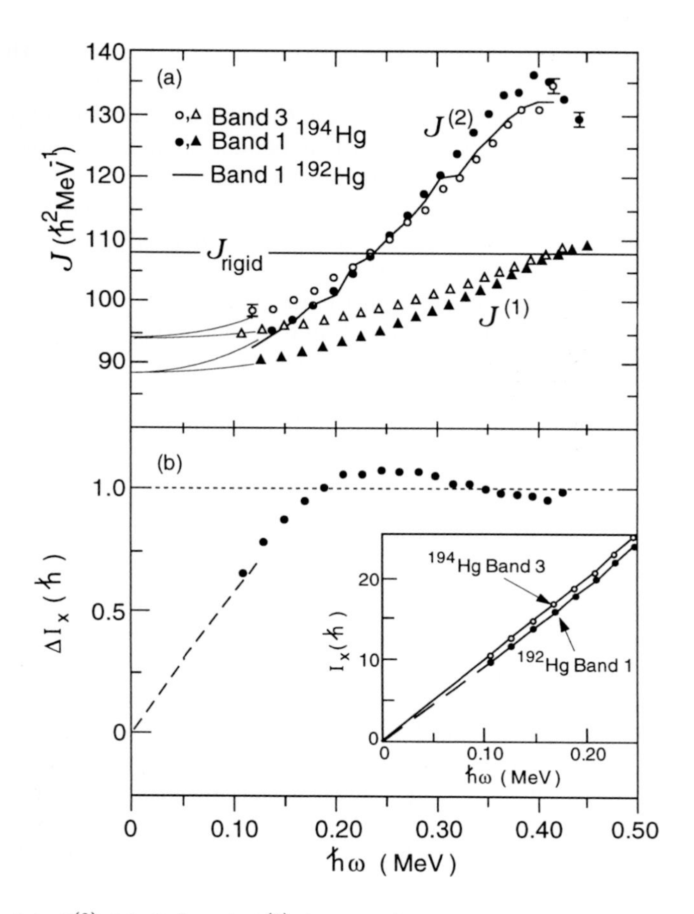

Figure 22: (a) $\mathcal{J}^{(2)}$ (circles) and $\mathcal{J}^{(1)}$ (triangles) moments of inertia for bands 1 (filled symbols) and 3 (open symbols) in ^{194}Hg. The extrapolations to zero frequency (thin lines) are from fits to Eq. 2 and Eq. 4. The solid line shows $\mathcal{J}^{(2)}$ for the vacuum SD band in ^{192}Hg. (b) ΔI_x *vs.* ω. $\Delta I_x = I_x(^{194}\text{Hg SD band 3}) - I_x(^{192}\text{Hg SD band 1})$ and $I_x = \sqrt{[I(I+1) - K^2]}$, where $K=$ 2 and 0, respectively, for the two bands, which have identical transition energies for $\omega > 0.17$ MeV. The dashed line shows extrapolations using Eq. 3. Inset: I_x for the two SD bands.

dependent on the density of states. The latter is controlled by pair correlations and by the damping of these correlations with excitation energy. Hence, the decay spectrum provides a probe of pairing[45]. This unanticipated application of the study of the decay process is explained pictorially in Fig. 23. The figure also contrasts how the pair gap Δ quenches with temperature in a macroscopic

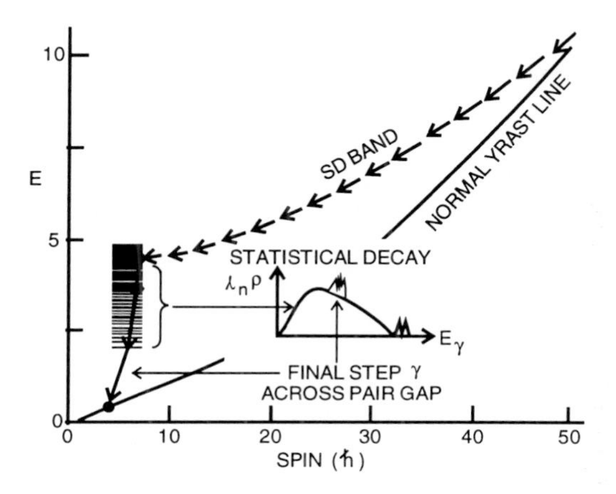

Decay from "single sharp" excited state a sensitive probe of:

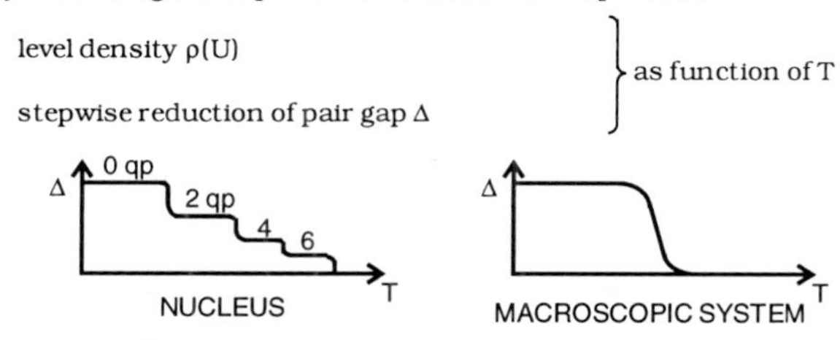

Calculations of - Døssing et al.

self-consistent 0,2,4,6... quasiparticle energies, particle no, projectile diagonalization

ρ (U)

γ cascades

spectrum

Figure 23: Illustration of how the decay from an excited state can probe properties of lower-lying states [45]. The decay spectrum is governed by the level densities, which are controlled by pair correlations. The decrease in pair gap Δ with temperature T in a nucleus and in a macroscopic superfluid are also compared.

or mesoscopic superconductor. In the former, the decrease in Δ is rather abrupt. However, one would expect the drop to be smeared out in the latter due to pairing fluctuations. The predicted[45] decrease in Δ is even more interesting, with stepwise drops corresponding to increasing numbers of quasiparticle pairs.

Døssing et al. [45] have used a self-consistent BCS treatment of pairing, with particle-number projection and diagonalization, to calculate the energies of quasiparticle levels. Based on these levels, the γ spectrum in statistical decay from a sharply defined state was also calculated and compared with the experimental decay spectrum (Fig. 24a). The theoretical spectrum with $\Delta = 0.7$ MeV reproduces the main features of the experimental spectrum,

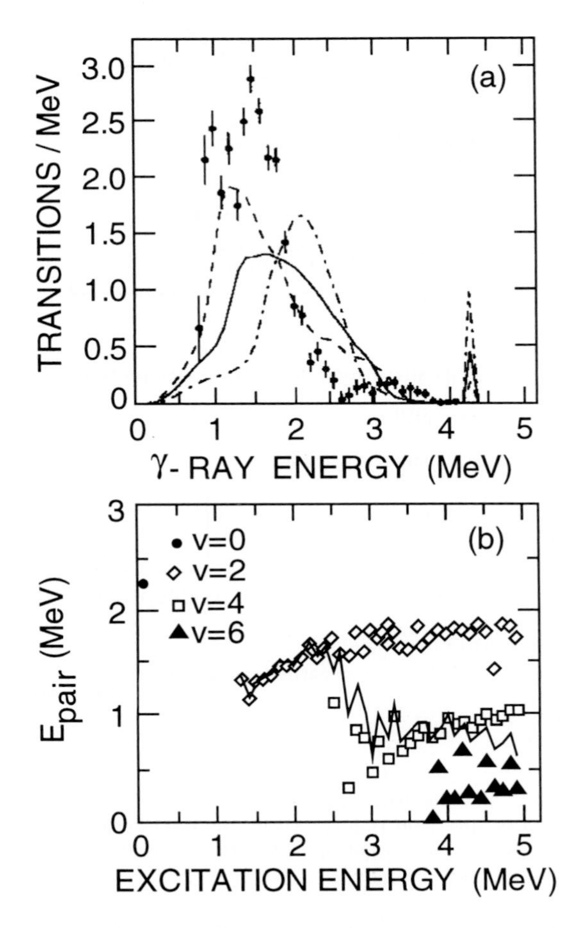

Figure 24: (a) Statistical spectra of an even-even nucleus from an excited state at U = 4.3. Lines are calculated with levels obtained from BCS theory, with particle-number projection and diagonalization. Dashed, solid and dot-dashed lines correspond to different pairing strengths, parameterized by gaps $\Delta = 0.7$, 0.9 and 1.1 MeV. The spike at 4.3 MeV represents a one-step transition. Data points with errors represent the spectrum from decay out of the ^{192}Hg SD band, which could be extracted only above 0.8 MeV. (b) Pair correlation energy for $\Delta = 0.9$ MeV. The points are for 0-, 2-, 4- and 6- quasiparticle states; the line is the average value. From Ref. [45].

namely the "squeezing" of the spectrum around 1.6 MeV, the depletion in yield at low energy, and the occurrence of one-step high-energy transitions. The pair gap has a significant consequence: the last step in the γ cascade has to vault the gap (see Fig. 23), leading to a concentration of strength around 1.6 MeV. Another distinctive feature is also predicted, namely a depletion in yield extending ~1.6 MeV below the one-step lines, from the lack of primary decays into the pair gap. However, this depletion is difficult to observe.

The creditable reproduction of the experimental spectrum (Fig. 23a) lends confidence to the theoretical prediction about the decrease in pairing energy E_{pair} as a function of excitation energy. The stepwise decrease in E_{pair} from the onset of 2- and 4- quasiparticle excitations is evident in Fig. 24b (circle and solid line). At higher energy, remnant pairing energy persists and any further stepwise decrease is washed out. The combination of theory and experiment provides the first plausible look at the quenching of pairing with temperature in a nucleus - many decades after the first recognition of the occurrence of nuclear pairing. This example also illustrates the synergistic collaboration of experiment and theory. An important theoretical step for comprehending the properties of excited nuclei is the calculation of the same quantity as is measured, namely the spectrum.

3.7 Onset of Chaos with Excitation Energy

Another unexpected application of the decay of SD states is as a probe of the onset of chaos (in ND states) with excitation energy. Cold states near the yrast line are characterized by order, quantum numbers and selection rules. In contrast, highly-excited states near the neutron separation energy, ~8 MeV, are known to be chaotic from the distribution of nearest-neighbor spacings [9] and from the fluctuation properties [36] of neutron- and γ- decay widths. [46,47] How does the transition from order to chaos occur with increasing energy? Is chaos present at intermediate energies, e.g. between 2.5 and 5 MeV? In this range, the decaying SD state provides a sharp probe state for determining chaoticity. (It is otherwise very difficult to obtain a highly-excited sharp state.)

One signature of chaos is a breakdown of selection rules in the electromagnetic decay of excited states, leading to a fluctuation and, hence, a distribution in transition strength. A well-known example is the Porter-Thomas distribution, [36]

$$P_S = (2\pi S\overline{S})^{-1/2} exp(-S/2\overline{S}), \tag{5}$$

which describes the fluctuations in neutron- and γ- decay widths S near the neutron threshold [46,47]. (The fluctuations provide a more direct measure of the complexity of the wave function than the nearest neighbor spacings.) The contrast between ordered decay, governed by selection rules imposed by quantum numbers, and chaotic decay, governed by level densities and fluctuating strengths, is illustrated for the different decay modes of SD states in Fig. 25.

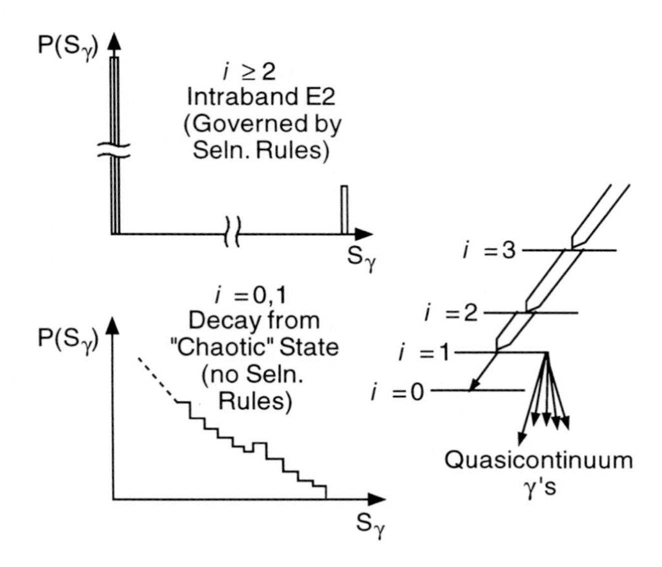

Figure 25: Decay modes of SD levels. Levels with i $\leq$ 1 decay to ND states, those with larger i deexcite to band members. The distribution of reduced transition strengths S_γ for the two cases are illustrated. Decays from states with $i \geq 2$ are governed by selection rules, with population of only one state, with the same K quantum number. Primary-transition strengths for i = 0, 1 display fluctuations, which has a Porter-Thomas form [36] in the chaotic limit.

The statistical-like form of the decay spectrum out of SD states suggests that the admixed ND excited state is quite chaotic. Further evidence is provided by the fluctuating strengths of primary transitions from the SD state. A qualitative illustration of fluctuations is the contrast in the intensities of the one-step lines (Fig. 17) in $^{192,194}Hg$; Fig. 17 shows that these are strong in ^{194}Hg, but are undetectably weak in ^{192}Hg. Quantitative evidence is provided [48] in Fig. 26, which shows the distribution in the reduced transition strengths of primary transitions from the component $|ND^* >$ in the eigenfunction of the SD state. This distribution is consistent [48] with the Porter-Thomas form, shown as a dashed line in Fig. 26. In the fully chaotic limit, the fluctuations should indeed have a Porter-Thomas distribution. However, we cannot yet draw a definite conclusion that this limit is attained, since the primary transitions are so weak that we can detect only strong transitions in the tail of the

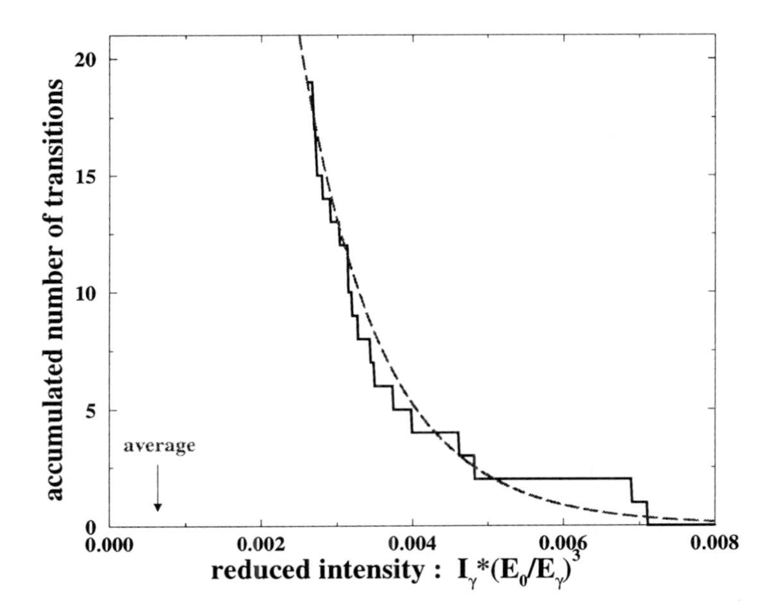

Figure 26: Cumulative number of primary transitions *vs.* reduced transition strengths, $I_\gamma(E_0/E_\gamma)^3$, from ^{194}Hg SD band 1. (The counting starts from the most intense transition.) The dashed line shows the cumulative distribution expected from a Porter-Thomas distribution that fits the data; its average value is indicated. From Ref. [48].

distribution. Nevertheless, it is already clear that ND states ~4.3 MeV above the yrast line are quite chaotic - they are certainly not ordered - but we are still in the process of trying to quantify the degree of chaoticity. Furthermore, some open questions remain, e.g. it is not understood why ^{194}Hg is so "cooperative" in revealing the one-step lines from two of its three SD bands.

The long-term goal is to understand how quantum numbers damp out with excitation energy, as reflected in Γ_Q/D, where Γ_Q is the spreading width of states with a particular quantum number, e.g. K, and D is the average level spacing. Another aim is to understand how the good quantum numbers (K, σ) of the SD state dissolve upon tunneling from the false into the true vacuum.

4 Some Other Examples of Tunneling in Nuclear Physics

The nucleus offers other opportunities for investigating tunneling. A few examples from recent work will be briefly mentioned here.

4.1 ΔI=4 Bifurcation in SD Bands

SD intraband transitions have energies that smoothly increase with spin and have almost equal spacings. This high degree of regularity makes it possible to inspect for small excursions from a perfectly smooth behavior. The first example of a small irregularity was found [49] in the SD band of ^{149}Gd. Transitions from states with ΔI=4, i.e. every other transition, were found to be slightly displaced by ~500 eV with respect to the other set with intermediate spins. The small oscillatory behavior [49] in ^{149}Gd is displayed in Fig. 27 (from Ref. [50]). Two other examples have been found in ^{148}Eu (yrast) and ^{148}Gd (excited SD band 6)- see Fig. 27. There are only very few examples of such oscillations which have been confirmed; the vast majority of bands do not exhibit this behavior, as typified by the yrast SD band in ^{148}Gd (yrast). It is interesting and significant that the three bands that show ΔI=4 bifurcation have either identical transition energies or moments of inertia $\mathcal{J}^{(2)}$. Perhaps the Hamiltonians for the three bands all share some common feature (possibly a symmetry).

Several explanations for the $\Delta I = 4$ bifurcation have been proposed. An interesting suggestion, in which tunneling is responsible for the observed oscillations, was advanced by Hamamoto and Mottelson. [51] The rotational Hamiltonian has the general form:

$$H_{rot} = A_0\overline{I}^2 + AI_3{}^2 + B_1(I_1{}^2 - I_2^2) + B_2(I_1{}^2 + I_2{}^2) \tag{6}$$

$\overline{I}$ is the angular momentum vector, I_i denote its three components in the body-fixed coordinates of the deformed nucleus, and the A and B terms are parameters. Hamamoto and Mottelson pointed out that under special conditions, namely $A{\sim}100B_1$, $B_2{\sim}0$, regular oscillations of the type observed in ^{149}Gd can be obtained. These conditions imply a small symmetry breaking of Y_{44} type around the symmetry axis. This gives rise to a four-fold symmetry about this axis, so that $\overline{I}$ is trapped in one of four minima. Tunneling between adjacent-minima splits the degenerate states. If the tunneling matrix element changes sign regularly each time $\mid \overline{I} \mid$ increases by 2, alternate shifts in energy can be obtained. While it has been recognized that one particular form of the Hamiltonian can give rise to the observed oscillations, the conditions for obtaining this form are not understood. In other words, the underlying reason for getting the oscillations in the particular three cases is still not known.

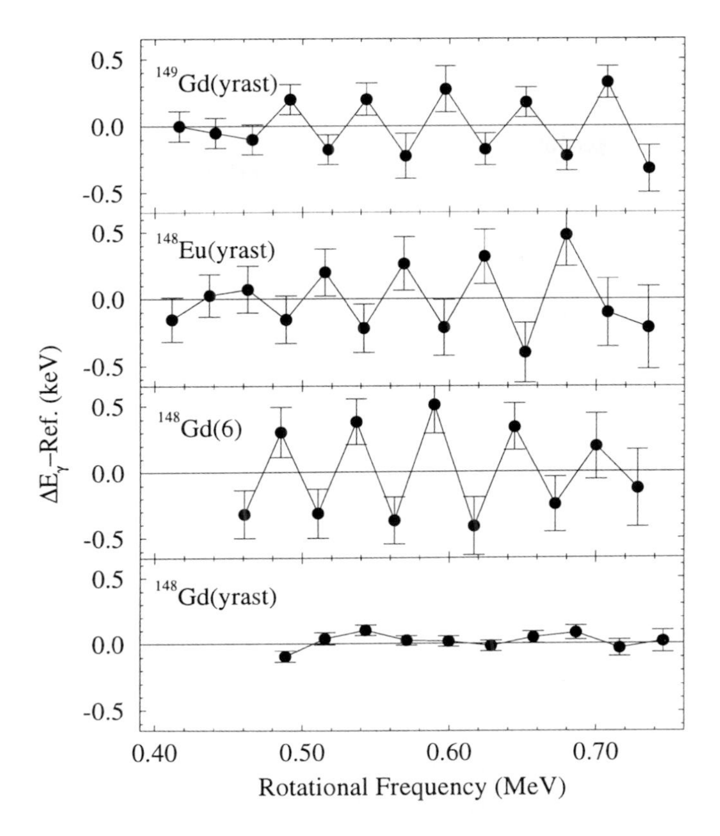

Figure 27: Deviations of transition energies from a smooth "fit" of the measured transition energies in four SD bands. Three bands, which have identical $\mathcal{J}^{(2)}$, exhibit $\Delta I = 4$ bifurcation, but most bands are smooth, as exemplified by the spectrum in the bottom panel. From Ref. [50]

4.2 Sub-barrier Fusion

It has been known since the 1970's that nuclei can fuse even though there is insufficient energy to surmount the Coulomb barrier. A particularly interesting example is the production [52] of element 110 with the reaction ^{64}Ni + ^{208}Pb; at the optimal energy, the surfaces of the fusing nuclei are still separated by ~2 fm at the point of closest approach. Nevertheless, in this case, and many others, fusion below the barrier is observed. The fusion is enabled, in one view, by fluctuations of the barrier height brought about by coupled-channel effects from inelastic excitation in the target and/or projectile. An alternative view is that surface vibrations, induced by the polarizing effects of the two ions, provides the pathway to fusion. In other words, not only is the radial separation r between the ions important, deformation degrees of freedom α also play a role - giving an example of tunneling in a multidimensional (r, α) space. This subject is discussed in recent reviews of sub-barrier fusion [53,54].

4.3 Proton Emitters

Proton-rich nuclei beyond the line of stability can exhibit proton radioactivity [55]. The ground states of these nuclei are proton unbound, and decay by emission of protons. Recent work at Argonne [56] has added many new proton emitters - see Fig. 28. Using powerful new experimental techniques in conjunction with a Fragment Mass Analyzer for unambiguously identifying

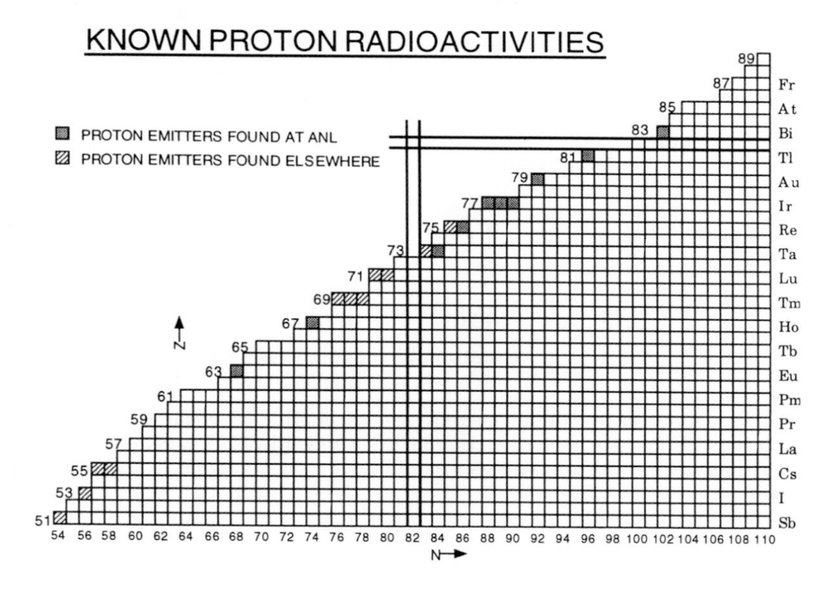

Figure 28: Known radioactive proton emitters [56], which are found *beyond* the proton-drip line. They decay when a proton tunnels across its Coulomb barrier.

evaporation residues, very clean decay proton spectra can be observed, [57] e.g. from ^{167}Ir, as shown in Fig. 29(a). Proton decay is a simple but interesting case of tunneling of an elementary particle through its Coulomb barrier (Fig. 29(b). It is analogous to α- decay, but is simpler to calculate since there is no need to evaluate a pre-formation factor. In general the proton tunneling rates can be calculated [56,58] if the orbital angular momentum ℓ of the proton

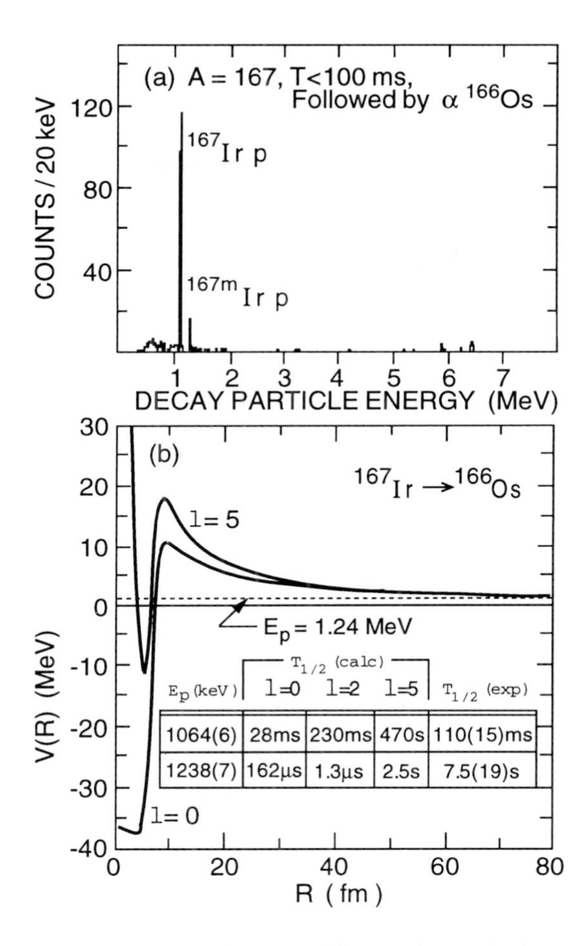

Ep (keV)	T1/2 (calc) l=0	l=2	l=5	T1/2 (exp)
1064(6)	28ms	230ms	470s	110(15)ms
1238(7)	162μs	1.3μs	2.5s	7.5(19)s

Figure 29: (a) Protons from the ground state and from an isomer in the proton emitter ^{167}Ir – from Ref. [57]. (b) Proton tunneling through Coulomb and centrifugal (l = 0, 5) barriers – from Ref. [56].

is taken into account. Tunneling through a deformed potential is an interesting example of multi-dimensional tunneling, and examples of this phenomenon have recently been found [59].

Tunneling from excited states of a proton emitter provides an opportunity to study cases where the initial state can change character from ordered to chaotic. One can also investigate the interplay between the proton energy and angular momentum: an excited high-spin state can take advantage of

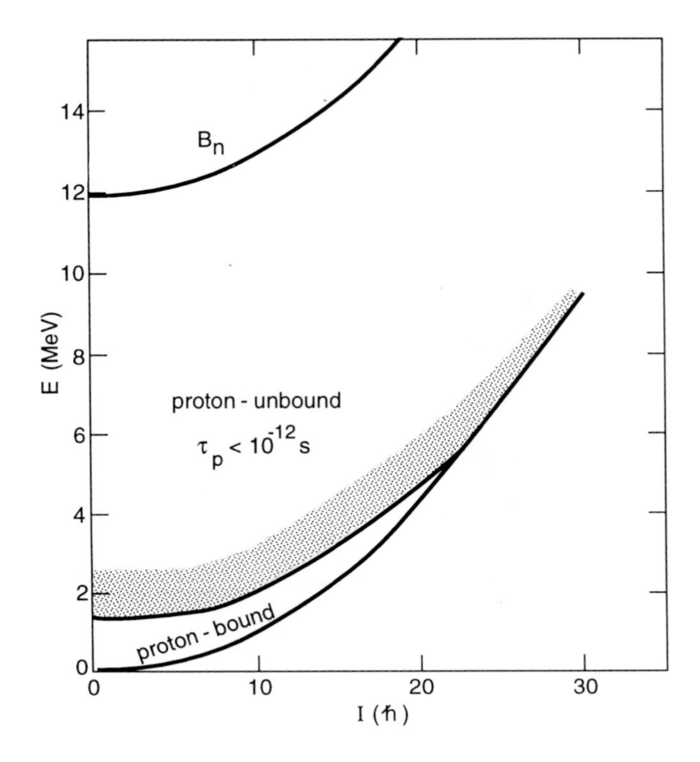

Figure 30: In a proton emitter, such as ^{167}Ir, only a small spin-energy region is proton bound. The region above that, as well as the ground state, decay by proton tunneling. Protons from the dotted region are expected show a new phenomenon – sub-Coulomb peaks and structures. Levels above the line marked B_n are neutron-unbound.

the increased energy for decay to lower-spin states to overcome the additional centrifugal barrier in decay with $\ell > 0$.

Although the ground state decays by proton emission, states immediately above that, or along the yrast line, preferentially decay by γ emission. However, $\Gamma_p \sim \Gamma_\gamma$ at some excitation energy, estimated to be > 1.2 MeV above the yrast line in the proton emitter ^{167}Ir; thereafter $\Gamma_p > \Gamma_\gamma$. As a consequence, it is expected that the proton-bound region of a proton-emitter is unusually small (Fig. 30). In contrast, in stable nuclei the particle-bound region is much larger, defined typically by the neutron separation energy of ~8 MeV. Excited states in the dotted region of Fig. 30 should exhibit a new type of proton emission, namely *sub-Coulomb protons that exhibit peaks and structures.* It would be interesting to search for this new form of proton emission. As only the proton-bound region of Fig. 30 can lead to population of the proton-emitter ground state via γ decay, there is another distinctive signature for the small bound region. The γ ray spectrum, which normally extends to ~8 MeV, will end at a much lower energy of ~1.5 MeV in a proton emitter. The prospective coupling of Gammasphere to the Fragment Mass Analyzer at Argonne will make it possible to examine the γ spectra leading to the ground state of proton emitters.

5 Comments on tunneling in nuclei

The nucleus provides examples of tunneling where the initial and final states can be either ordered or disordered states. In the feeding of SD bands, tunneling between hot SD and ND states plays a role. In decay out of SD bands, an ordered state tunnels to a chaotic one. The tunneling probability is very small in the latter. However, it would have been even smaller if the tunneling were to another ordered state. In other words, the chaotic nature of the final state facilitates the tunneling rate [60]. This situation is related to chaos- assisted [11] tunneling, where the transition between two ordered regions of phase space can be amplified by coupling with an intervening chaotic region. In a Poincaré surface of section, one visualizes a diffuse chaotic region surrounding the ordered regions and abetting the tunneling between them.

Restated in the conventional language of nuclear physics, one would say that the coupling between cold SD and ND states would be negligible because the states have different structures and quantum numbers. However, the compound nature of the hot ND state aids the coupling between SD and ND states, since compound states are able to couple to all states. One may also view this problem in terms of the shape parameters defined in Fig. 7. The SD state, well localized in β and γ, can have some overlap with a hot ND state, which is spread out in the $\beta - \gamma$ plane. On the other hand, the overlap with a cold state, localized elsewhere in the plane, is considerably smaller and is also further reduced by a barrier.

For nuclei which lie beyond the proton drip line, proton decay from radioactive ground states and isomers connect cold initial and final nuclear states (although the proton is in the continuum representing an unbound final state). However, the character of the initial state can be changed from ordered to chaotic by increasing its excitation energy. This provides another incentive for studying proton emission from excited states of a proton emitter.

5.1 Analogies with other systems

Tunneling from a SD state to the ND well can occur only when there is a nearby excited ND state. In other words, this is a resonant tunneling process. Some interesting analogies can be made with other processes where resonances are responsible for large-amplitude motion. In *neutron- induced fission*, intermediate resonances are observed –see Fig. 7 in Ref. [3] – which represent enhanced fission when the neutron energy corresponds to a vibrational state in the secondary SD well of actinide nuclei. *Molecular pre-dissociation*, a process equivalent to nuclear fission, occurs when a bound level crosses an unbound one at some radial separation[61]. A third example comes from theoretical work [62] on a *hydrogen atom in a polarized microwave field.* A stable non-spreading electronic wave packet (Floquet state) is predicted. The crossing of the energy level of the wave packet with that of chaotic states (as a function of, say, the electric field strength) induces ionization of the stable wave packet, and spikes in the photoabsorption spectrum are predicted at these crossings.

6 Summary

The feeding and decay of SD bands are summarized in Fig. 31. Following neutron evaporation, there is an initial cooling by statistical γ rays (component 1). This is the hot chaotic phase, where the nuclear shape is not well-defined because of thermal fluctuations. When the nucleus is well within the SD false vacuum, shape fluctuations begin to diminish, the elongated shape becomes established and collective E2 transitions from excited SD bands form a large E2 peak (component 2). Cooling into the SD minimum is brought about by M1/E2 transitions (component 3). Once trapped into the bottom of the SD vacuum, the nucleus undergoes cold, ordered rotation. The resultant spectrum, with characteristic equally-spaced sharp lines (component 4), is distinctly different from the broad, unresolved, precursor spectrum. A sudden tunneling to ND states signals a return to chaos, with a re-emergence of a broad statistical spectrum (component 5) and fluctuating primary-line strengths. Finally, the nucleus settles towards the ground state along the cold ND yrast line (component 6) by emission of sharp lines. Hence, SD nuclei provide examples of order embedded in chaos, display a unique double cycle of chaos-to-order transition, and exhibit tunneling from ordered to chaotic states, as well as between disordered SD and ND states. The decay spectra from SD bands reveal that the decay mechanism is a tunneling from an ordered state in the false vacuum to a chaotic one in the true vacuum. The high-energy, one-step decay transitions yield excitation energies and spin/parity quantum numbers, which characterize the SD bands as ordered states; reveal their microscopic structure; and sug-

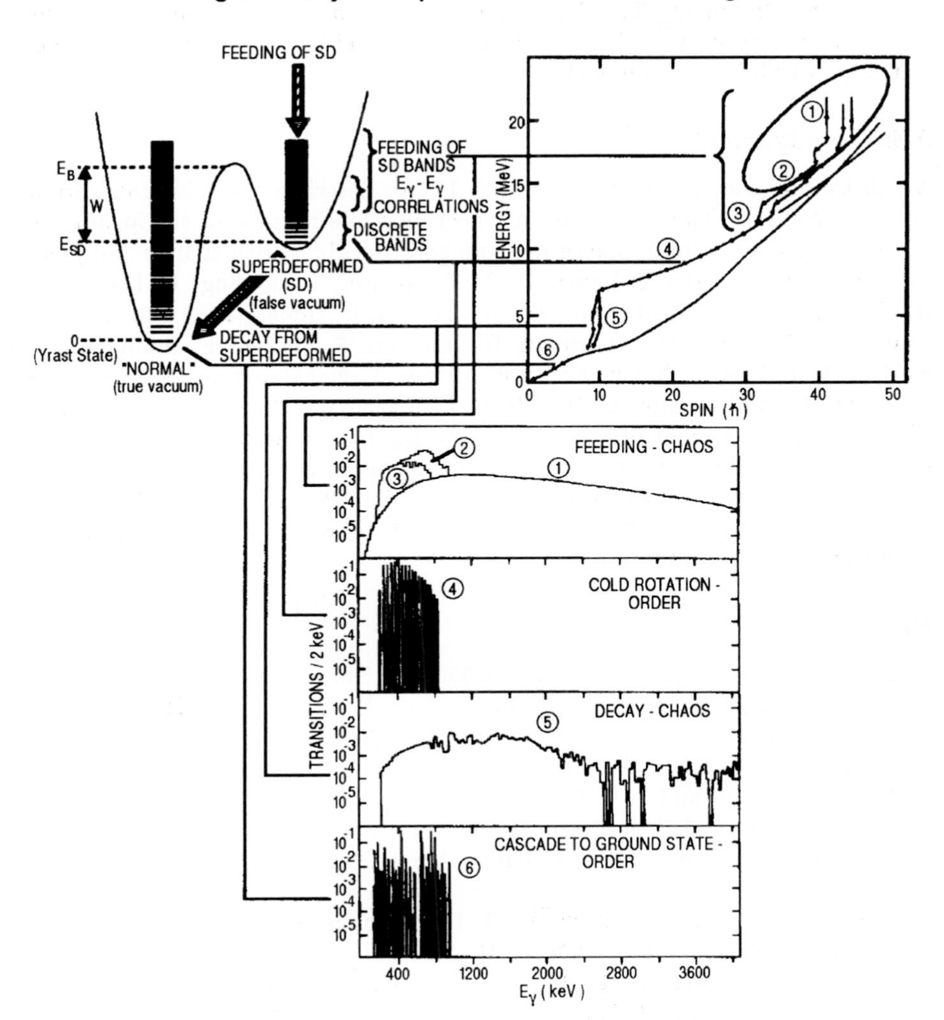

Figure 31: Gamma spectra from different stages in the feeding and decay of a SD band. The different characters of the spectra reveal a unique cycle of chaos-to-order-to-chaos-to-order transition.

gest that identical bands arise from accidental cancellations between pairing and particle alignment. The decay spectrum also provides information about the quenching of pairing with excitation energy and indicates that excited ND states at ~4 MeV are largely chaotic.

Acknowledgements

A substantial portion of the work that I have described here is drawn from my own research. This work has only been possible because of the invaluable contributions and stimulation of my collaborators. It is with thanks and pleasure that I acknowledge T. Lauritsen, M. Carpenter, T. Døssing , G. Hackman, R. G. Henry, A. Lopez-Martens, D. Ackermann, I. Ahmad, H. Amro, D. Blumenthal, I. Calderin, D. Gassmann, R. Janssens, S. Fischer, F. Hannachi, A. Korichi, I. Y. Lee, A. Macchiavelli, F. Moore, T. Nakatsukasa, D. Nisius and P. Reiter. I wish to thank: C. Davids, R. Firestone and D. Haslip for providing some figures; R. Janssens for carefully reading this manuscript; and the Institute of Nuclear Theory for its hospitality during the program "Tunneling in Complex Systems", where these lectures were delivered. The work at Argonne National Laboratory is supported by the U. S. Department of Energy, under Contract No. W-31-109-ENG-38.

References

1. V. M. Strutinsky, Nucl. Phys. A 95, 420 (1967).
2. S. Hofmann, V. Ninov, F.P. Hessberger, P. Armbruster, H. Folger, G. Mnzenberg, H.J. Schtt, A.G. Popeko, A.V. Yeremin, S. Saro, R. Janik, M. Leino, Z. Phys. A 354, 229 (1996).
3. S. Bjørnholm and J. E. Lynn, Rev. Mod. Phys. 52, 725 (1980).
4. P. J. Nolan and P. J. Twin, Annu. Rev. Nucl. Part. Sci. 38, 533(1988).
5. R. V. F. Janssens and T. L. Khoo, Annu. Rev. Nucl. Part. Sci. 41, 321 (1991).
6. R. Krücken , G. Hackman, M. A. Deleplanque, R. V. F. Janssens, I. Y. Lee, D. Ackermann, I. Ahmad, H. Amro, S. Asztalos, D. J. Blumenthal, M. P. Carpenter, R. M. Clark, R. M. Diamond, P. Fallon, S. M. Fischer, B. Herskind, T. L. Khoo, T. Lauritsen, A. O. Macchiavelli, R. W. MacLeod, D. Nisius, G. Schmid, F. S. Stephens and K. Vetter, Phys. Rev. C 54, R2109 (1996).
7. A. Bohr and B. Mottelson, *Nuclear Structure*, vol. 2, p. 592 (W. A. Benjamin, Inc., Reading, 1975).
8. B. Singh, R. B. Firestone and S. Y. F. Chu, Table of Superdeformed Nuclear Bands and Fission Isomers, WWW Edition (1997);

http://csa5.lbl.gov/~fchu/sdband/sdbindex.html.
9. R. U. Haq, A. Pandey and O. Bohigas, Phys. Rev. Lett. 48 1086 (1982).
10. B. Mottelson in *The Frontier of Nuclear Spectroscopy*, p. 7, ed. Y. Yoshizawa, H. Kusakari and T. Otsuka (World Scientific, Singapore, 1993).
11. O. Bohigas, D. Boose, R. Egydio de Carvalho and V. Marvulle, Nucl. Phys. A560, 197 (1993); S. Tomsovic and D. Ullmo, Phys. Rev. E 50, 145 (1994).
12. I. Y. Lee, Nucl. Phys. A520, 641c (1990).
13. T. Lauritsen, Ph. Benet, T. L. Khoo, K. B. Beard, I. Ahmad, M. P. Carpenter, P. J. Daly, M. W. Drigert, U. Garg, P. B. Fernandez, R. V. F. Janssens, E. F. Moore, F. L. H. Wolfs and D. Ye, Phys. Rev. Lett. 69, 2479 (1992).
14. K. Schiffer and B. Herskind, Nucl. Phys. A520, 521c (1990).
15. T. L. Khoo, M. P. Carpenter, T. Lauritsen, D. Ackermann, I. Ahmad, D. Blumenthal, S. Fischer, R. V. F. Janssens, D. Nisius, E. F. Moore, A. Lopez-Martens, T. Døssing , R. Krücken , S. J. Asztalos, J. A. Becker, L. Bernstein, R. M. Clark, M. A.Deleplanque, R. M. Diamond, P. Fallon, L. P. Farris, F. Hannachi, E. A. Henry, A. Korichi, I. Y. Lee, A. O. Macchiavelli and F. S. Stephens, Phys. Rev. Lett. 76, 1583 (1996).
16. B. Herskind, T. Døssing , S. Leoni, M. Matsuo, N. Nica, D. C. Radford and P. Rasmussen, Nucl. Phys. A557, 191c (1993).
17. S. Leoni, B. Herskind, T. Døssing , A. Atac, I. G. Bearden, M. Bergstrom, C. Fahlander, G. B. Hagemann, A. Holm, D.T. Joss, M. Lipoglavsek, A. Maj, P. J. Nolan, J. Nyberg, M. Palacz, E. S. Paul, J. Persson, M. J. Piiparinen, N. Redon, A. T. Semple, G. Sletten and J.P. Vivien, Phys. Lett. B 409, 71 (1997).
18. E. F. Moore, T. Lauritsen, R. V. F. Janssens, T. L. Khoo, D. Ackermann, I. Ahmad, H. Amro, D. Blumenthal, M. P. Carpenter, S. M. Fischer, G. Hackman, D. Nisius, F. Hannachi, A. Lopez-Martens, A. Korichi, S. J. Asztalos, R. M. Clark, M. A.Deleplanque, R. M. Diamond, P. Fallon, I. Y. Lee, A. O. Macchiavelli F. S. Stephens, J. A. Becker, L. Bernstein, L. P. Farris and E. A. Henry, Phys. Rev. C 55, R2150 (1997).
19. C. Baktash, B. Haas and W. Nazarewicz, Annu. Rev. Nucl. Part. Sci. 45, 485 (1995).
20. E. Vigezzi, R. A. Broglia and T. Døssing , Phys. Lett. B 249, 163 (1990).
21. Y. R. Shimizu, E. Vigezzi, T. Døssing and R. A. Broglia, Nucl. Phys. A 557, 99c (1993).
22. F. Barranco, G. F. Bertsch, R. A. Broglia and E. Vigezzi, Nucl. Phys. A512, 253 (1990).

23. T. L. Khoo, T. Lauritsen, I. Ahmad, M. P. Carpenter, P. B. Fernandez, R. V. F. Janssens, E. F. Moore, F. L. H. Wolfs, Ph. Benet, P. J. Daly, K. B. Beard, U. Garg, D. Ye and M. W. Drigert, Nucl. Phys. A 557, 83c (1993).
24. F. Soramel, T. L. Khoo et al, to be published.
25. R. G. Henry, T. Lauritsen, T. L. Khoo, I. Ahmad, M. P. Carpenter, B. Crowell, T. Døssing , R. V. F. Janssens, F. Hannachi, A. Korichi, C. Schück, F. Azaiez, C. W. Beausang, R. Beraud, C. Bourgeois, R. M. Clark, I. Deloncle, J. Duprat, B. Gall, H. Hübel, M. J. Joyce, M. Kaci, Y. Lecoz, M. Meyer, E. S. Paul, N. Perrin, N. Poffe, M. G. Poquet, N. Redon, H. Sergolle, J. F. Sharpey-Schafer, J. Simpson, A. G.Smith, R. Wadsworth and P. Willsau, Phys. Rev. Lett. 73, 777 (1994).
26. B. Crowell, M. P. Carpenter, R. G. Henry, R. V. F. Janssens, T. L. Khoo, T. Lauritsen and D. Nisius, Nucl. Inst. Meth. A 355, 575 (1995).
27. D. C. Radford, I. Ahmad, R. Holzmann, R. V. F. Janssens and T. L. Khoo, Nucl. Inst. Meth. A 258, 111 (1987).
28. R. Holzmann, I. Ahmad, R. V. F. Janssens, T. L. Khoo and D. C. Radford, Nucl. Inst. Meth. A260, 153 (1987).
29. A. Lopez-Martens, F. Hannachi, T. Døssing , C. Schück, R. Collatz, E. Gueorguieva, Ch. Vieu, S. Leoni, B. Herskind, T. L. Khoo, T. Lauritsen, I. Ahmad, D. J. Blumenthal, M. P. Carpenter, D. Gassmann, R. V. F. Janssens, D. Nisius, A. Korichi, C. Bourgeois, A. Astier, L. Ducroux, Y. Le Coz, M. Meyer, N. Redon, J. F. Sharpey-Schafer, A. N. Wilson, W. Korten, A. Bracco and R. Lucas, Phys. Rev. Lett. 77, 1707 (1996).
30. T. Lauritsen, A. Lopez-Martens, T. L. Khoo, R. V. F. Janssens, M. P. Carpenter, G. Hackman, D. Ackermann, I. Ahmad, H. Amro, D. J. Blumenthal, S. M. Fischer, F. Hannachi, A. Korichi and D. Nisius in *Proc. Conf. On Nuclear Structure at the Limits*, Argonne, 1996, ANL/PHY-97/1, p. 35.
31. G. Hackman, T. L. Khoo, M. P. Carpenter, T. Lauritsen, A. Lopez-Martens, I.V. Calderin, R. V. F. Janssens, D. Ackermann, I. Ahmad, S. Agarwala, D. J. Blumenthal, S. M. Fischer, D. Nisius, P. Reiter, J. Young, H. Amro,, E. F. Moore, F. Hannachi, A. Korichi, I. Y. Lee, A. O. Macchiavelli, T. Døssing and T. Nakatsukasa, Phys. Rev. Lett. 79, 4100 (1997).
32. A. Lopez-Martens, F. Hannachi, A. Korichi, C. Schück, E. Gueorguieva, Ch. Vieu, B. Haas, R. Lucas, A. Astier, G. Baldsiefen, M. Carpenter, G. de France, R. Duffait, L. Ducroux, Y. Le Coz, Ch. Fink, A. Gorgen, H. Hübel, T. L. Khoo, T. Lauritsen, M. Meyer, D. Pevost, N. Redon, C. Rigollet, H. Savajols, J. F. Sharpey-Schafer, O. Stezowki, Ch. Theisen,

U. Van Severen, J.P. Vivien and A. N. Wilson, Phys. Lett. B 380, 18 (1996).
33. K. Hauschild, L. A. Bernstein, J. A. Becker, D. E. Archer, R. W. Bauer, D. P. McNabb, J. A. Cizewski, K.-Y. Ding, W. Younes, R. Krücken , R. M. Diamond, R. M. Clark, P. Fallon, I.-Y. Lee, A. O. Macchiavelli, R. MacLeod, G. J. Schmid, M. A. Deleplanque, F. S. Stephens and W. H. Kelly, Phys. Rev. C 55, 2819 (1997).
34. J. Schirmer, J. Gerl, D. Habs and D. Schwalm, Phys. Rev. Lett. 63, 2196 (1989).
35. S. Lunardi, R. Venturelli, D. Bazzaco, C. M. Petrache, C. Rossi- Alvarez, G. de Angelis, G. Vedovato, D. Bucurescu and C. Ur, Phys. Rev. C 52, R6 (1995), and references therein.
36. C. E. Porter and R. G. Thomas, Phys. Rev. 104, 483 (1956).
37. T. Nakatsukasa, K. Matsuyanagi, S. Mizutori and Y. R. Shimizu, Phys. Rev. C 53, 2213 (1996).
38. B. Crowell, R. V. F. Janssens, M. P. Carpenter, I. Ahmad, S. Harfenist, R. G. Henry, T. L. Khoo, T. Lauritsen, D. Nisius, A. N. Wilson, J. Sharpey- Schafer and J. Skalski, Phys. Lett. B 333, 320 (1994).
39. S. Bouneau, F. Azaiez, J. Duprat, I. Deloncle, M-G. Porquet, U. J. van Severen, T. Nakatsukasa, M. M. Aleonard, A. Astier, S. Baldsiefen, C. W. Beausang, F. A. Beck, C. Bourgeois, D. Curien, N. Dozie, L. Ducroix, B. Gall, H. Hübel, M. Kaci, W. Korten, M. Meyer, N. Redon, H. Sergolle, and J. F. Sharpey-Schafer, in *Proc. Conf. On Nuclear Structure at the Limits*, Argonne, 1996, ANL/PHY-97/1, p. 105.
40. P. H. Heenen, J. Dobaczewski, W. Nazarewicz, P. Bonche and T. L. Khoo, preprint (1997).
41. J. E. Draper, F. S. Stephens, M. A. Deleplanque, W. Korten, R. M. Diamond, W. H. Kelly, F. Azaiez, A. O. Macchiavelli, C. W. Beausang and E. C. Rubel, Phys. Rev. C 42, R1791 (1990); J. Becker, E. A. Henry, A. Kuhnert, T. F. Wang, S. W. Yates, R. M. Diamond, F. S. Stephens, J. E. Draper, W. Korten, M. A. Deleplanque, A. O. Macchiavelli, F. Azaiez, W. H. Kelly, J. A. Cizewski and M. J. Brinkman, ibid. C 46, 889 (1992).
42. S. M. Harris, Phys. Rev. 138, 508B (1965).
43. P. Fallon, W. Nazarewicz, M. Riley and R. Wyss, Phys. Lett. B 276, 427 (1992).
44. R. Wyss and W. Satuła, Phys. Lett. B 351, 393 (1995).
45. T. Døssing , T. L. Khoo, T. Lauritsen, I. Ahmad, D. Blumenthal, M. P. Carpenter, B. Crowell, D. Gassmann, R. G. Henry, R. V. F. Janssens and D. Nisius, Phys. Rev. Lett. 75, 1276 (1995).

46. S. S. M. Wong, *Nuclear Statistical Spectroscopy*, p. 32 (Oxford Univ. Press, Oxford 1986).
47. H. E. Jackson, J. Julien, S. Samour, A. Bloch, C. Lopata, J. Morgenstern, H. Mann and G. E. Thomas, Phys. Rev. Lett. 17, 656 (1966).
48. A. Lopez-Martens et al, to be published.
49. S. Flibotte, H. R. Andrews, G. C. Ball, C. W. Beausang, F. A. Beck, G. Belier, T. Byrski, D. Curien, P. J. Dagnall, G. de France, D. Disdier, G. Duchene, Ch. Finck, B. Haas, G. Hackman, D. S. Haslip, V. P. Janzen, B. Kharraja, J. C. Lisle, J. C. Merdinger, S. M. Mullins, W. Nazarewicz, D. C. Radford, V. Rauch, H. Savajols, J. Styczen, Ch. Theisen, P.J. Twin, J. P. Vivien, J. C. Waddington, D. Ward, K. Zuber and S. Åberg, Phys. Rev. Lett. 71, 4299 (1993).
50. D. S. Haslip, S. Flibotte, G. de France, M. Devlin, A. Galindo- Uribarri, G. Gervais, G. Hackman, D. R. LaFosse, I. Y. Lee, F. Lerma, A. O. Macchiavelli, R. W. MacLeod, S. M. Mullins, J. M. Nieminen, D. G. Sarantites, C. E. Svensson, J. C. Waddington and J. N. Wilson, Phys. Rev. Lett. 78, 3447 (1997).
51. I. Hamamoto and B. Mottelson, Phys. Lett. B 333, 294 (1994); Phys. Scr. T 56, 27 (1995).
52. S. Hofmann, V. Ninov, F.P. Hessberger, P. Armbruster, H. Folger, G. Mnzenberg, H.J. Schtt, A.G. Popeko, A.V. Yeremin, A.N. Andreyev, S. Saro, R. Janik, M. Leino, Z. Phys. A 350, 277 (1995)
53. R. Vandenbosch, Annu. Rev. Nucl. Part. Sci. 42, 447 (1992).
54. A. Winter, Nucl. Phys. A594, 203 (1995).
55. S. Hofmann in *Nuclear Decay Modes* p.143, ed. D. N. Poenaru, (Institute of Physics Publishing, 1996).
56. P. J. Woods and C. N. Davids, Annu. Rev. Nucl. Part. Sci. 47, 541 (1997).
57. C. N. Davids, P. J. Woods, J. C. Batchelder, C. R. Bingham, D. J. Blumenthal, L. T. Brown, B. C. Busse, L. F. Conticchio, T. Davinson, S. J. Freeman, D. J. Henderson, R. J. Irvine, R. D. Page, H. T. Penttila, D. Seweryniak, K. S. Toth, W. B. Walters and B. E. Zimmerman, Phys. Rev. C 55, 2255 (1997).
58. S. Åberg, P. B. Semmes and W. Nazarewicz, Phys. Rev. C56, 1762 (1997).
59. C. N. Davids, priv. com. (1997).
60. O. Bohigas, priv. com. (1997).
61. L. D. Landau and E. M. Lifshitz, *Quantum Mechanics, Non- Relativistic Theory*, p. 310 (Pergamon Press, London, 1958).

62. Z. Zakrewski, D. Delande and A. Buchleitner, Phys. Rev. Lett. 75, 4105 (1995).

Solitons in the Bose Condensate

William P. Reinhardt
Department of Chemistry and School of Chemistry
University of Washington University of Melbourne
Seattle, Washington, 98195-1700 Parkville 3052
USA Victoria, Australia

e-mail: rein@chem.washington.edu

Abstract

The simplest mean field description of a dilute gaseous Bose condensate is reviewed. Special attention is paid to the case of atoms with repulsive interactions in strongly anisotropic, pseudo-one-dimensnional traps. For this special case the Gross-Pitaevskii, or Non-Linear Schrodinger Equation (NLSE), is approximately separable. We then focus on finding analytical solutions for the stationary states of the NLSE for the simplest possible one dimensional trap: the familiar one dimensional "particle in a box". Solving the NLSE for this simple stationary state case is accomplished via Jacobi elliptic functions, whose introduction and properties form, perhaps surprisingly, an essential part of the tutorial. This follows from the fact that starting from the stationary state solutions, the natural question of collisions between condensates, each in its own stationary ground state, leads us immediately to discussion of solitons whose elementary analytical theory involves these same Jacobi functions. Following the introduction of zero energy condensate collisions, and a review of elementary soliton theory, an analogy with the Josephson tunneling effect is made, in that the velocity of an isolated soliton separating two coherent pieces of condensate is a simple sinesoidal function of the relative phase of the two condensates. Solitons are thus a potentially useful "relative phase meter" for interacting condensates which otherwise dipslay a remarkable phase rigidity in the mean field approximation. The discussion is then extended to multi-soliton dynamics in 1, 2, and 3 dimensions, then to solitons in a harmonic trap, and concludes with a brief discussion of the possible experimental consequences of observation and caerful monitoring of such solitons.

1. Introduction

Background: The recent experimental realization of Bose condensation in magnetically trapped dilute alkali gases with repulsive[1,2], and attractive[3,4] atomic interactions is the culmination of decades[5] of intensive experimental work, and opens an entirely new window into the behavior of macroscopic quantum systems, these formerly being "only" the condensed phase superfluids and superconductors. It is now experimentally possible to make wide variations in density and interaction strength (including even sign), given the many atomic species which are candidates for condensation, and the fact that densities are controlled (up to limits set by three body collisions[6]) by the trap, rather than the intrinsic, and only slightly variable, densities of solids and liquids. Also, the quantum properties of these systems are "directly" observable[1,2], in that they are not hidden in, for example, a crystal lattice. The fact that low density gaseous condensed systems are at hand is a gift to theory, as many of the long standing issues, for example those related to the phase of macroscopic quantum systems[7–11], can be quantitatively *initially* approached using quite simple and well established theoretical models. The simplest mean-field theory, the Gross – Pitaevskii[12,13], or Non-Linear Schrodinger Equation (NLSE), and its generalization(s) to treat "RPA" level excitations[14], has been shown to have quantitative predictive power, and when used to analyze experiment has produced satisfying agreement and qualitative physical understanding[15 a,b]. Namely, densities and low energy collective excitations, both in terms of amplitude and frequency, are well modelled in mean field theory for both the strongly anisotropic traps used at MIT[2], and the more nearly isotropic Boulder traps[1].

An obvious next question is: "what happens when condensates collide"? In an extraordinary experiment[16] the MIT group verified that independently created condensates, separated by a light sheet potential, and then ballistically expanded, showed dramatic and robust de-Broglie like interferences in regions of overlap. The issue of such interacting condensates has a long theoretical history, beginning with questions originally posed by Anderson, as neutral superfluid analogs of the Josephson effect were sought, and the question of the possible meaning of the (relative) phase of superconductors/superfluids "which have never met" probed. In weakly coupled superconductors the Josephson current is driven by this phase difference. The difficulties of observing the analog of this effect in neutral superfluids are reviewed in Tilley and Tilley[17]. Other, more recent, theoretical modelling[18] has looked at interferences expected in the overlap of freely expanding condensates (ie. the direct analog of the MIT experiment[16]) modelled, using the NLSE, but now interpreted in the framework of theory originally developed in quantum optics[19,20].

Outline of the Tutorial: Tony Leggett has, in his tutorial presentaions in this INT workshop, reviewed the basic physics of the magnetic trapping of dilute alkali atomic gases, and the formation, and general properties of Bose-Einstein condenstates, BECs in what follows. His discussion in the present volume[21] thus provides a readily available introduction to the field. We move more directly into the development of a detailed analytically soluble model of a trapped BEC. Advantage is taken of the fact that the trapped BECs at MIT are highly anisotropic. Such anisotropy gives approximate separabilities of the relevant non-linear PDEs, allowing development of simple models which nonetheless maintain some contact with actual experiments.

The tutorial disucssions to follow consist of:

- preliminary discussion of mean field theory for the condensate
- development of a one-dimensional model for the BEC stationary states in a "box" trap
- solution of the one-dimensional NLSE for "many bodies in a box": the Jacobi SN functions
- stationary states of the NLSE and collisions between initially separated condensates.....a "surprise"
- an introduction to solitons via the KdV equation: Jacobi CN functions
- solitons in the BEC

Apologia: The material which follows is based quite closely on the actual pedagogical Tutorial Lectures presented at the INT workshop. It is a mixture of review, old pedagogy, new pedagogy, and with new results and applications mixed in. Almost none of the workshop attendees, as it happened, were initially familiar with Jacobi functions, solitons, or even the BEC itself, many having arrived after the conclusion of Leggett's actual lecture series introducing the BEC. Such is the nature of Workshops! So, the intent of the Tutorial was to introduce the basic ideas in a genuinely elementary, and even oversimplified fashion, but with the hope that enough of the mathematics, and physics behind such, could be included to allow interested participants easy access to further study. Such being the case, no further apology will be made for the fact that much of what is presented is in no way original.

2. Preliminaries: mean fields, the condensate wave function, and the psuedo-potential.

Mean Field Theory for a Zero Tempearture Bose Gas: Boson mean field theories may be derived variationally from conventional quantum mechanics by assuming that the N-body wave function is a symmetrized product of one body wave-functions. This is the Boson analog of the usual approach of chemists, looking for an "orbital" picture of the stationary states of electrons in molcules. Alternatively, field theoretic derivations[14] proceed via factorizaion of two body Green's functions in terms of an appropriately symmetrized or antisymmetrized (for fermions) products of one body Green's funcions. Both approaches lead to "Hartree" or "Hartree-Fock" type pictures, the former if exchange is simply neglected. We proceed here heuristically: suppose that (before any symmetrization) particle "i" is in orbital φ_i. Further assume that these identical particles, of mass m, are to be considered elementary, and interact via a pairwise two body potential V(r_{ij}), where r_{ij} = | $\bar{r}_i$ - $\bar{r}_j$|, and also with the externally implosed trap potential $V^{trap}(\bar{r}_i)$. The mean field idea, that the jth particle moves in the average and static potenital generated by the remaining i = 1,2...,N, (i ≠ j) particles each with its one body wave function φ_i, suggests, intuitively, that the one particle Schrodinger equation for the determination of the orbital φ_j may be written as

$$- \frac{\hbar^2}{2m} \nabla^2 \varphi_j(\bar{r}_j) + V^{trap} \varphi_j(\bar{r}_j)$$

$$+ \left(\sum_{i=1\,(i \neq j)}^{N} \int \varphi_i(\bar{r}_i)^* V(r_{ij}) \varphi_i(\bar{r}_i) d^3 \bar{r}_i \right) \varphi_j(\bar{r}_j) = \mu \varphi_j(\bar{r}_j), \qquad (1)$$

where μ is the chemical potential, or "orbital energy". μ is referred to as a chemical potential, as it gives the energy for removal of one particle from the system. For large N in homogeneous systems this is also the energy of addition of a single particle.

The fact that μ is related to an energy difference between systems containing N and (N-1) particles is known to theoretical chemists as Koopman's Theorem[22]. This same result follows immediately from the field theoretic derivation of the mean field approximation via an equation of motion for $< \psi(\bar{r}, t)^\dagger \psi(\bar{r}', t')>$. Here the ψ 's are Heisenberg representation Bose field operators[14], and <,> an N body ground state expectaion value, wherein | > = |N,0>. The ground state expectation value may be interpreted by insertion of the identity, resolved in terms of a complete set of (N-1) body states:

$$\mathrm{I} = \sum_{i=0}^{\infty} |N-1,\ i> < N-1,\ i\,|. \qquad (2)$$

The ket "|N-1,i>" denoting the ith excited state of the N-1 body system. Such an insertion gives, for example, an expansion of the one body density matrix:

$$\gamma(\bar{r}, \bar{r}') \equiv \ < \psi(\bar{r}, t)^\dagger \psi(\bar{r}', t')> \qquad (3)$$

$$= \sum_{i=0}^{\infty} < N, 0 \left| \psi(\bar{r}, t)^{\dagger} \right| N-1, i > < N-1, i \mid \psi(\bar{r}', t') \mid N, 0 >, \qquad (4)$$

for $t = t'$. The $< N-1, i \mid \psi(\bar{r}', t') \mid N, 0 >$'s are Dyson amplitudes[14], whose time dependence involves energy differences of the form $E_{N-1,i}$ - $E_{N,0}$ as follows from the Heisenberg time evolution of the field operators:

$$\psi(\bar{r}, t) = e^{-\mathrm{iHt}/\hbar} \psi(\bar{r})\, e^{\mathrm{iHt}/\hbar}. \qquad (5)$$

This representaion of the density matrix, $\gamma(\bar{r},\bar{r}')$, has been introduced at this point not only to give some indication of why the μ of Eqn. (1) is an energy difference, but also to provide a foundation for some qualitative remarks made later in this section.

Returning to equation (1), as the temperature $T \to 0$ the essential idea of the Bose condensation is that all of the N identical Bose particles end up in the single (and lowest energy) single particle state φ_o, which we will simply refer to as φ . In this case Eqn (1) reduces to

$$- \frac{\hbar^2}{2m} \nabla^2 \varphi(\bar{r}) + V^{\mathrm{trap}} \varphi(\bar{r})$$

$$+ N \int \varphi(\bar{r}')^* \mathrm{V}(|r - r'|) \varphi(\bar{r}') d^3 \bar{r}' \varphi(\bar{r}) = \mu \varphi(\bar{r}). \qquad (6)$$

As it is assumed that N>>1, the approximation $N \approx N-1$ has been made in setting the prefactor to the non-local effective potential. Eqn (6) is a Schrodinger-like integro-differential equation for the wavefunction $\varphi(\bar{r})$, which must be determined self-constently from this Hartree-type equation. $\varphi(\bar{r})$ may be thought of as a one-particle condensate wavefunction. Eqn. (6), albeit a non-linear Schrodinger equation, is not *the* NLSE of BEC theory. This latter will follow from an additional approximation appropriate to the physical parameters and length scales typical of actual trapped BECs in the late 1990s. But first a more detailed discussion of "the condensate wavefunction" is in order.

The Condensate Wavefunction: The term *condensate wavefunction* is both mathematically and historically far more complex than simply assuming that the ground state solution of Eqn.(6) defines the condensate, and that $\varphi(\bar{r})$ is the "condensate wavefunction". Mathematically, the derivation of Eqn.(6) needs to be placed within the context of an exact many body theory, and both the nature of the approximations made in deriving Eqn.(6), and prescriptions for the resulting higher order corrections determined. This will not concern us here, as Eqn.(6), itself, will be seen to be adequate for the description of many aspects of currently available gaseous condensates, and, simple as it might seem, will be found to contain surprising physical consequences. Readers interested in the full many-body aspects of BECs are referred to an enormous literature[14,23,29–32].

The term condensate wavefunction may be traced back to the work of London[24] and Laudau[25] in the earliest days of the phenomenological treatment of superfluids. Especially for the novice, great care must be taken in understanding various usages in this area. For example, Landau introduced a complex order parameter $\eta(\bar{r})$ such that the local superfluid density, $\rho(\bar{r})$, could be written as $|\eta(\bar{r})|^2$. Then

$$\eta(\bar{r}) = \sqrt{\rho(\bar{r})}\, e^{i\phi(\bar{r})}, \tag{7}$$

where $\eta(\bar{r})$ is also often called the condensate wave function. All of the quantities in Eqn (7) may also depend on time. The spatially dependent phase $\phi(\bar{r})$ will be seen to play a crucial role in what follows. What is the relationship between the $\varphi(\bar{r})$ of Eqn. (6) and Landau's $\eta(\bar{r})$? The latter of these might well be defined for a strongly correlated system, while Eqn (6) ignores even the possibility of correlations.

The exploration of the relationship betweeen the phenomenological and mathematical descriptions of the superfulid ground state, in terms of a one body condensate wavefunction, was begun by Onsager and Penrose[26]. These workers defined the superfluid state in terms of the possibility of factorization of the one particle density matrix as

$$\gamma(\bar{r}, \bar{r}') = \eta(\bar{r})^* \, \eta(\bar{r}'). \tag{8}$$

Namely, superfluidity is characterized by the property of *off-diagonal long range order*, or ODLRO, as implied by Eqn (8), and containing the idea that $\gamma(\bar{r}, \bar{r}')$ does not vanish as $|\bar{r} - \bar{r}'|$ becomes large, as long as both $\bar{r}$ and $\bar{r}'$ are within the superfluid. That ODLRO is indeed a special property may be immediatly understood by examination of the more general diagonal expansion of the density matrix, long known in theoretical chemistry as the "natural orbital expansion",

$$\gamma(\bar{r}, \bar{r}') = \sum_{i=0}^{\infty} \nu_i \, \eta_i \, (\bar{r})^* \, \eta_i(\bar{r}'), \qquad (9)$$

where the η_i's are eigenfunctions and the ν_i 's the eigenvalues of $\gamma(\bar{r}, \bar{r}')$ considered as an integral kernel[27]. ODLRO thus is equivalent to assuming that the sum in Eqn. (9) is dominated by a single term, and the value of "ν_o" then absorbed into the normalizaion of $\eta_o \equiv \eta$, consistent with the idea that in the Bose condensed state all of the particles are in an identical single particle state. What is the mathematical representaion of this state? Here is where there are divergences in the approaches of different workers. Anderson[8,9], for example, combinies Equations (3) and (8) to the give the factorization

$$< \psi\,(\bar{r})^{\dagger}\, \psi(\bar{r}')> \;\;\approx\;\; < \psi(\bar{r})^{\dagger} > \, <\psi(\bar{r}')> \qquad (10)$$

suggesting the identifications

$$\eta(\bar{r})^* \;=\; < \psi(\bar{r})^{\dagger} > \quad \text{and} \quad \eta(\bar{r}') \;=\; <\psi(\bar{r}')> \qquad (11)$$

Thus a connection between an expectation value of an Heisenberg Bose field operator and the phenomenolgical London-Landau order parameter seems at hand. However, the expectation value <,> of Eqns (10,11) is certainly *not* that of Eqn (3), as if $<\psi(\bar{r}')>$ is to be non-vanishing, the expectation value cannot be taken with respect to a number conserving state |>. Anderson takes this expectation value with respect to a Glauber-type coherent state[19,28] with an average particle number $\overline{N}$ and a mean phase $\overline{\phi}$. These then satisfy an "uncertainty" relation of the form $\Delta N \Delta \phi \approx 1$. Hohenberg and Martin[29] use a somewhat different, but also coherent, definition of a non-number conserving |>, in making an analogous identification. In the large N (thermodynamic) limit either of these provides a suitable formalism. However for the finite (and fixed) N systems at hand in magentic traps, while it may still be a useful approximation to use the coherent state formalism as an approximation, even for N as "small" as 10^3 or 10^4, it is important to realize that this does not imply that a given system actually is *in* such a coherent state. This is made clear by noting the simple alternative, employed by Pitaevskii[30] and Nozieres and Pines[31],

$$\eta(\bar{r}') \;=\; < N-1, 0\, |\psi(\bar{r}')\, |\, N,0> \qquad (12)$$

which is, again assuming that a single term dominates the sum in Eqn (4), consistent with ODLRO. This interpretation has the further advantages that there is then no "phase-particle number uncertainty relation" and, in the simplest mean field limit, the Dyson amplitude of Eqn.(12) is easily shown to be identified with the $\varphi(\bar{r})$ of Eqn (6). Finally, it should be noted that this is a controversial area and both the literature and individuals associated with it should be approached cautiously!

The Pseudopotenital, *the* NLSE, and a First Approximate Solution: The Gaseous Bose condensates of current experimental interest are very "dilute". What this actually means is seen from inspection of typical length scales and number densities in currently attainable traps. Some approximate (these do differ widely from experiment to experiment) relevant length scales are 1) the range of an interatomic atomic potenetial ≈ 5 to $10 \times 10^{-4}\mu$ (μ=microns); 2) the trap size 10 to 50 μ; 3) number of trapped particles 10^3 to 10^5. A typical particle density might then be $10^{12-14}/\mathrm{cm}^3$ giving a mean interatomic spacing in the micron range. In the condensate, the local thermal (typically $\lesssim 10$ nanoKelvin) deBroglie wavelength becomes large compared to the interparticle spacing, and thus very large compared to atomic length scales. In this long wavelength limit, a very convenient low energy approximation may be made[32]. Assuming only s-wave scattering between individual isolated pairs of condensate atoms, in the k $\rightarrow 0$ limit ($p = \hbar k$ is the momentum) the cross section $\sigma(\mathrm{k})$ reduces to

$$\sigma(\mathrm{k}) = \frac{4\pi}{k}\sin^2(\delta_s(\mathrm{k})) \;\rightarrow\; 4\pi {a_s}^2 . \tag{13}$$

In Eqn. (13), $\delta_s(\mathrm{k})$ is the s-wave scattering phase shift and a_s is the scattering length, which will play an essential role in what follows. The scattering length is not necessarily "just" the range of the interatomic potential. a_s can actually diverge, and can also be negative. For example, for Rb^{87} pairs $a_s \approx 0.2\mu$, while for Li^7 $a_s \approx -.0015\mu$. As the magnetically trapped alkalis are kept in spin-parallel states (otherwise alkali clusters or microcrystals would form at 10 nK....the trapped BEC is highly metastable) these are not the ordinary ground state scattering lengths, and their accurate deduction from experimental spectroscopic observation is a cutting edge activity[33]. In the limit of validity of Eqn.(13) the interparticle potential in Eqn.(6) may be replaced by the pseudopotential (which gives the identical s-wave scattering cross section as $k \rightarrow 0$),

$$\mathrm{V}(|r - r'|) = \frac{4\pi\hbar^2 a_s}{m}\delta(\bar{r} - \bar{r}'), \tag{14}$$

where m is the particle mass, and $\delta(\bar{r} - \bar{r}')$ is the Dirac δ-function. This approximation is well suited for use in the mean field approximation of Eqn (6), and, as the experimental values of m and a_s are then used therein, results in an effective summation of an infinite class of pair interaction ladder diagrams. The implicit or explicit substitution of the pseudopotential of Eqn.(14) into expressions already containing higher levels of approximation must thus be done with great care[34]. With the use of Eqn (14) the mean field equation (6) takes on a very simple form:

$$-\frac{\hbar^2}{2m}\nabla^2\varphi(\bar{r}) + V^{\text{trap}}\varphi(\bar{r}) + \frac{4\pi\hbar^2 a_s}{m} N\,|\varphi(\bar{r})|^2\ \varphi(\bar{r}) = \mu\,\varphi(\bar{r})\,. \tag{15}$$

This, at last, is the NLSE, or Gross – Pitaevskii[12,13], equation, which will occupy us for the ramainder of the tutorial. We will, in what follows, focus almost entriely on the case where a_s is positive: the factor $\frac{4\pi\hbar^2 a_s}{m} N\,|\varphi(\bar{r})|^2$ then acts as a repulsive effective potential, which causes, for large enough N, the wavefucntion to approach a constant were it not for V^{trap}. This is seen by allowing N to grow to the point where the kinetic energy may be neglected (the Thomas Fermi limit). Then, if $V^{\text{trap}} = 0$,

$$\varrho(r) \approx |\varphi(\bar{r})|^2 \approx m\mu/(4\pi\hbar^2 a_s N) \tag{16}$$

In actuality, as a constant $\varphi(\bar{r})$ gives zero kinetic energy in any case, this uniform density limit is quickly reached in the interior of a replusive condensate in a constant potential trap. This easy preliminary result will set the stage for the variable separations to follow.

3. Separation of Variables in a Box, Jacobi Solutions for the 1-dimensional NLSE

An Approximate Separation of Variables: A First Reduction to a 1-Dimensional NLSE: To take advantage of the result of Eqn.(16) in a way that will allow development of an anlytically soluble model of a trapped BEC, we will choose our trap to be a three dimensnional box, with $V^{\text{trap}} = 0$ within the box, and $V^{\text{trap}} = \infty$ otherwise. As development of a natural "condensate" length scale will be key to the understaing of solitons in an appropriately trapped condensate we begin that task for the box BEC. The appropriate intrinsic condensate length scale is the healing length, or skin depth, ξ, of the condensate; ξ is also called the condensate correlation length. To motivate this, consider confining a condensate in a box $0 \leq x \leq L$, and running between $-L/2$ and $+L/2$ (with L very large) in the y and z directions, and with $V^{\text{trap}} = 0$ in this volume. The result of Eqn (16) makes it perfectly plausible to take $\varphi(\bar{r})$ as a constant away from $x = 0$, and away from the walls at $y,z = \pm L/2$. Writing

$$\varphi(\bar{r}) = f(x)g(y)g(z)/L^{\frac{3}{2}}\,, \tag{17}$$

and taking $g(y) = g(z) = 1$ near $y=z=0$ (i.e. away from the boundaries in y and z) , we examine the behavior of $f(x)$ near $x = 0$, namely at the boundary in x. Substitution of this ansatz into Eqn (15) gives

$$-\frac{\hbar^2}{2m} f''(x) + \frac{4\pi\hbar^2 a_s}{mL^3} N\,|f(x)|^2\ f(x) = \mu\, f(x). \tag{18}$$

Noting that N/L^3 is the bulk number density, ϱ, and dividing through by the factor $4\pi\hbar^2 a_s\varrho/m$ we rewrite Eqn (18), assuming $f(x)$ to be real, as

$$-\xi^2 f''(x) + f(x)^3 = \tilde{\mu}\, f(x), \tag{19}$$

where $\xi^2 = 1/(8\pi a_s\varrho)$, and where $\tilde{\mu}$ is now dimensionless. As a_s has units of length, it is evident that so does ξ. To find a solution such that $f(x) \rightarrow 1$ away from the boundary, where the kinetic energy term may be neglected, [see Eqn (16)] we need only set $\tilde{\mu} = 1$, giving

$$-\xi^2 f''(x) + f(x)^3 - f(x) = 0 \tag{20}$$

Eqn (20), as will be the case in all of the examples which follow, has a simple integrating factor: namely, multiplication of Eqn (20) by $f'(x)$ followed by integration with respect to x gives:

$$-2\xi^2\,(f'(x))^2 + f(x)^4 - 2\,f(x)^2 + \mathbb{C} = 0, \tag{21}$$

where the constant of integration, $\mathbb{C}$, will be chosen to ensure satisfaction of the boundary conditions $f(x) \rightarrow 1$ away from $x = 0$, and $f(x=0) = 0$. The proper choice is $\mathbb{C} = 1$, as we then find:

$$dx/(\sqrt{2}\,\xi) = df/\sqrt{[(1-f^2)(1-f^2)]} \tag{22a}$$

$$= df/(1-f^2). \tag{22b}$$

Integrating theleft side over x, and right over f we then find

$$\int^x 1/(\sqrt{2}\,\xi)\, d\tilde{x} = \int^f 1/(1-\tilde{f}^2)\, d\tilde{f}\,, \tag{23}$$

which may be immediatly integrated in terms of elementary functions to give

$$x/(\sqrt{2}\,\xi) = tanh^{-1}(f). \tag{24}$$

Or, at last, actually taking the inverse,

$$f(x) = tanh[x/(\sqrt{2}\,\xi)]. \tag{25}$$

The solution, Eqn.(25), indeed satisfies the boundary conditions $f(x) = 0$ at $x = 0$, and $f(x) \rightarrow 1$ as $x \rightarrow \infty$. The range of x over which this transition takes place is indeed $\approx\xi$. Figure 1 shows a graph of this solution for two values of ξ, and the origin of the term "skin depth" is clearly illustrated.

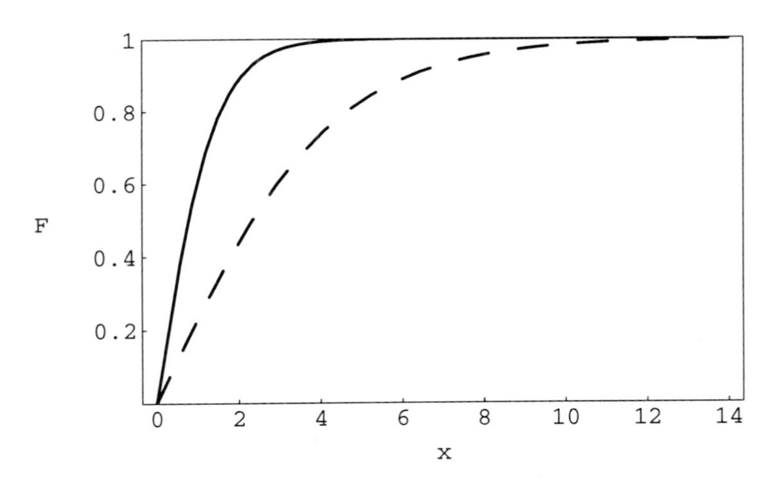

Figure 1. The function F = $tanh[x/(\sqrt{2}\,\xi)]$ for $\xi = 1$ (solid line) and $\xi = 3$ (dashed line).

It may well seem that we be have belabored a rather simple example. Why didn't we just announce that the solution was $tanh[x/(\sqrt{2}\,\xi)]$ and verify this by substitution? Can the choice $\mathbb{C} = 1$ be understood by reasoning other than *ex post facto* checking of the result?

The reasons, both for the logical development undertaken and for the choice of $\mathbb{C}$, will become clear in the next sections. Precisely the same series of techniques will be utilized to first solve for the stationary states of a condensate in a finite box, and then to develope the theory of solitons in repulsive condensates. Namely, use of $f'(x)$ as an integrating factor, followed by representation of the solution in terms of an integration exactly in the spirit of Eqns (23) and (24) in that the solution will be found as the inverse of a function with a fairly simple algebriac integral representation. However, in both of the next cases, rather than the familiar $tanh^{-1}$ integral representation, we will introduce the inverse Jacobi elliptic functions, and then the Jacobi functions themselves via an integral representation which is a seemingly small generalization of that of Eqn 22a. The motivation for leaving Eqn (22a) in the factored form shown will become clear. Further, and very important for developing a physical, as well as a mathematical, understanding of the results, we will develop and utilize a "Hamiltonian" interpretation of Eqn.(21) and its relatives which will allow developemnt of an intiutive understanding of some of the unusual properties of the Jacobi funcitons, both as solutions of the stationary state NLSE, and as the soliton solutions of its time dependent counterpart.

Satisfying the "Other" Boundary Condition, Stationary State Solutions for a Finite Box Trap: The alert reader will have noted that the solution

$$f(x) = tanh[x(/ \sqrt{2}\ \xi)]$$

is actually not a suitable solution over the entire range $0 \leq x \leq L$. The natural particle-in-a-box boundary conditions are $f(x) = 0$ at both $x = 0$ and $x = L$. The solution obtained at this point only satisfies the first of these. If $L >> \xi$ we can correctly guess that the desired ground state solution will be fall to zero in a "tanh-like" manner on a length scale ξ both near $x = 0$ and $x = L$, while being essentialy constant away from either boundary. The "special" funcion which has these properties is the Jacobi elliptic function[35] $f(x|m)$ defined implicitly by an integral representation which generalizes the differential of Eqn (22a)

$$x = \int^{x} d\tilde{x} = \int^{f} \frac{1}{\sqrt{[(1-\tilde{f}^2)(1-m\tilde{f}^2)]}}\, d\tilde{f} \tag{26}$$

This form arrises immediately if $\mathbb{C} < 1$ in Eqn (21), rather than being equal to 1 which implies *"m" = 1* and taking us immediatly back to Eqn 22a,b, and the *tanh[x]* solution. Jacobi introduced this generalized represenation as a way of studying doubly periodic functions for x complex, a point which we will not develop, except to note that even "single" periodicity is lost for real x, if $m = 1$. The *tanh[x]* solution is thus actually a singular limit of the *"f"* of Eqn (26). That this inegral representation can indeed define a function periodic in x, is seen by considering the $m = 0$ limit, where again an elementary function arises:

$$x = \int^{x} d\tilde{x} = \int^{f} \frac{1}{\sqrt{[(1-\tilde{f}^2)]}}\, d\tilde{f} = sin^{-1}[f]\,, \tag{27}$$

and thus *f(x)= sin[x]*. As the function *"f(x)"* defined implicitely in Eqn.(26) reduces to the sine function when $m = 0$, it is referred to as the Jacobi SN function, and we will denote it as *SN[x|m]*, where "x" is the *argument* and "m" the *parameter*.{ In the literature m is often written as $m = k^2$, and k is then called the *modulus*. Great care should be taken in noting and employing the many and variable notations[35] for Jacobi functions! }

Figure 2 shows the function *SN[x|m]* as a functnion of x for several values of the parameter. The period is a slowly incleasing function of m as $m \rightarrow 1^-$.

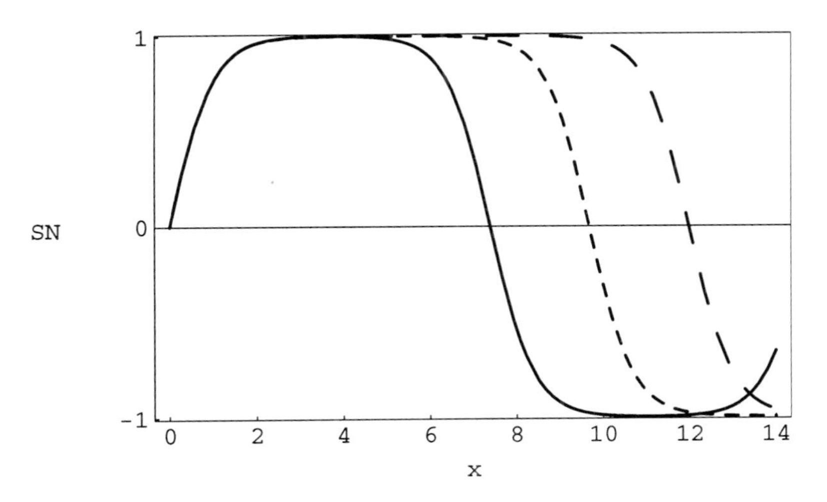

Figure 2. The Jacobi elliptic function *SN[x|m]* plotted as a function of x for $m = 0.99$ (solid line), 0.999 (short dashes), and 0.9999 (long dashes) .

The period, *Π[m]*, of *SN[x|m]* is given explicitly in terms of a *complete elliptic integral*[35]

$$\Pi[m] = 4 \int_0^1 \frac{1}{\sqrt{[(1-\tilde{f}^2)(1-m\,\tilde{f}^2)]}}\, d\tilde{f} \tag{28}$$

The integral in Eqn (28) is "*complete*" as its integration range is fixed. $\Pi[m]$ actuallly diverges as[35]

$$\sim ln[16/(1-m)] \tag{29}$$

as $m \rightarrow 1^-$. At $m = 1$ the period is indeed infinite and the original *tanh[x]* solution reflects this. Figure 3 illustrates this logarityhmic divergence as a function of *m*.

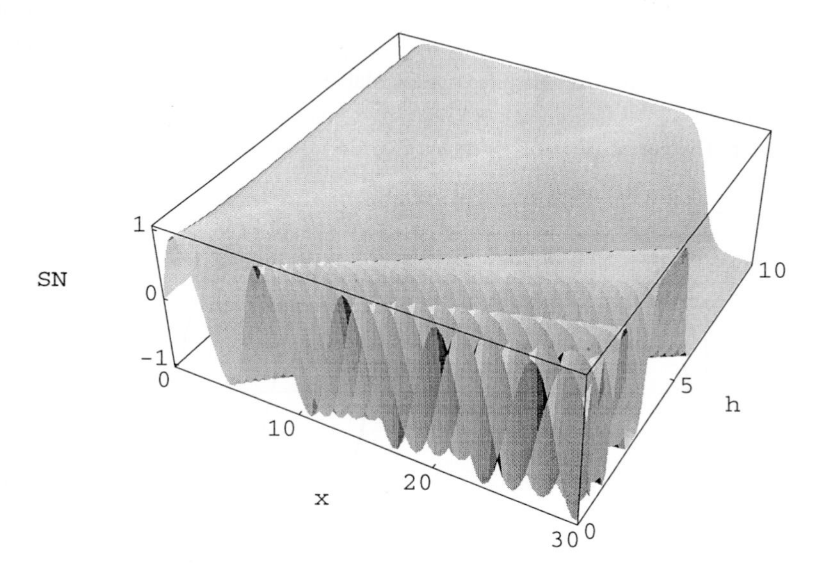

Figure 3. Plot of SN[x | m] as a function of m = 1 - $10^{-\eta}$ for η = 0 through 10. the logarithmic nature of the period as a function of m is clearly illustrated.

Solving Hamilton's Equations to get Jacobi Functions: Can we make "physical" sense of these Jacobi solutions of Eqn (20,21) which arise by simply changing the integration constant $\mathbb{C}$, and thus *m?* In particular, can we understand their simgular behavior as *m* $\rightarrow 1$? Fortunately, there is a very simple physical picture underlying the initially rather abstruce Jacobi functions. This same physical picture will also be of use in understanding the origins and nature of solitons in Section 5.

This is a case where a simple notational change can make a great difference in ease of understanding. Suppose Eqn (21) is rewritten as

$$\mathbb{C} = 2\xi^2 \dot{q}^2 + 2q^2 - q^4 \tag{30}$$

$$= 2\xi^2 \dot{q}^2 + V(q) \tag{31}$$

where we identify t = *"time"* with the *"position"* x, the velocity $\dot{q} = \dot{q}(t)$ with $f'(x)$ and position $q = q(t)$ with $f(x)$; and where $V(q) = 2q^2 - q^4$. If we further identify $2\xi^2 \dot{q}^2$ as a kinetic energy, and $V(q)$ as a potential energy, is it immediatly evident that the "constant" $\mathbb{C}$ is a conserved total energy for a Hamiltonian system of mass $4\xi^2$, and momentum $p = 4\xi^2 \dot{q}$. Visualization of the properties of "trajectories" q(t) now follows from our well developed experience in relating the properties of such a trajectory to the total and potential energy. The potential $V(q) = 2q^2 - q^4$ is shown in Figure 4. Trajectories run on the potential curve of Figure 4 are oscillatory near q = 0, with longer and longer periods as the "energy" , $\mathbb{C}$, approaches 1^-. Trajectories started at energy 1, exactly, for $|q| < 1$ will simply run up the potential and stop at $q=\pm 1$ in unstable equilibrium. Such solutions correspond to the *tanh(x)* limit of *SN[x|m]* as $m \to 1^-$, and indeed have infinte period. An actual trajectory run on this potential, starting with zero velocity at $q(0) = 0.999$, corresponding to $\mathbb{C} = 0.999996$ is shown in Figure 5. The resemblance to the SNs of Figure 3 is immediatly evident.

Trajectories run on the potential of Fig 4 oscillate (maximally) between -1 and 1, with oscillatory solutions near energy *0* having quite small amplitudes, as determined from the classical turning points of $V(q)$. This follows from the choice $\mu = 1$. Should we wish solutions with larger (or smaller) amplitudes, the chemcial potential μ of Eqn (19) must be adjusted accordingly, resulting in a family, V_μ(q), of μ dependent potentials. Thus the "effective" potential whose Hamiltonian (or Newtonian) trajectories are equivalent to solution of Eqn (19), depends on the specific boundary conditions imposed in addition to the structure of the equation itself. This coupling of boundary conditions and the "dynamics" will persist in our discussion of solitons. Note also that as all of these equations are non-linear, simple multiples of a solution, be it $f(x)$ or $q(t)$, are no longer solutions of the given equation. Thus, unlike solutions of the more familiar linear Schrodinger equation, normalization of the solutions of Eqn (19) is non-trivial, and is equivalent to determination of the self-consistent solution of this Hartree-type equation. This is the topic of the following section.

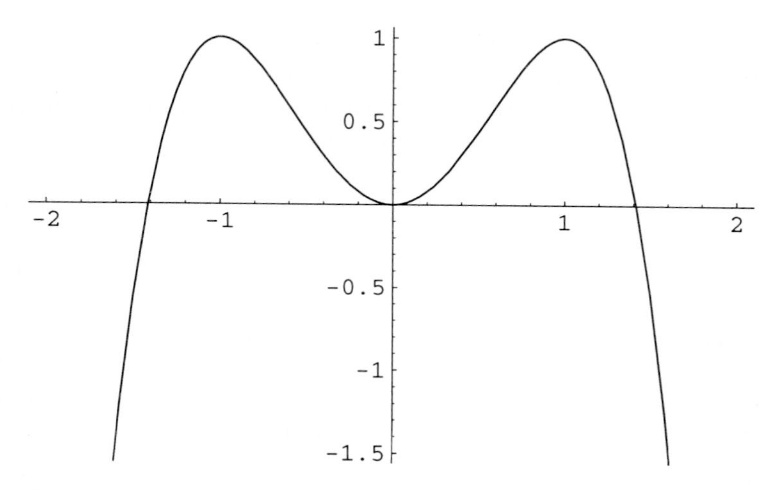

Figure 4. The effective potential corresponding to the Hamiltonian dynamics equivalent to Eqn (21) for $\mu = 1$. The grid lines run through the points (1,-1) and (1,1) indicating maxima of $V(q)$ at $q = \pm 1$, with $V(\pm 1) = 1$. Motion near $q = 0$ is thus bounded and oscillatory for energies $\mathbb{C}<1$, but non-oscillatory and unbounded for $\mathbb{C} > 1$. The energy $\mathbb{C}= 1$ is the separating (or separatrix) energy, which gives rise to trajectories which may "stick" in unstable equilibirum at $q = \pm 1$.

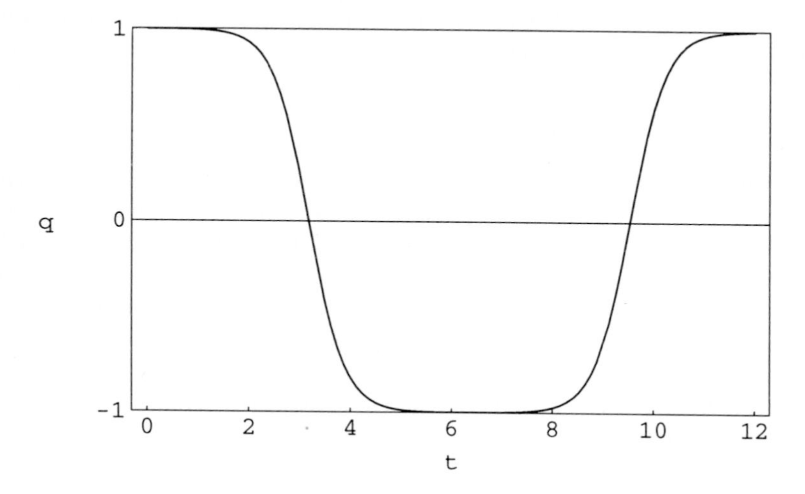

Figure 5. A trajectrory started in the convex region of the potential $V(q)$ of Fig 4, and with an energy $\mathbb{C} = 0.999996$. An immediate identification with the $SN[x|m]$ of Figure 2 may be made.

4. Stationary State Solutions of the NLSE and Condensate Collisions

Solving the NLSE for particles in a 1-dimensional box: The requirements for finding stationary state solutions of the one dimensional NLSE for a box of lengfth L is now clear. We need to find *SN[x|m]*, automatically satisfing the boundary condition $\varphi(0)=0$, but which now also must satisfy the two additional boundary condisitons $\varphi(L)=0$, and $\int\varphi^2(x)dx=1$, where the intergral is over $(0,L)$. As the solutions are non-degenerale, φ(x) has been take to be real.. Satisfaction of these two additional conditions may be met by appropriate choice of the two free parameters μ and $\mathbb{C}$, the former of which determines the chemical potential of the system. This new normalization of φ(x) may be enforced by writing $\varphi(x) = f(x)/\sqrt{L}$, as was done in Eqns (17,18), and normalizing $f(x)$ as

$$\int_0^L f^2(x)\, dx = L, \tag{32a}$$

or, as we will do in what follows, working with φ(x) itself, with norm

$$\int_0^L \varphi^2(x)\, dx = 1, \tag{32b}$$

in which case Eqn(19) is replaced by

$$-\xi^2 \varphi''(x) + L\,\varphi(x)^3 = \tilde{\mu}\,\varphi(x). \tag{33}$$

As in Eqn. (19), $\xi^2 = 1/(8\pi a_s \varrho)$, as it is still useful to have the correlation length ξ defined in terms of the bulk desnity, ϱ.

Real solutions of Eqn (33) vanishing at $x = 0,L$, and normalized as in Eqn (32b) are then given by appropriate determination of μ and $\mathbb{C}$.. This is accomplished, for states labled by quantum number n = 1,2,3..., by solution of transcendental equation[36]

$$\left(\frac{L}{2\,n\xi}\right)^2 = 2K(m)(K(m) - \mathbb{R}(K,m)) \tag{34}$$

for the unknown parameter m. In Eqn. (34) $K(m)$ is the complete elliptic integral *of the first kind*, as appears in Eqn (28),

$$K(m) = \int_0^1 \frac{1}{\sqrt{[(1-\tilde{f}^2)(1-m\tilde{f}^2)]}}\, d\tilde{f}, \tag{35}$$

and

$$\mathbb{R}(\mathrm{K,m}) \equiv \int_0^{K(m)} DN^2[x\,|\,m]\, dx \tag{36a}$$

$$= E[\,sin^{-1}\{SN[\,K(m)|m]\}]\,, \tag{36b}$$

where *DN[x|m]* is the Jacobi *DN* function, (to be defined in Section 5) which has no simple trigonometirc interpretation, as *DN[x|m]* $\rightarrow$1, as $m \rightarrow 0$. *E[m]* is the complete elliptic integral *of the second kind*[35]

$$E[m] = \int_0^1 \frac{\sqrt{(1-m\tilde{f}^2)}}{\sqrt{1-\tilde{f}^2)}}\, d\tilde{f}\,. \tag{37}$$

Eqn (34) may be solved iteratively for each desired n. However, near $m = 1$, which will be the actual region of interest, as $L >> \xi$ in trypical traps, the solutions are very much *tanh*-like, and rather great numerical care must be taken. This follows as *K(m)* diverges logarmthmly at $m = 1$, see Eqn (29), and both approprite expansions valid near $m = 1$, and very high numerical precision are needed to get sensible results. This is evident on noting that for the ground state $m \approx 1 - e^{-L/\xi}$, and thus for L/ξ the order of 30 to 50, m will be equal to 1 to a rather large number of significant figures.

Once m is determined, for a given n, the dimensionless chemical potential of Eqn (33) is, for the nth state with (n-1) internal nodes, then give simply as[36]

$$\tilde{\mu}_n = \frac{2K^2(m)}{\{1+(1-m)/(1+m)\}\,(L/2\,n\xi)^2}\,. \tag{38}$$

As $m \rightarrow 0$ it may be checked that usual particle in a box energy levels μ_n are recovered from $\tilde{\mu}_n$. In the opposite limit, $m \rightarrow 1$, the chemical potentials (stationary state energies) are evenly spaced with spacing [36]

$$\Delta\mu = \frac{\sqrt{2}\,\hbar^2}{m\mathrm{L}\xi}\,. \tag{39}$$

Eqn. (39) is valid in the limit $L/(n\xi) >> 1$. In this limit the energy simply increases linearly as each node is added to the wave function, suggesting a similarity, especially as the density vanishes at each node, to adding non-interatcing vortecies to a 2-dimensional superfluid. The energy of adding such a node, as given in Eqn. (39), may be reinterpred by noting that in the regions where φ(x) is constant, and away from other nodes, $(|\varphi(x)|)^2 \simeq 1/L$, and so Eqn.(39) may be re-written as

$$\Delta\mu = \frac{\sqrt{2}\,\hbar^{2}\,(|\varphi(x)|)^{2}}{m\xi} \quad , \tag{40}$$

noting carefully that $\varphi(x)$ is the wavefunctnion at point x *before* the node is added. Thus, when nodes are placed in a region of otherwise constant density the energy cost is sensibly proportional to the density, and inversely proportional to the correlation length.

Figure 6a,b,c shows φ_n, for $L/\xi = 50$, the first three stationary states n =1,2,3 of the NSLE in a box, confirming the notion that these nodes are "added" relatively independently to the *tanh*-like solution of the ground state, and that each has a scale size $\approx 2\xi$, and a local shape least roughly independent of how many other nodes are in the system. Note also that the solutions, illustrated in Fig 6., are non-orthogonal, as is possible as Eqn (33) is non-linear. Put a different way, in each solution of Eqn.(33) all of the particles are collectively fully excited, and thus in each state "see" a completely different effective potential. They are thus eigenstates of different Hamiltonians, and as such, there is no reason to expect orthogonality. Of course if $L/(n\xi)$ finally becomes comparable to, or less than, 1 as n increasaes, m will then decrease rapidly and the solutions become *sine*-like, see Figure 6d where the $n = 20$ state is shown. These are still not orthogonal to the low energy *tanh*-like states, excpet by symmetry, as all of these 1-dimensional particle in a box solutions are non-degenerate and either even or odd. In the low density limit ξ diverges, m then vanishes and the usual particle in a box states are obtained. This is as expected. However the particle in a box limit is also achieved if $L/\xi \lesssim 1$, i.e. as L simply becomes small or comparable to ξ, a remark which will be of importance later on.

Physical Interpretation of NLSE Stationary Solutions: It is important to realize that the excited states illustrated in Figure 6, and whose energy separation is given for large L/ξ by Eqns. (39 or 40), are states very different from the low lying excitations observed expermientally in Refs. (15a) and modelled by RPA like extensions of the NLSE in Refs. (15b,23). In the RPA collective excitation picture, the ground state φ(x), is mixed with a

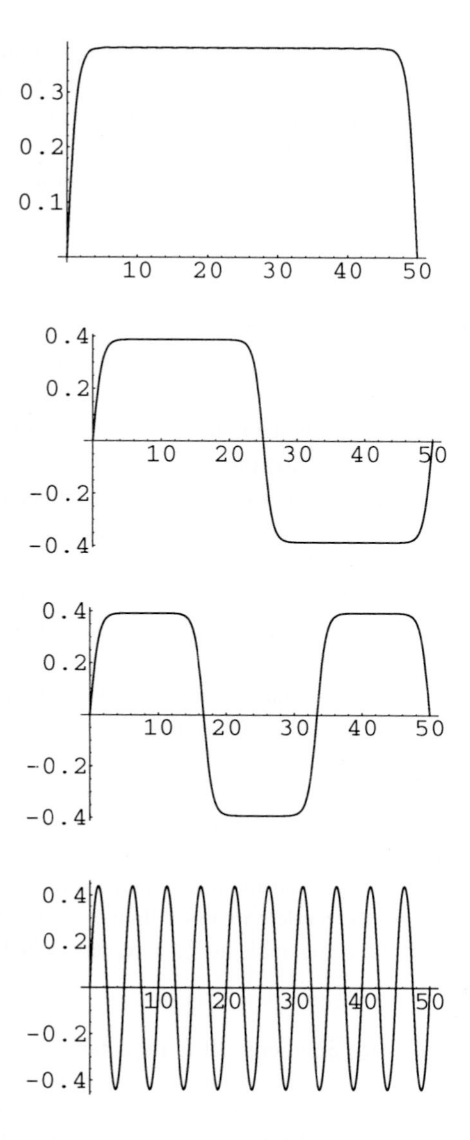

Figure 6. Stationary states of the NSLE for $L/\xi = 50$, for n = 1,2,3 and 20, respectively.

time dependent correction, χ(x,t) ,which is assumed small:

$$\phi^{\mathrm{RPA}}(\mathrm{x,t}) = \varphi(\mathrm{x})e^{-i\mu t} + \chi(\mathrm{x,t}) \qquad (41)$$

The trial funcion of Eqn. (41) is then subject to a perturbatrion of the form Asin(ωt), and only linear terms in χ in the then time dependent quantum equaitons of motion are retained. Excitation frequencies are deduced in the linear response regime. Conisistent formulation of this picture and its corrections is non-trivial for finite systems, and is discussed in Refs. (23). As noted above, the RPA gives a good accounting of the observed small amplitude ocillations of observed trapped BECs.

Where do the states of Figure 6, then, enter this low energy excitation picture? Baldly stated, they simply don't. The energies of Eqn (39) may be far higher than those of the usual RPA excitations. This is not surprising: the excited states shown in Figure 6 require that at every particle be excited to a first excited state and thus every particle has an actual node in its single particle wavefunction. An example from atomic physics may make this difference clear. Suppose the Ne atom, with nominal electron configuration $(1s)^2(2s)^2(2p)^6$ is put in a weak oscillatory electric field. Further, as we wish to discuss a Bosonic ground state, assume that the "electrons" in Neon are Bosons, and so the ground state is then $(1s)^{10}$. In the RPA the effect of a low intensity field will be to mix in some "2p" character with each of 1s orbitals. If we then, as the in RPA, linearize in the "2p" amplitude, the dominant configurations coupled to the Bosonic ground state are of the form $(1s)^9(2p)^1$. Such real or virtual single particle excitations are a far cry of those envisaged above: the excitations illustratred in Figure 6b would corespond to wholesale excitations of the type $(1s)^{10} \rightarrow (2p)^{10}$.

A Simple Condensate Collision: How might nodes such as those illustrated in Figure 6 be created? Consider two condensates, each with N/2 particles, and each in a box of length $L/2$. The first box runs form 0 to $L/2$, the second from $L/2$ to L. Imagine these one dimensional boxes are initially separated by an δ-function potential, $A\delta(x - L/2)$, with a very large prefactor A, effectively forcing the boundary condition $\varphi(x = L/2) = 0$ in both boxes, and preventing tunnelling. Further, assume that in each box we have a ground state condensate with wave functions localized on the left and right hand sides of the extnded box of length L. Thus particles on the left are in the one paricle ground strate $\varphi_o^{left}(x)$ which is non-zero only between 0 and $L/2$, and those on the right are in the ground state $\varphi_o^{right}(x)$ which is non-zero between $L/2$ and L. At time $t = 0$, we remove the δ-functinon and follow the subsequent dynamics *of this zero energy collision between two ground state condenstaes*. This is easily done in terms of the "delocalized" basis functions[18,11]

$$\psi(x) = a\,\varphi_o^{left}(x) \quad + b\,\varphi_o^{right}(x). \tag{42}$$

If the left and right hand functions are individually normalized, and as they are certainly orthogonal, we may, without loss of generality, rewrite this in the from

$$\psi(x)_{\chi} = [\,\varphi_o^{left}(x) \quad + e^{i\chi}\,\varphi_o^{right}(x)]/\sqrt{2}\ . \tag{43}$$

How does this initial state propagate in time, and how is such propagation affected by the "phase" χ ?

For definiteness we will assume that $\varphi_o^{left}(x)$ and $\varphi_o^{right}(x)$ are both real and positive nodeless Jacobi SN functions in their respective half boxes. Further, as the mean number density is the same on both sides of the box, and in fact does not change if the δ-function barrier is removed, a single correlation length ξ characterizes both the left, right, and total systems. Then for phase $\chi = \pi$ the wave funcion

$$\psi(x)_{\pi} = [\,\varphi_o^{left}(x) \quad - \quad \varphi_o^{right}(x)]/\sqrt{2}\ , \tag{44}$$

is precisely the first excited Jacobi *SN* state for all N particles in the box of length *L*, and thus a stationary state. If $\psi(x)_{\pi}$ is indeed a stationary state of the full NLSE, its time evolution should not change the denstiy, as this will consist of only a time dependent phase factor of the form exp(-iμt/$\hbar$). To check this, the initial state $\psi(x)_{\pi}$ may be time propagared according to the time dependent NLSE (tdNLSE in what follows)

$$i\partial_t\,\psi(x,t) = -\xi^2\,\partial_{x,x}\,\psi(x,t) + L|\,\psi(x,t)\,|^2\,\psi(x,t)\,, \tag{45}$$

where Eqn, (45) is integrated with initial conditinon $\psi(x,t{=}0) = \psi(x)_{\pi}$.

The result of such an integration is shown in frames a) and b) of Figure 7. The integration of the tdNLSE was carried out via a pseudo-spectral expansion, of the form

$$\psi(\hat{x},t) = \sum_{j=1}^{M} C_j(\mathrm{t})\Theta_j(\hat{x})\,, \tag{46}$$

where the orthogonal pseudo-spectral functions $\Theta_j(\hat{x})$ are defined only on an evenly spaced grid $\hat{x}$, and the time dependence of the coefficients C_j(t) propagaed from their initial $t = 0$ values by a fourth order Runge-Kutta integrator. Typical converged

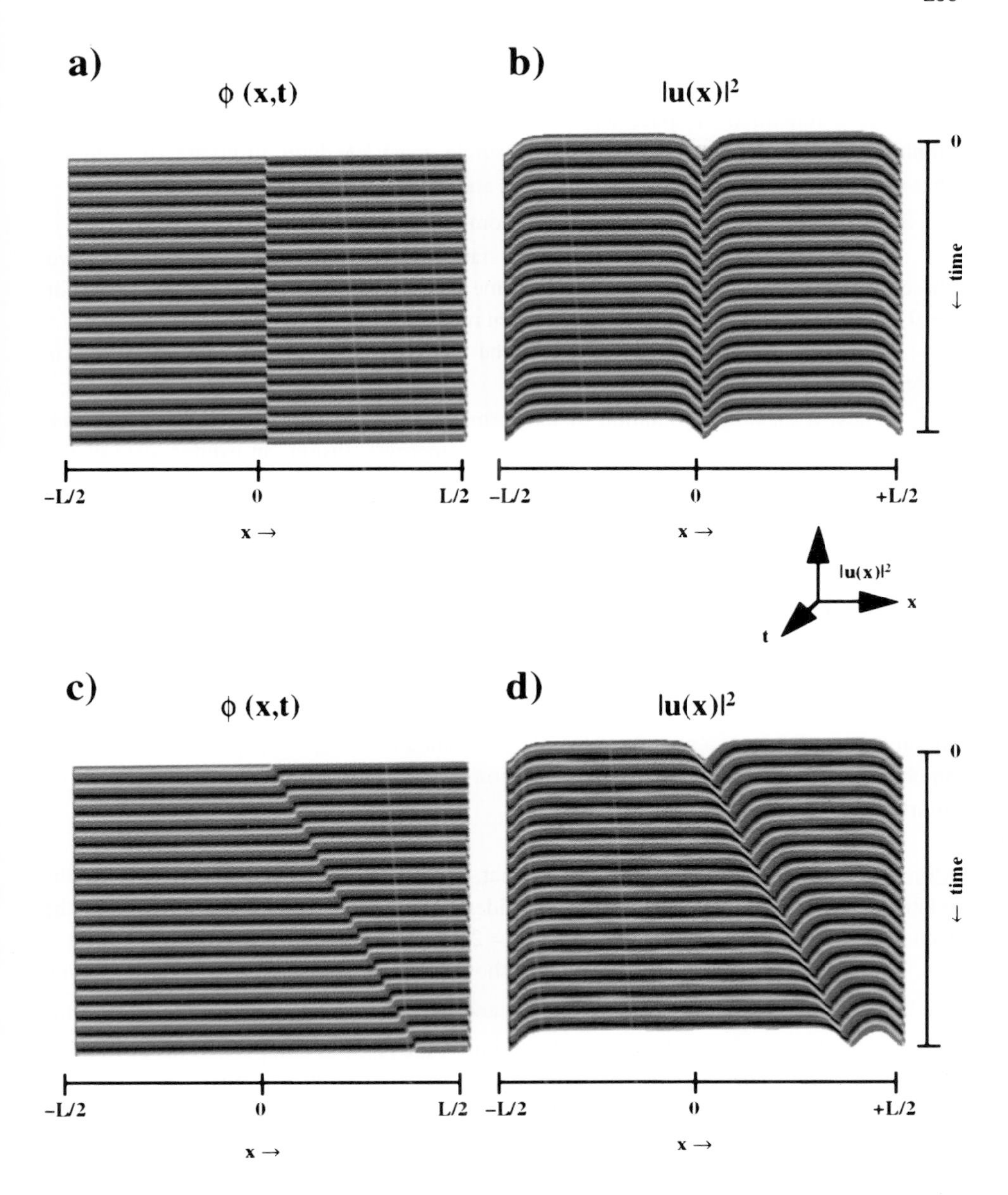

Figure 7. a) and b) the time evolution of the phase (a) color coded; and the phase and magnitude (b) for the zero energy collision of two condensates in a box trap, with an initial phase offset of π. Figures c) d) show the same for an initial offset of $\pi + \pi/8$. A soliton has been launched.

expansions for dynamics in one spatial dimension in a box with $L/\xi = 40$ involved the order of 80 to 120 grid points and up to $M = 60$. In Figure 7a) the phase of $\psi(\hat{x}, t)$ is plotted as a functinon of space and time by a color representation in which the phase running from 0 to 2π is mapped onto a color circle in a 1:1 fashion. In Figure 7b) both the phase information (in the idential color code), and the density of $\psi(\hat{x}, t)$ are shown.

Several things are immediately clear[37] from this propagaion of $\psi(x)_\pi$:

1) $\psi(x)_\pi$ is, as predicted, a stationary state. Namely, only its overall multiplicative phase factor time-evolves; its magnitude is time independent. The density notch present at $t = 0$ is indeed time invariant following sudden removal of the spearating δ-potential..

2) The phase discontinuity of π between the left and right hand sides of $\psi(x)_\pi$ is preserved exactly.

Thus, if an initial condition of the form $\psi(x)_\pi \sim \varphi_o^{left}(x) - \varphi_o^{right}(x)$ can be experimentally arranged via an appropriate zero energy collision, an excited state of the form of Fig. 6b) has been created. If control of this type is experimentally available, a suitable triple condensate collision at zero energy might well produce the state of Fig 6c)!

However, for better or worse, such an inital state preparation is not simple. In fact, for the zero energy collision described abovre there is equal probability of finding (in a single measuarement) a state of the form

$$\psi(x)_\chi = [\,\varphi_o^{left}(x) \;+\; e^{i\chi}\,\varphi_o^{right}(x)]/\sqrt{2}\;, \tag{47}$$

for any $\chi \in [0,2\pi)$. Is a specific value of χ, obtained in a single experimental realization, an observable? This seemingly simple question takes us, at last, to the main subject of the tutorial!

<u>More General Condensate Collisions:</u> What, indeed, is the effect of changing χ, the relative phase offset of the left and right sides of the initial condensate away from the value $\chi = \pi$ which give rise to the excited ($n = 2$) stationary state of Figure 6b? Setting $\chi = \pi + \frac{\pi}{8}$ in Eqn (47) gives the time evolution shown in Figure 7c) and d). Similarly, setting $\chi = \pi + \frac{\pi}{4}$ gives the evolution of Fig 8a) and b); an initial $\chi = 2\pi$ results in the time evolution of Figure 8c),d). In each of these frames the phase is color coded (as in 7a,b) on the left side of each pair, and the color-coded phase and the wavefucntion magnitude shown on the right. These results[37] were (initially) quite a surprise to us, as we had had no expection of what was to come. What is seen is that the density notch, for χ not too far from π propagates linearly in time, and if followed long enough, as in Figure 8a,b) reflects off the "wall" at $x \sim L$, *but with an exactly reversed velocity..* Further as χ nears 2π the original notch bifurcates, and exactly at 2π gives the perfectly symmertic two notch

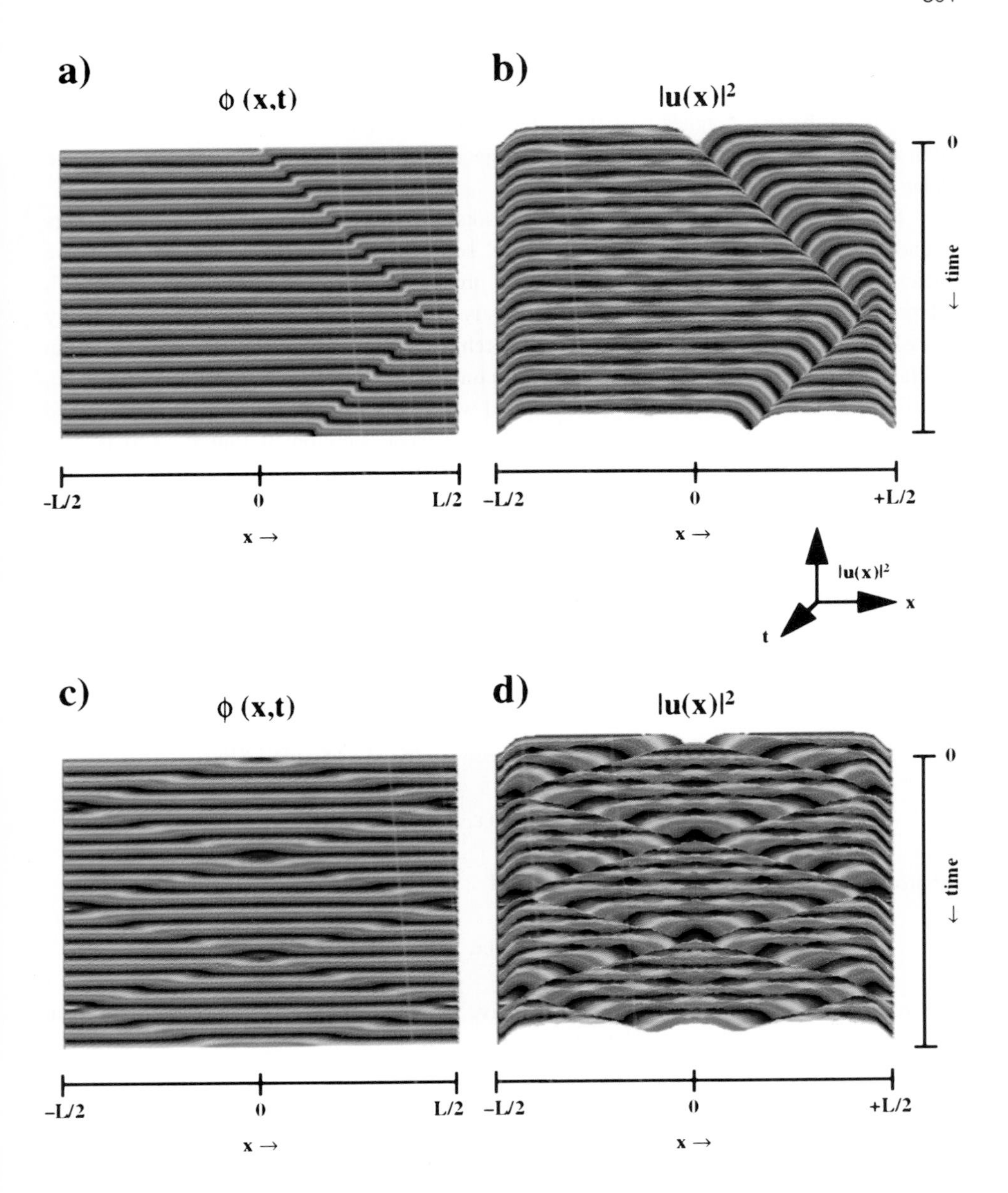

Figure 8. a) and b) as in Figure 7, but with an initial phase offset of $\pi + \pi/4$. The soliton now is seen to "bounce" off the boundary, reversing both direction and phase discontinuity. In c) and d) the initial phase offset is $0 = 2\pi$, and a bi-soliton is launched.

dynamics of Figure 8c,d). Further, it appears that the (sharp) phase discontinuity across each moving isolated density notch, while no longer π is a consant of motion. Also, in Figure 8c,d) the two notches pass through one another, and emerge completely intact, but with a slight "phase shift" in their linear trajectories, as if there had been a slight time delay imposed by the collision.

So, what have we? All of the above remarks, and the evolution of the notches in Figs 7 and 8 are characteristic of "dark" or "gray" solitons[37,38,39]. By *"dark"* or *"gray"* we mean distrubances in the density which propagate as highly localized "notches", "depressions" or "dips" in the ambient denstiy. *Dark* implies that the density actually vanishes at the minimum, *gray* implies the notch is "partial". By the term *soliton* we mean solitary waves which travel coherently with unchanged profile as a function of $\zeta \equiv x - ct$, c being the velocity of an isolated soliton, and which, after interacting, emerge unscathed except for a possible time offset. The stationary state solution of Figure 7a,b) is now interpred as a zero velocity dark soliton! Likewise, the boundary conditions at $x = 0$ and $x = L$, are now thought of pinned dark solitons, and the reflection of the single moving soliton from the wall in Figure 8a,b is a soliton-soliton collision[38]. Figure 8c,d is then dominated by two moving gray solitons and two stationary dark solitons, all of which perfectly maintain their shape and speed (but may change velocity in sign) "in between" collisions. This is a multi-soliton system.

5. An Elementary Introduction to Solitons.

In retrospect, we should not have been surprised. The tdNLSEs

$$i\partial_t\, \psi(x, t) \;=\; -\xi^2\, \partial_{x,x}\, \psi(x, t) \;+\; \mathrm{L}|\,\psi(x, t)\,|^2\, \psi(\mathrm{x,t}), \tag{48}$$

appropriate to a positive scattering length, and

$$i\partial_t\, \psi(x, t) \;=\; -\xi^2\, \partial_{x,x}\, \psi(x, t) -\; \mathrm{L}|\,\psi(x, t)\,|^2\, \psi(\mathrm{x,t})\,, \tag{49}$$

where now the scattering length a_s is negative, which changes the sign of the non-linear term in Eqn (48), and forces redefinition of ξ as

$$\xi^2 = 1/(8\pi\,|\,a_s|\varrho), \tag{50}$$

are both known to support single and multi-soliton solutions[38–41]. Equation(48) corresonding to a repulsive atomic interaction supports localized waves of depression of a specific form and velocity, while rapidly dispersing localized peaks in the density. Eqn

(49), on the other hand, corresponds to an attractive atomic interation, will disperse density depressions or notches, while allowing localized positive density waves (waves of elevation) of specific forms and velocity to propagate with unchanged profiles. The tdNLSE of Eqn (49), also describes, in appropriate circumstances, the propagation of localized light pulses in dispersive non-linear optical fibers which are the optical solitons[40] used to propagate "bits" and "bytes" over the World Wide Web, and allow us to make trans- and interconental phone calls via fiber-optical cables. What is shown in Figures 7 and 8 are thus families of solitons, well known in contexts outside of atomic physics, and the BEC.

In the following subsections we introduce, following a brief historical discussion, the theory of solitons as localized waves of elevation or depression. This is done via an equation rather less complex to analyze than the tdNLSE, the Korteweg - de Vries, or "KdV", equation. The solutions of Eqn, (48) and (49) are then presented, in section 6., for the simplest single soliton cases. Analytic solution of the tdNLSE Eqn. (48) will confirm much of the qualitative discussion of Figures 7 and 8, and allow derivation of an exact relationship between the *phase discontinuity*, χ , and the soliton velocity c. It will emerge, through this relationship between phase and soliton velocity, that the currents implicit in the motion of a soliton are a "barrierless" analog of the familiar Josephson[7–9,17] effect of superconductivity.

Introduction to Solitons via the KdV Equation: Solitary waves and solitons were first described in the scientific literature by Russell in a series of papers beginning in 1844, in which translational waves of elevation are described. The actual term *soliton,* with its modern emphasis on stability following multiple soliton collisons and on the recurrances in these *integrable* systems, first appeared in the work of Zabusky and Kruskal[42] in 1965. Historical soliton "lore" appears in many texts[39], and is as charming as it is fascinating. What follows is strongly influenced by the delightful elementary presentation of Drazin and Johnson, and makes full use of the theory of Jacobi elliptic functions develped for solution of the stationary state NLSE problem in Section 3, and even adds yet another Jacobi function, the *CN* to our repertoire. We present only the elementary theory of single soliton solutions of either the KdV (introduced below) or tdNLSE equations. By elementary we mean those solitons described as Jacobi elliptic *SN* and *CN* functions in their limits as *tanh*- and *sech*-like functions. We will thus avoid, at all times, the far more complex *inverse scatering* formalism[39,43] needed for exact elucidation of the multi-soliton solutions of the very same equations. There is no shortage of introductions to this more advanced material, and its inclusion here would take us too far from the essential physics. There is actually an interesting recent backlash[44] related to the complexity of the mathematical formalism needed to obtain such exact multi-soliton solutions.

Solitons arrise when non-linearity and dispsersion *exactly* balance allowing waves to propagate with an unchanging profile. The importance of the KdV equation is that it is the simplest *PDE* illustrating the possibility of such a balance. To see what is ment by dispersion, and to introduce the third derivative term appearing in the KdV equation, which seems quite an oddity at first sight, consider the simple wave equation

$$c f_x + f_t = 0 \tag{51}$$

where in general $f = f(x,t)$ and $f_x \equiv \partial_x f(x,t)$ and $f_t \equiv \partial_t f(x,t)$. Solutions which simply translate in time at speed c, will be of the form $f(x,t) = f(x - ct)$. It is readily seen that any differentiable funciton of $\zeta = x\text{-}ct$ satisfies Eqn. (51). These are linear and non-dispersive waves, non dispersive meaning that they retain their original wave form for all times. These, however, are not solitons: they are unstable with respect to any small perturbation of Eqn. (51). The effects of dispersion, dissipation, and non-linearity are now systematically introduced.

A Dispersive Wave Equation: the simplest generalization of Eqn. (51) illustrating dispersion is

$$f_{xxx} + f_x + f_t = 0, \tag{52}$$

where f_{xxx} is the 3rd partial, and time has been rescaled so that $c = 1$. The reason for introduction of f_{xxx} instead of f_{xx} will become clear in the following subsection. Eqn. (52) supports oscillatory solutions as seem by substitution of the trial form

$$f(x,t) = exp[\ i(kx - \omega t)], \tag{53}$$

which yields a solution if $\omega(k) = k - k^{3.}$ Substituting this back into Eqn. (53) gives

$$f(x,t) = exp[\ i(kx - (1 - k^2)t)]. \tag{54}$$

Waves with different wave numbers, k, thus move at differning velocities $(1 - k^2)$, and thus initially localized packets, of the form

$$f(x,0) = \int_0^\infty A(k)\, e^{ikx}\, dk, \tag{55}$$

will disperse. The relation $\omega(k) = k - k^3$ is the *dispersion relation* for this system..

A Dissapative Wave Equation: if instead of adding a third derivative to Eqn (51) we had chosen to investigate

$$-f_{xx} + f_x + f_t = 0, \tag{56}$$

the dispersion relation from the ansatz of Eqn.(53) is then $\omega(k) = k - i\,k^2$,which gives a solution of the form

$$f(x,t) = e^{-k^2 t} exp[\, ik(x-t)]. \tag{57}$$

This solution is dissapative, in that rather than spreading it translates at unit speed, and decays rapidly. This is not a good beginning if solitons are desired as an outcome.

Introduction of Non-linearity: the KdV and Burgers Equations: A non-linear term may be added to either Eqn.(52) or (56). This will yield far more interesting solutions. A canonical, and often physically realistic, way of doing this is to assume that the speed, c, of the wave depends on the amplitude : $c = c(1+f(x,t))$, via the multiplicative factor $(1 + f)$. Again, in time units such that the constant c is 1, Eqn (52) becomes

$$f_{xxx} + (1+f)\, f_x + f_t = 0, \tag{58}$$

which is one form of the KdV equation. Eqn. (56) becomes

$$-f_{xx} + (1+f)\, f_x + f_t = 0, \tag{59}$$

which is the Burgers equation. Both are justly famous.

The KdV equation is essentially the simplest non-linear wave equation with dispersion. Note that the additive constant "1" may be surpressed: taking $u(x,t) = 1 + f(x,t)$,

$$u_{xxx} + u\, u_x + u_t = 0\,, \tag{60}$$

or, changing the sign of u, we have

$$u_{xxx} - u\,u_x + u_t = 0\,, \tag{61}$$

We will see that the KdV equation can, depending on the sign of the non-linear term, support dark/grey solitons, or solitons as waves of elevation above a constant or zero background. For historical reasons, and eventual algebraic convenience[39], the KdV equation is usually written in the form (obtained by appropriate scaling of u, x, and t)

$$u_{xxx} - 6\,u\,u_x + u_t = 0. \tag{62}$$

It is in this standard form that we will undertake its solution.

Solution of the KdV Equation for a Single Soliton, Hamiltonian Approach: An ioslated soliton, propagating at speed c, is a function only of the variable $\zeta = x - ct$. Substituting $f(x,t) = f(\zeta)$ into Eqn (62) gives the non-linear ordinary differential equation

$$f''' - 6ff' - cf' = 0, \tag{63}$$

where now the f' denotes $\partial_\zeta f$. This immediately integrates to

$$f'' - 3f^2 - cf = A, \tag{64}$$

which in turn has the integrating factor f', giving the non-linear ODE[39]

$$\frac{1}{2}(f')^2 = f^3 + \frac{1}{2}cf^2 + Af + B, \tag{65}$$

where A and B are the two integration constants. We are now on familiar gound! Eqn. (65) is of precisely the same form as the time independent equation (21) derived from the time independent NLSE in Section 3, the only difference being that the polynomial is a cubic rather than a quartic. Nonetheless, the soutions of Eqn. (65) are Jacobi elliptic functions, and the Hamiltonian equation of motion analysis of Section 3 still applicable.

Let us undertake the qualitative trajectory analysis first. The analysis of section 3 suggested that the quartic polynomial in Eqn. (21) functioned as a potential, see Figure 4, and that trajectories run on this potential could give us a flavor of the expected solutions. Figure 9a,b shows two possible cubic potentials, $V(q)$, of interest. In each case there is a region of the co-ordinate *"q"* over which the potential *V(q)* allows stable oscillatory classical motion, and point of unstable equilibrium at $q = 0$ *or 1, and* where $V(q) = 0$, in each case. In both cases, this period of this oscillatory motion will become infinite, as the "energy" of the system approaches *0* from below, resulting in trajectories which "start" and/or "stop" at the unstable maximum. These special infinite period trajectories correspond to the soliton solutions of the KdV equation. Comparison of the potentials 9a,b, which differ only by a sign and re-setting of constants, also immediatly indicates that, as the point of unstable equilibrium may be to the lefr *or* right of the other turning point, a sign choice in the KdV equation determines the existence of solitons of elevation or depression.

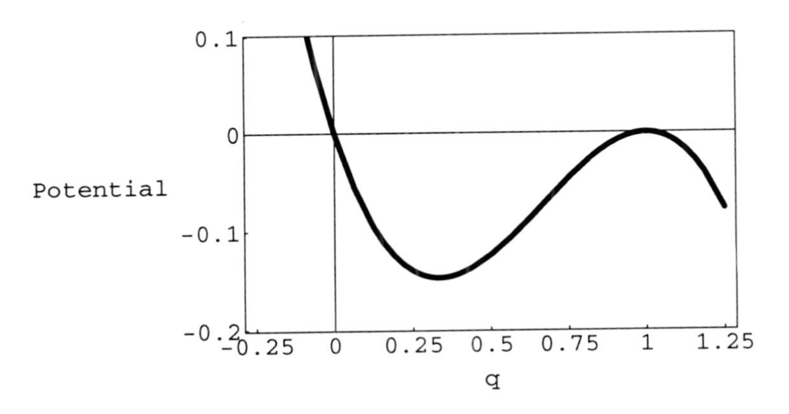

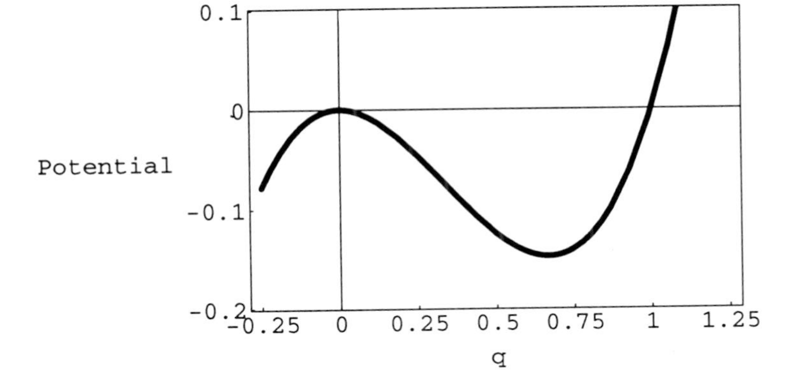

Figure 9a,b.. Effective potentials, V(q), for the q(t) time dynamics equivalent to the solution of the KdV equation. The top figure, 9a, with its local maximum on the "left" will give rise to solitons of depresson (dark or grey); the bottom potential, 9b, with its local maximum on the "right" will give rise to solitons of elevation. See Fig. 10a,b. These cubic potentials, may, via adjustment of constants, be translated and rescaled, giving rise to a variety of possible periodic and soliton type sof solutions.

Such a sign change corresponds to the difference between Eqns. (60) and (61). To make this more definite, Figure 10a,b shows trajectories run, starting very near $V(q) = 1$, on each of the potentials of Figure 9a,b. Identifying these trajectories $q(t)$ with $f(\zeta)$ and then noting that $\zeta = x - ct$ gives propagating waves of speed c and fixed waveform in both cases. Thus, for appropriate constituative constants in the equation itself combined with a proper choice of boundary conditions, the KdV equations (60,61) may give rise to solitons of elevation and depression. Other choices of these parameters may well also give rise to oscillatory solutions, or, if the potential well in the cubic disapears, only unbounded solutions.

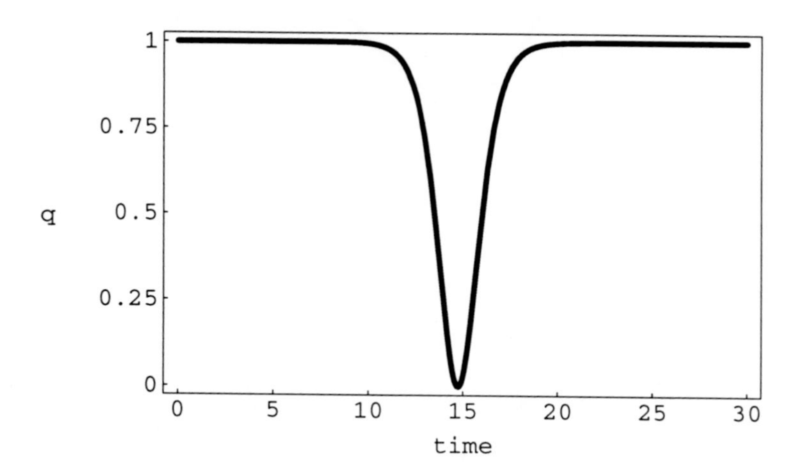

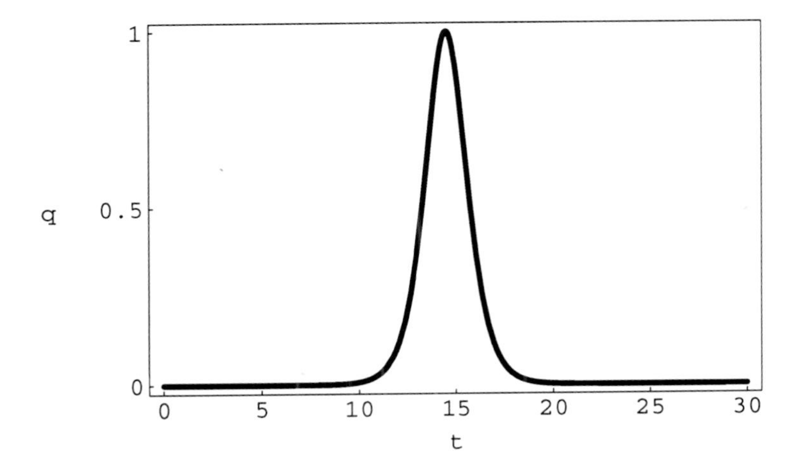

Figure 10a,b. Actual trajectories q(t) run at energies "slightly less than 0" on the potentials of Fig. 9a,b, respectively. In each case the trajectories were started quite near the point of local unstable equilibrium, but within the potential well. These Newtonian trajectories clearly illustrate what will emerge to be solitons of depression (a) and elevation (b). Note that the cubic, rather than quartic (as in Figure 4), potential gives "one sided" oscillations appropriate to the fact that physical densities are positive.

Analytical Solution of the KdV Equation for Single Solitons: Eqn. (65) may be immediately formally integrated, following the pattern of Eqns. (21-23) of Section 3:

$$\zeta = \int^{\zeta} d\tilde{\zeta} = \int^{f} \frac{1}{\sqrt{[2(f^3 + \frac{1}{2} c f^2 + Af + B)]}} d\tilde{f}. \tag{66}$$

Perhaps surprisingly, as the radicand is now a cubic, rather than a quartic as in Eqn.(26), the defining equation for the Jacobi *SN*, the function *f* defined implicitly in Eqn. (66) is still a Jacobi elliptic function. This cubic[45] may be thought of as a quartic with one root moved to infinity, and Eqn (66) may be put into a standard Jacobi *SN[x|m]* form, which for appropriate *A,B,c* and boudnary conditions (needed to allow for the trapped classical motion of Figs. 9a,b) will yield oscillatory *SN* solutions and finally analytical soliton solutions in the $m \rightarrow 1^-$ limit.

This is most directly done via an alternative form of Eqn. (26), often itself used to define the SN function itself. Under the variable change[35] $f \rightarrow \sin(\theta)$

$$\int^{f} \frac{1}{\sqrt{[(1-\tilde{f}^2)(1-m\,\tilde{f}^2)]}}\, d\tilde{f}$$

$$\rightarrow \int^{\theta=\arcsin(f)} \frac{\cos(\tilde{\theta})}{\sqrt{[(1-\sin^2(\tilde{\theta}))\,(1-m\sin^2(\tilde{\theta}))]}}\, d\tilde{\theta}\,. \qquad (67)$$

Following cancellation of the $cos(\theta)$ in the numerator and denominator of (67), we find that if

$$x \equiv \int_0^{\theta} \frac{1}{\sqrt{[\,1-m\,sin^2(\tilde{\theta})]}}\, d\tilde{\theta}, \qquad (68)$$

comparison of Eqns. (67) and (68) then gives an alternative definition of the Jacobi *SN* function as

$$\theta = sin^{-1}\ (\ SN[x|m]\ ) \qquad (69a)$$

or

$$sin(\theta) = SN[x|m]. \qquad (69b)$$

This representation is particularly useful as it also allows direct definition of the Jacobi *CN* and *DN* (see Eqn. (36a)) funcitons as (again, implicitely via Eqn.(68))

$$CN[x|m] \equiv cos(\theta) \qquad (69c)$$

$$DN[xc|m] \equiv \sqrt{(1\ -\ m\,sin^2(\theta))} \qquad (69d)$$

The *CN* and *DN* are also peiordic functions of x for $0 \leq m < 1$, and satisfy the trig-like identity $SN^2 + CN^2 = 1$, but are not simply related by an argument shift, except in the $m=0$ limit, where $CN[x|0] = cos(x)$.

With this new armament we now approach Eqn.(66): Assume that the cubic polynomial $F(f) = f^3 + \frac{1}{2} c f^2 + Af + B$ has the three distinct roots $f_1 > f_2 > f_3$ and that sign is such that the situation for the eqivalent $V(q)$ is that of Figure 9a, so that there is a classically allowed region between f_2 and f_3. Following Drazin and Johnson[39], $F(f) = (f - f_1)(f - f_2)(f - f_3)$, and thus

$$\zeta = \int^{\zeta} d\tilde{\zeta} = \int^{f} \frac{1}{\sqrt{[2((\tilde{f} - f_1)(\tilde{f} - f_2)(\tilde{f} - f_3))]}} d\tilde{f}. \tag{70}$$

The variable change

$$\tilde{f} = f_2 - (f_2 - f_3)\cos^2(\tilde{\theta}), \tag{71}$$

and insertion of lower limits corresponding to f_3 gives

$$\zeta = \zeta_3 \pm \sqrt{\frac{2}{(f_1 - f_3)}} \int_0^{\theta} \frac{1}{\sqrt{[1 - m \sin(\tilde{\theta})^2]}} d\tilde{\theta}, \tag{72}$$

where the parameter m is $(f_2 - f_3)/(f_1 - f_3)$, which is indeed between 0 and 1. Relations (69c) and (71) now imply[39]

$$f(x - ct) = f(\zeta) = f_2 - (f_2 - f_3)\, CN^2[(\zeta - \zeta_3)\{\frac{2}{(f_1 - f_3)}\}^{-1/2} \mid m] \tag{73}$$

Equation (73) brings us to the end of our formal development, as it contains most of the behavior of interest for both oscillatory and soliton solutions of both the KdV and (surprisingly!) the tdNLSE. The presence of the Jacobi *CN* function in Eqn. (73) leads to the name *cnoidal waves* for the oscillatory solutions of the KdV equation. Figure 11 shows the development of such waves as a function of x and t, for m = 0.99.

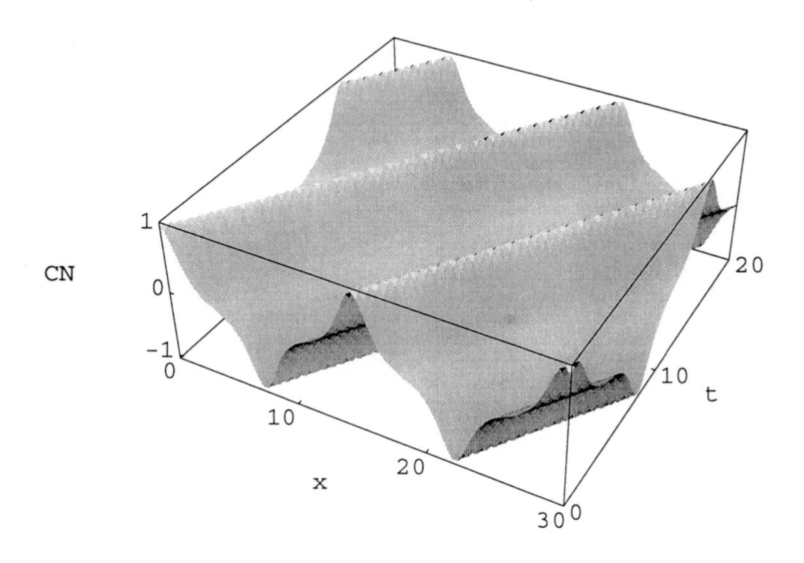

Figure 11. A Cnoidal wave as a function of = x - ct, plotted as a function of x and t for c = 1, and m = 0.99,

KdV Solitons at Last: Solitons, rather than oscillaory functions, *arise as* $m \to 1^-$. In this limit, $CN[x|m] \to sech[x]$. In Eqn.(73) this limit is attained as the roots f_1 and f_2 coalesce, namely the Hamiltonain "energy" is such that the unstable equilibrium point may be attained exactly. In this case (73) becomes, setting $\zeta_3 = 0$, and $\beta = (f_2 - f_3)$

$$f(x - ct) = f(\zeta) = f_2 - \beta\, sech^2 [\ \sqrt{\beta/2}\ \zeta\ ] \tag{74}$$

This is plotted as a fucntion of x and t in Figure 12. The dynamics is eerily reminiscent of the plots of $|\varphi(x, t)|^2$ in Figures 7 and 8. This is not surprising: if f_2 and β are , for example, both 1, the identity $tanh^2 [\zeta] = 1 - sech^2 [\zeta]$ is evidently at play. KdV solitons of depression, solutions of Eqn.(62), are closely related to those for the repulsive BEC. If the " -6" in Eqn.(62) is changed to "+6", solitons of elevation of the form $\alpha + \beta \mathrm{sech}^2[$

ζ] result. A simple example where the backgound or ambient wave amplitude, α, is zero, is shown in Figure 13. This type of solution corresponds to solitons arising in the solution of the tdNLSE for the attractive BEC, a subject of interest, but beyond the scope of the present tutorial..

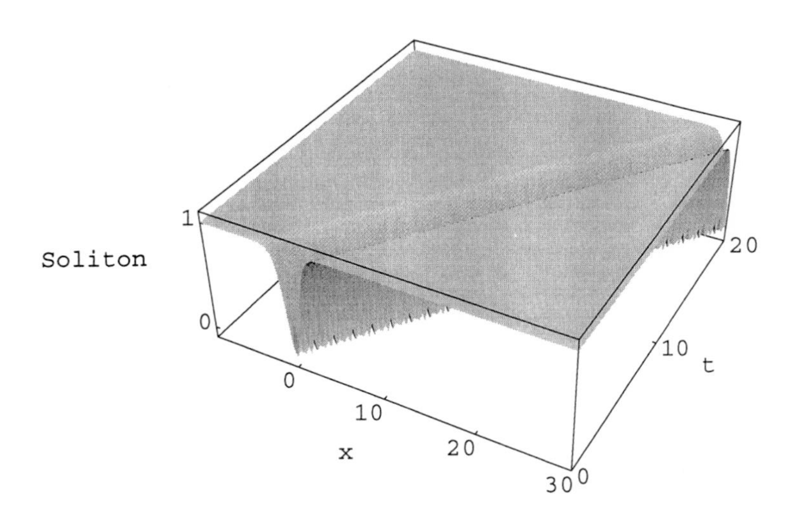

Figure 12. A KdV soliton of depression, closely related to the BEC NSLE notch solitons of Figs 7,8. This is plotted as a fuction of x, and t for c = 1.5, and m = 1.

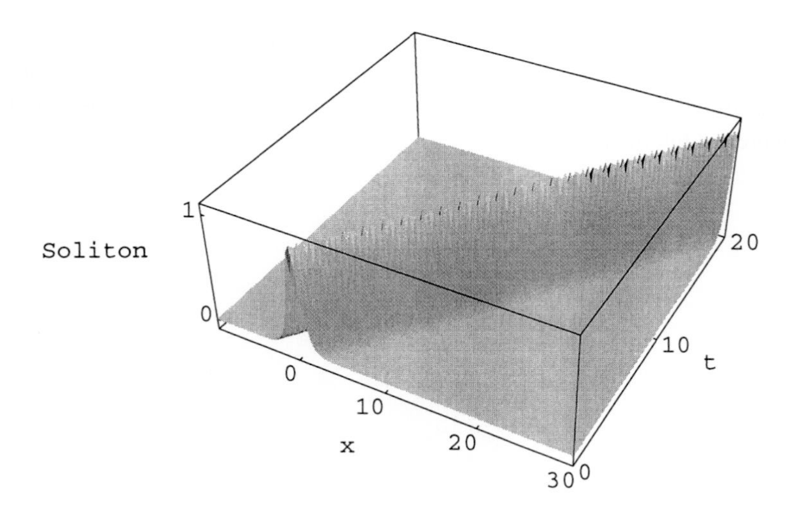

Figure 13. A KdV soliton of elevation (bright soliton), plotted as a function of x and t for c = 1.5.

6. Solitons in the BEC

Solitons for the 1-Dimensional tdNLSE: The stage has now been reached where all that needs to be done is to cast Eqn(45) into a familiar form. In a small extension of the notation of Drazin and Johnson[39] (making the two important lengths scales ξ and L explicit)

$$i\partial_t\, \psi(x, t) \;=\; -\xi^2\, \partial_{x,x}\, \psi(x, t) + L|\, \psi(x, t)\,|^2\, \psi(x,t) \tag{75}$$

is easily put into a now familiar form by writing

$$\psi(x,t) = r(x - ct)exp[\, i\, (\phi(x - c\, t) + n)t\,], \tag{76}$$

where r and ϕ are real functions of $x - ct$. Substitution of this ansatz into (75) gives, after two initial integrations, as in the KdV case, an equation of the form

$$\frac{1}{2}(S')^2 = -F(S), \tag{77}$$

where F is a cubic polynomial in S, and $S = r^2$. We are right back to the KdV type system familiar from the previous sections. Boundary conditions are somewhat different as the tdNLSE is not invariant to shifts of the type "$a + \psi$", as was the KdV. But solutions for $r(x\text{-}ct)$ will include cnoidal wave behavior and solitons of elevation for the attractive condensate, and notch solitons, as we have seen numerically already, for the repulsive case. The algebra is more complex, and will not be shown here. For single soliton solutions in the repulsive condensate r and ϕ are given by

$$r^2(x - ct) = \left(\mu - 2k^2[\mu,c,\xi]\mathrm{sech}^2[k[\mu,c,\xi](x - ct)/\xi]\right)/L, \tag{78a}$$

$$\phi(x\text{-}ct) = \arctan[\,2(k[\mu,c,\xi]/(c/\xi))\tanh[\,k[\mu,c,\xi](x - ct)/\xi]], \tag{78b}$$

where

$$k[\mu, c, \xi] = \sqrt{(2\mu - (c/\xi))^2}\,/\,2, \tag{78c}$$

for

$$(2\mu - (c/\xi))^2 \geq 0. \tag{78d}$$

The constant n in Eqn.(76) is given in terms of the chemical potential as $-\mu$. Note that "far away" from the soliton center $\zeta = 0$, $(|\,\psi(x, t)\,|)^2 \sim \mu/L$, which is the Thomas-Fermi result. The chemical potential thus sets the density (or *vice versa*!) and also imposes a bound on the maximum velocity of the soliton through the crucial restriction of Eqn.(78d). The attractive well in the cubic "potential" vanishes at equality in (78d), and is non-existant if it is violated. This soliton velocity maximum is $c^{\max} = \xi\sqrt{2\mu}$, which reduces to[37]

$$c^{\max} = \sqrt{4\pi\hbar^2 a_s\, \varrho / m^2}\;, \tag{79}$$

on insertion of all units. This is the Bogoliubov sound speed[46] in the condensate. At this maximum velocity the "depth of the soliton" below the ambient μ / L vanishes.. The maximal depth soliton has speed $c = 0$, where $r(x\text{-}ct)$ actually vanishes at $x = 0$. This corresponds exactly to the case shown in figure 7a,b where the initial collision partners were phased to give an actual node in a real wave functions at $x= L/2$. What was observed is clearly, now, a zero velocity (dark) soliton, of maximal depth. In fact the stationary state and zero velocity soliton solutions are identical, as follows from our second use of the identify $tanh^2 [\zeta]= 1 - sech^2 [\zeta]$. All solutions may be translated by adding a co-ordinate shift, x_0 to x to give the appropirate position at $t=0$. Similarily, the moving grey solitons in Figs. 7c,d and 8a,b,c,d are easily reprsented via Eqns(78), except in regions where they are interacting with one another.

Although $r(x\text{-}ct)$ behaves very much like a KdV solution, ψ itself has far richer behavior: the phase $\phi(x - ct)$, which is simply not present in the KdV analysis, shows remarkable behavior, and allows a novel connection with the Josephson effect of superconductivy to be made. The phase ϕ of Eqn.(78b) is shown as a function of x and t in Figure 14a,b for two different speeds, c. It is evident that the analytical solution displays the same sharp and conserved phase discontinuities seen in Figures 7 and 8.

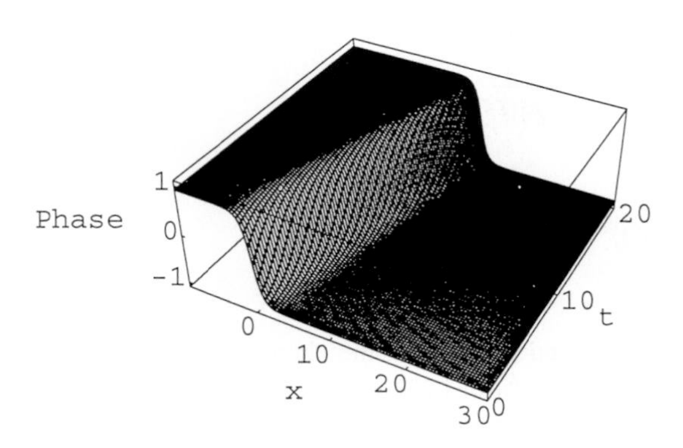

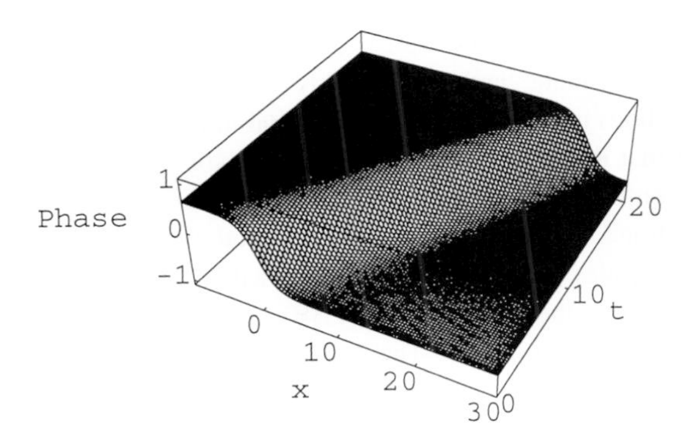

Figure 14. Phase of a NLSE grey soliton as a function of x and t for c = 0.5 (top), and c = 1.2(bottom), this latter being near the maximum speed of $\sqrt{2}$ in these dimensionless units. The sharp, conserved phase discontinuity is clearly illustrated, although the discontinuity becomes smaller as the speed approaches $\sqrt{2}$.

A phase "discontinuity" may be defined as

$$\Delta = \tan^{-1}(2k/(c/\xi)) - \tan^{-1}(- 2k/(c/\xi)), \tag{80}$$

and is usefully written as $\Delta = \pi + \delta$. This relationship may be inverted, using also Eqn. (78d), to give[37]

$$c = c^{max} \sin(\delta/2), \tag{81}$$

a "new" result[47] which relates the speed of a soliton to the Bogoliubov sound speed, and its phase discontinuity. The phase discontinuinty thus drives the soliton, and determines its depth. Away from the soliton center the phase discontinuity is a constant of the motion, as envisaged by Anderson in his qualitative discussions of phase rigidity[9].

Validity of the Separation of Variables for Highly Anisotropic Traps: Both the 1-dimensional NLSE and tdNLSE discussed above have followed from a primitive separation of variables, Section 3, in which it is assumed that the condensate is constant in the transverse y, and z, directions. This may well be the case if the systems is infinte in the y,z directions. But then it is, in fact, well known[41.48] that the 2 and 3 solitons of the KdV and tdNLSE are unstable, and will quickly dissipate, most likely forming vortecies in 2D and vortex lines in 3D through "snake" instabilities[48].

It is thus no accident that Russell's observations in the 1840s were of water waves in narrow canals, not the open ocean, and these were then well described by solution of a 1-dimensional KdV equation. This illustrates a fundamental fact:: if and only if the transverse confining length scales be on the order of, or smaller than, the natural soliton length scale, which is $\sim 2\xi$ in the repulsive tdNLSE case, the system will behave essentially as if it were one dimensional.. The bright optical solitons propagaing in fibers[40] are examples of stability in this type of pseudo-1-dimensional environment.

A simple adiabatic scale separation (in time/energy) allows an understanding of this result, and is well illustrated by a brief disucssion of a condensate in a highly anisotropic box trap. Suppose the condensate is in a box of dimensions a by a by L. Further assume, that a is of the order of a few healing lengths, or smaller, but that as before, $L >> \xi$. The transverse one dimensional wave functions will be usual particle in a box functions, as for $a \approx \xi$, the m parameter of the Jacobi funcitons is small. But the energy spacing of the transverse levels $\Delta \mho$ is >>> than the $\Delta\mu$ of Eqn.(39). Thus transverse excitations are of very high energy compared to that of longitudional soliton propatation, and we can expect an equivalently long term stability. While a nominal density may still be defined as $\varrho = N/(La^2)$ the actual shape of the transverse functions may require use of a slightly modified value of ξ. However, this is a small effect as well illustrated in Figures 15 and 16, in the next section.

Higher Energy Intra-trap Condensate Collisions: The above remarks on separability suggest that the solitons generated by removing a "thin" barrier between two pieces of ground state condensate, as illustrated in Figures 7 and 8, should be observable if the system has appropriate transverse confinement. The strongly anisotropic traps in current use at MIT[2,46] suggest that this is already nearly attainable. However, it is currently not realistic to assume that condensates may be easily initially separated by a length scale on the order of ξ itself. The MIT group has very recently created ground state condensates separated by a "light sheet", but the beam waist creating the potential barrier is minimally of the order of the light wavelength, 0,6μ, and in actuality much larger $\approx$ μ, while the healing length is the order 0.2μ.

What happens if two zero energy condensates separated by a length large compared to ξ are alowed to colllide within a trap[37,46], rather than in a state of low density ballistic expansion, as in Refs. (16,18)? Results of a one dimensional simulation of such a collison are shown in Figure 15. Here the initial separation is ξ. To check the stability with respect to the pseudo-one-dimensional confinement, the same system,

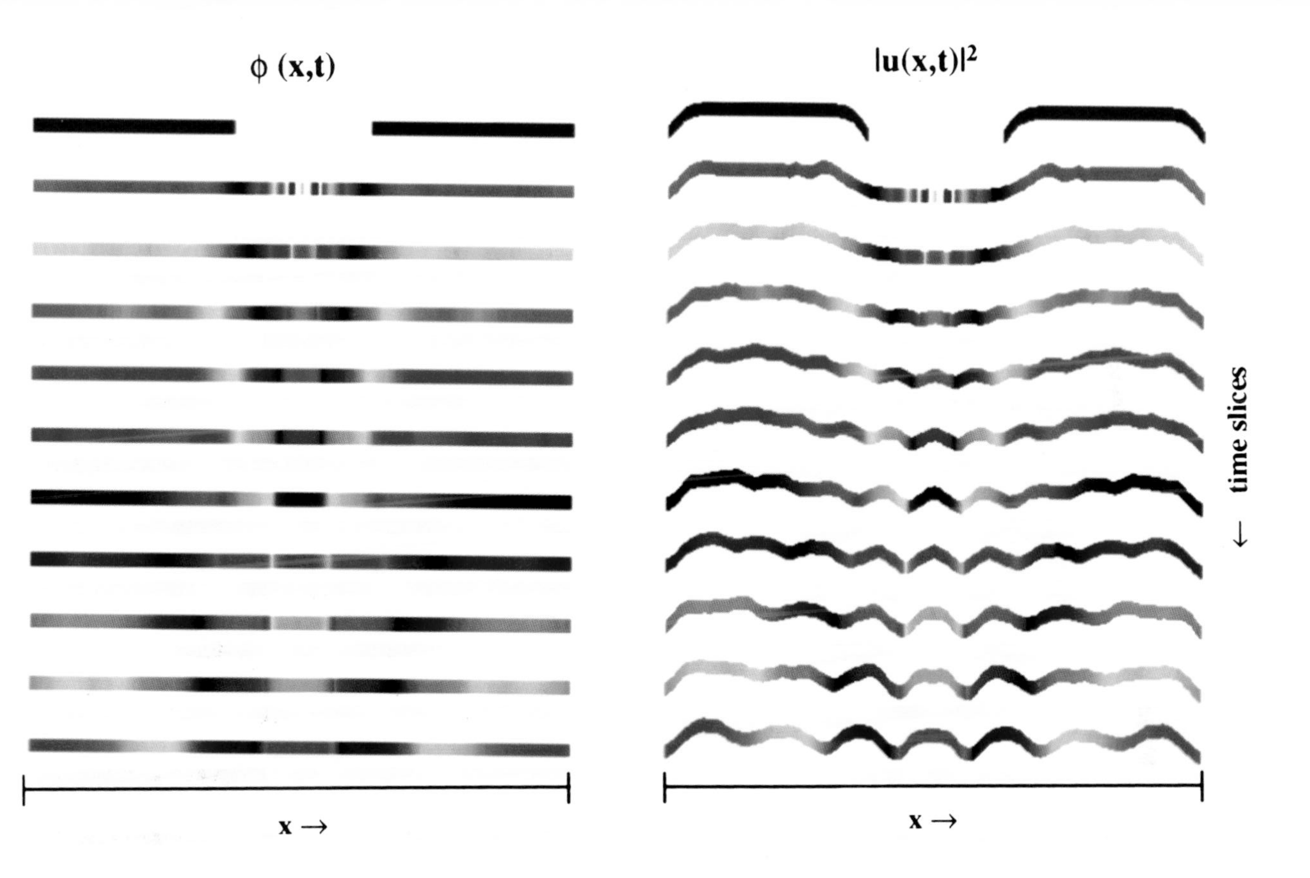

Figure 15. Intra-trap collision of initially separated condensates in one dimension, as shown in a series of time slices. Phase evolution is shown on left, phase and magnitude on the right. The dynamics is "multi-solitonic".

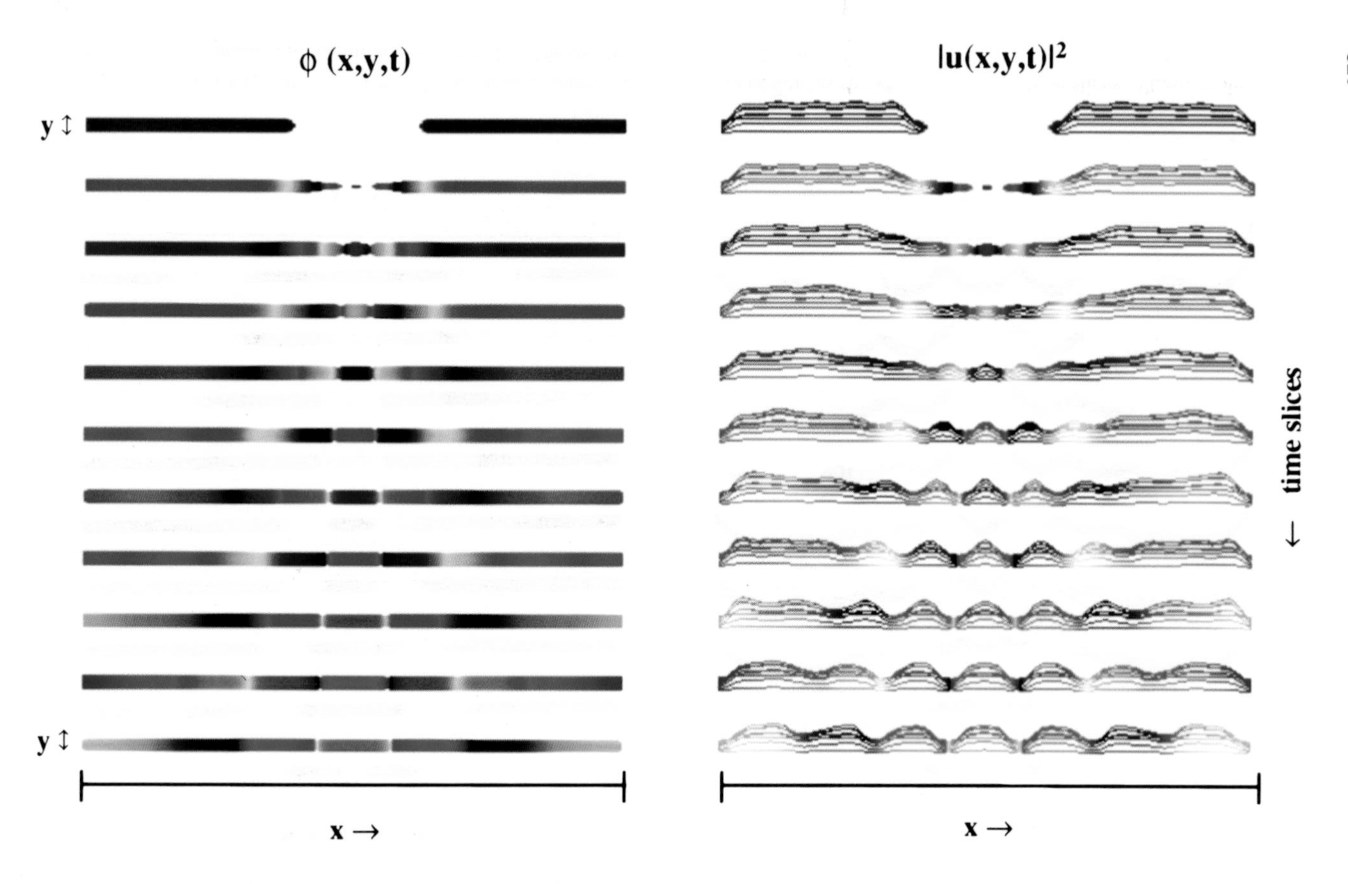

Fig 16. As in Figure 15, except that the NLSE is now solved in two spatial dimenstions. The transverse spatial confinement ($\approx 4\xi$)is seen to have little if any affect on the dynamics, or on the evident phase rigidity.

assumed to be separable at $t = 0$, but then allowed to fully couple in the subsequent time evolution, was numericaly time evolved in 2 and 3 dimensions with transverse confinement lengths of 4ξ, 8ξ and 12ξ. Results for a such a two dimensional collision with transverse length scale 4ξ are shown in Figure 16, in a manner which allows easy compairson with the one-dimensional results of Figure 15. Results in 3 dimensions for this short y,z confinement are indistinguishable from those in shown in 2-dimensions. Confinements as large as 12ξ in 2-dimensions show strong early time soliton stability, but not quite the sharp transverse phase rigidity shown in Fig.16, perhaps presaging eventual long term instability.

Collisons in Harmonic Traps: As indicated above, traps currently available at MIT have strong enough anisotropy to allow pseudo-one-dimensional condensate confinement, and thus potentially stable solitons. However these traps produce harmonic rather than "box" confinement. A crucial point thus revolves around the issue as to whether solitons of the type expressed in Eqn. (78) exist and are stable in harmonic traps. Understanding of this issue has progressed well beyond the stage of development at the time of the tutorial lectures[37]. We content ourselves here with the remark that a theory of such solitons is now well developed, and that they are in fact quite stable, and propagate as harmonically confined classical particles under the conditions of the MIT experiments[46]. Figure 17 contains phase and density profiles for the early time evolution of zero energy collisions of a harmonically trapped condenstses, initially separated by a narrow ($\leq \xi$) light sheet[37]. The solitons shown are stable at long times (many oscillator periods), and the propagation of single solitons is perfectly harmonic, and isochronous. This series of Figures stands in anology to those of Figs 7,8 for "box" trapped condensates, in that as the phase difference, χ, is evolved from π, solitons of larger velocities are created, and eventually, when $\chi = 2\pi$, a bi-soliton results. The solitons shown are thus again a distinctive measure of the phase difference χ.

Conclusions and Final Remarks on the Condensate Phase: Through numerical and analytical studies of a mean field model of the repulsive BEC, it is suggested that dark and gray solitons should be ubiquitous. These will be of scale size ξ, and thus not necessarily easy to observe. Should their observation be possible in intra-trap dynamics it will provide a sensitive diagnostic of the coherence of the BEC. The phase offset, χ, which we have used as if it were a control parameter, is, as mentioned earlier, not a deterministic variable in the collision of two uncorrelated condensates. This is simply because the probability deduced from the projection of the full N-body wave fucntion, φ^N, corresponding to the delocalized state φ of Eqn.(43), onto the initial localized state $[\varphi^{\text{left}} \varphi^{\text{right}}]^{N/2}$ (with appropriate symmetrization) is independent of χ, and thus all χ

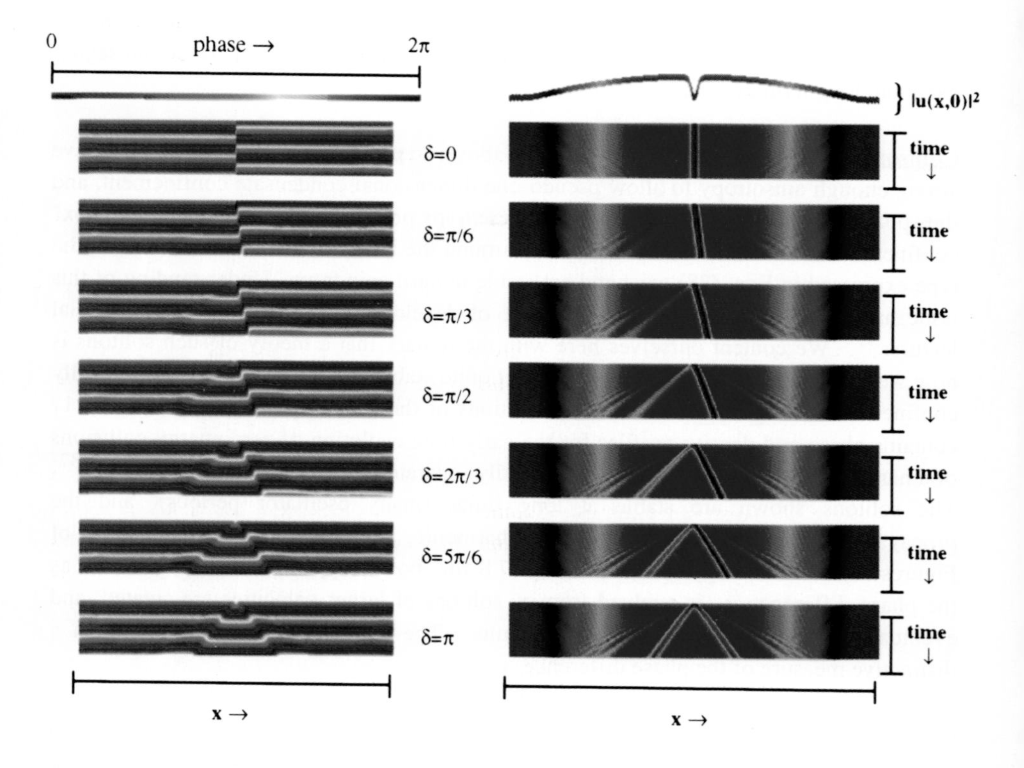

Figure 17. Solitons in an harmonic potential. On the left, a phase color codes the time evolution of a series zero energy collisions with phase offsets of the form $\pi + \delta$, for a series of values of δ. On the right, the color codes the density as shown in the top of the frame. The initially imposed phase shift generates solitons quite similar to those shown for the Box Trap, in Figs 7 and 8.

are equally probable in a given experimental observation. However, whatever the value of χ, as subsequent soliton evolution requires a stable phase discontinuity, the lack of such will lead to rapid soliton dispersion. Observation of solitons will thus provide a strong check on the applicability of Anderson's ideas of phase rigidity in real BEC systems. As individual dark or gray solitons following from the tdNLSE seem perfectly stable in both box and harmonic pseudo-one-dimensional confinement, their observation, and observations of their stability, will provide a sensitive test of the applicability of the tdNLSE, and allow subsequent checks on the development of more sophisticated theories.

Acknowledgments: It is especially important to note that the idea that it would be useful to have a "soluble" model of a trapped condensate arose in collaborative work with Charles Clark, begun at NIST in the Summer of 1996. The results of this collaboration appear elsewhere[36,37], and were supported by the Office of Naval Research. The creation and explorations of this model led to developments far beyond those originally anticipated. The author is most grateful to the organizers of the INT Workshop, Messrs. Tomsovic, Leggett and Bohigas for the invitation to present a tutorial, and to A. J. Leggett and D. Thouless for their enthusiastic, if sometimes skeptical, reception of the materials presented. Paul Hopkins at the University of Washington made active participation the Workshop possible; writing has been completed at the School of Chemistry, University of Melbourne, where the support of R. O. Watts, the Wilsmore Trust, and the US-Australia Fulbright Commission have been invaluable; all are most gratefully acknowledged. A "chance" meeting with Yuri Kivshar at Australian National University, Canberra, resulted in the availability of a pre-publication copy of Ref. 41, which was gratefully received.

References

1) M. H. Anderson, J. R. Ensher, M. R. Matthews, C. E. Wieman, and E. A. Cornell, Observation of Bose-Einstein Condensation in a Dilute Atomic Vapor, Science, 269,198(1995).

2) M. R. Andrews, M-O. Mewes, N. J. van Druten, D. S. Durfee, D. M. Kurn, and W. Ketterle, Direct Nondestructive Obvservation of a Bose Condensate, Science, 273,84(1996).

3) C. C. Bradley, C. A. Sackett and R. G. Hulet, Analysis of in situ Images of Bose -Einstein Condensates of Litium, Phys. Rev. A 55,3951(1997); and references therein.

4) C. C. Bradley, C. A. Sackett, and R. G. Hulet, Bose-Einstein Condensation of Lithium, Observation of Limited Condensate Number, Phys. Rev. Letts. 78,985(1997).

5) W. C. Stwalley and L.H. Nosanow, Possible "New" Quantum systems, Phys. Rev. Letts. 36,910(1976).

6) E. Tiesinga, A. J. Moerdijk, B. J. Verhaar, and H. T. C. Stoff, Conditions of Bose-

Einstein Condensation in Magnetically Trapped Atomic Cesium, Phys. Rev. A 46,R1167(1992).

7) B. D. Josephson,Possible New Effects in Superconductive Tunnelling, Phys. Letts. 1,251(1962).

8) P. W. Anderson,Considerations on the Flow of Superfluid Helium, Rev. Mod. Phys. 38,298(1966).

9) P. W. Anderson, Basic Notions of Condensed Matter Physics, Addison-Wesley, Reading, Mass, 1984.

10) A. J. Leggett and F. Sols, On the Concept of Spontaneously Broken Gauge Symmetry in Condensed Matter Physics, Found. Phys. 21,353(1991).

11) A. J. Leggett, Broken Gauge Symmetry in a Bose Condensate, in Bose -Einstein Condensation, A. Griffin, D. W. Snoke, and S. Stringari, eds., Cambride Univ. Press, 1995, pp 452.

12) E. P. Gross, Structure of a Quantized Vortex in a Bose System, Nuovo Cimento, 20,454(1961).

13) L. P. Pitaevskii, Vortex Lines in an Imperfect Bose Gas, Sov. Phys. JETP 13,451(1961).

14) A. I. Fetter and J. D. Walechka, Quantum Theory of Many-particle Systems, McGraw-Hill, New York, 1971; D. J. Thouless, The Quanatum Mechanics of Many Body Systems, Academic Press, New York, 2nd Ed. 1972.

15a) experiments are discussed in: D.S. Jin, J. R. Ensher, M. R. Mattehws, C. E. Wieman, and E. A., Cornell, Collective Excitation of a Bose-Einstein Condensate in a Dilute Gas, Phys. Rev. Letts. 77,420(1996); M-O. Mewes, M. R. Andrews, N. J. van Druten, D. M. Kurn, D. S. Durfee, C. G.Townsend, and W. Ketterle, Collective Excitations of a Bose-Einstein Condensate in a Magnetic Trap, Phys. Rev. Letts. 77,988(1996).

15b) theoretical explorations appear in : M. Edwards, R. J. Dodd, C. W. Clark, and K. Burnett, Properties of a Bose-Einstein Condensate in an Anisotropic Harmonic Potential, Phys. Rev. A 53,R1950(1996); M. Edwards, P.A. Ruprecht, K. Burnett, R. J. Dodd, C. W. Clark, Collective Excitations in Bose-Einstein Condensates, Phys. Rev. Letts. 77,1671(1996); K. G. Singh, and D. S. Rokhsar, Collective Excitations of a Confined Bose Condensate , Phys. Rev. Letts. 77,1667(1996); S. Stringari, Collective Excitations in a Trapped Bose Condensate, Phys. Rev. Letts. 77,2360(1996).

16) M. R. Andrews, C. G. Townsend, H.-J. Miesner, D. S. Durfee, D. M., Kurn, and W. Ketterle, Observation of Interference Between Two Bose Condensates, Science 275,637(1997).

17) D. R. Tilley, and J. Tilley, Superfluidity and Superconductivity, 3rd Ed. IOP, Bristol, 1990.

18) M. Naraschewski, H. Wallis, A. Schenzle, J. I. Cirac , and P. Zoller, Interference of Bose Condensates, Phys. Rev A. 54,2185(1996); W. Houston and L.You, Interference of Two Bose Condensates, Phys. Rev. A 53,4254(1966); H. Wallis, A. Rohrl, M. Naraschewski, and A. Schenzle, Phase Space Dynamcis of Bose Condensates, Interference Versus Interaction, Phys. Rev. A 55,2109 (1997); Rohrl M., A. Naraschewski, A. Schenzle, H. Wallis, Transition form Phase Locking to the Interference of Independent Bose Condensates: Theory versus Experiment, Phys. Rev. Letts. 78,4143(1997).

19) an overview is given in R. Louden, The Quantum Theory of Light, Oxford U. Press., Oxford,1973, Ch. 5.